Lecture Notes in Computer Science 1353

Edited by G. Goos, J. Hartmanis and J. van Leeuwen

T0223166

Springer

Berlin
Heidelberg
New York
Barcelona
Budapest
Hong Kong
London
Milan
Paris
Santa Clara
Singapore
Tokyo

Giuseppe DiBattista (Ed.)

Graph Drawing

5th International Symposium, GD '97
Rome, Italy, September 18-20, 1997
Proceedings

 Springer

Series Editors

Gerhard Goos, Karlsruhe University, Germany

Juris Hartmanis, Cornell University, NY, USA

Jan van Leeuwen, Utrecht University, The Netherlands

Volume Editor

Giuseppe DiBattista
Department of Computer Science
University of Rome III
Via della Vasca Navale 84, I-00146 Roma, Italy
E-mail: dibattista@iasi-rm.cnr.it

Cataloging-in-Publication data applied for

Die Deutsche Bibliothek - CIP-Einheitsaufnahme

Graph drawing : 5th international symposium ; proceedings / GD
'97, Rome, Italy, September 18 - 20, 1997 / Giuseppe DiBattista
(ed.). - Berlin ; Heidelberg ; New York ; Barcelona ; Budapest ; Hong
Kong ; London ; Milan ; Paris ; Santa Clara ; Singapore ; Tokyo :
Springer, 1998
  (Lecture notes in computer science ; Vol. 1353)
  ISBN 3-540-63938-1

CR Subject Classification (1991): G.2.2, F.2.2, D.2.2, I.3.5, J.6

ISSN 0302-9743
ISBN 3-540-63938-1 Springer-Verlag Berlin Heidelberg New York

© Springer-Verlag Berlin Heidelberg 1997
Printed in Germany

Typesetting: Camera-ready by author
SPIN 10661311   06/3142 – 5 4 3 2 1 0   Printed on acid-free paper

Preface

The 5th Symposium on Graph Drawing (Graph Drawing '97) was organized by the Università di Roma Tre and was held in Rome (Italy) at Villa Celimontana, September 18-20.

A strong technical program of 33 papers and 10 demos was selected from 80 submissions. All the presentations were of high quality. The invited talk of Jaroslav Nešetřil on the coloring of plane configurations was deep and interesting. About 90 participants attended from several countries and all of them contributed in creating a pleasant and friendly atmosphere.

Session topics included planarity, crossing theory, three-dimensional representations, orthogonal representations, clustering and labelling problems, packing problems, general methodologies, systems and applications. A report on the judging of the Graph Drawing Competition was, as usual, a highlight of the Symposium.

I wish to thank all the members of the Program Committee and of the Organizing Committee, the invited lecturer, all those who submitted papers for consideration, and all the referees.

Rome, October 1997   Giuseppe Di Battista

Organizing Committee:

Paola Bertolazzi (IASI-CNR, Italy)
Giuseppe Di Battista (Università di Roma Tre, Italy)
Giuseppe Liotta (Università di Roma "La Sapienza", Italy)
Maurizio Patrignani (Università di Roma Tre, Italy)
Francesco Vargiu (AIPA, Italy)

Program Committee:

Giuseppe Di Battista, chair (Università di Roma Tre, Italy)
Hubert de Fraysseix (CNRS, France)
Mike Goodrich (The John Hopkins University, USA)
Jan Kratochvíl (Charles University, Czech Republic)
Joe Marks (Mitsubishi Electric Research Laboratories, USA)
Petra Mutzel (Max Planck Institut für Informatik, Germany)
Stephen North (AT&T Bell Laboratories, USA)
Thomas Shermer (Simon Fraser University, Canada)
Kozo Sugiyama (Fujitsu Laboratories Ltd., Japan)
Roberto Tamassia (Brown University, USA)

Steering Committee:

Franz J. Brandenburg (University of Passau, Germany)
Giuseppe Di Battista (Università di Roma Tre, Italy)
Peter Eades (University of Newcastle, Australia)
Takao Nishizeki (Tohoku University, Japan)
Stephen North (AT&T Bell Laboratories, USA)
Pierre Rosenstiehl (EHESS, France)
Roberto Tamassia (Brown University, USA)
Ioannis G. Tollis (The University of Texas at Dallas, USA)
Sue Whitesides (McGill University, Canada)

Additional Referees:

James Abello
Gill Barequet
Ondra Čepek
Walter Didimo
Emden Gansner
Jan Hajič
Petr Hliněný
Christoph Hundack
Michael Jünger
Stephen Kobourov
Giuseppe Liotta
Vladan Majerech
Jaroslav Nešetřil

Patrice Ossona de Mendez
Maurizio Patrignani
Maurizio Pizzonia
Jack Snoeyink
Pavel Töpfer
Edward Tufte
Luca Vismara
Christopher Wagner

Sponsored by:

Università di Roma Tre, EC ESPRIT Long Term Research Project ALCOM-IT, Istituto di Analisi dei Sistemi ed Informatica (IASI) of the Italian National Research Council (CNR), Comitato per le Scienze e le Tecnologie dell'Informazione of the CNR, Comitato per le Scienze di Ingegneria ed Architettura of the CNR, Integra Sistemi, and CISCO Systems.

Contents

Systems II

Packing, Compressing, and Touching

Methodologies and Applications II

Graph-Drawing Contest

Drawable and Forbidden Minimum Weight Triangulations *

(Extended Abstract)

William Lenhart[1], Giuseppe Liotta[2]

[1] Department of Computer Science, Williams College, Williamstown, MA 01267.
lenhart@cs.williams.edu
[2] Dipartimento di Informatica e Sistemistica, Università di Roma 'La Sapienza', via
Salaria 113, I-00198 Roma, Italia.
liotta@dis.uniroma1.it

Abstract. A graph is *minimum weight drawable* if it admits a straight-line drawing that is a minimum weight triangulation of the set of points representing the vertices of the graph. In this paper we consider the problem of characterizing those graphs that are minimum weight drawable. Our contribution is twofold: We show that there exist infinitely many triangulations that are not minimum weight drawable. Furthermore, we present non-trivial classes of triangulations that are minimum weight drawable, along with corresponding linear time (real RAM) algorithms that take as input any graph from one of these classes and produce as output such a drawing. One consequence of our work is the construction of triangulations that are minimum weight drawable but none of which is Delaunay drawable—that is, drawable as a Delaunay triangulation.

1 Introduction and Overview

Recently much attention has been devoted to the study of combinatorial properties of *geometric graphs* such as Delaunay triangulations, minimum spanning trees, Gabriel graphs, relative neighborhood graphs, β-skeleton graphs, and rectangle of influence graphs. This interest has been motivated in part by the importance of these structures in numerous application areas, including computer graphics, pattern recognition, computational morphology, communication networks, numerical analysis, computational biology, and GIS.

Geometric graphs are 2-dimensional straight-line drawings that satisfy some additional geometric constraints. The problem of analyzing the combinatorial properties of a given type of geometric graph naturally raises the question of the characterization of those graphs which admit the given type of straight-line

---

* Research supported in part by by the EC ESPRIT Long Term Research Project
ALCOM-IT under contract 20244.

drawing. This, in turn, leads to the investigation of the design of efficient algorithms for computing such a drawing when one exists. Although these questions are far from being resolved in general, many partial answers have appeared in the literature. We give two such examples below.

A Gabriel graph (also called *Gabriel drawing* in the graph drawing literature) is a straight-line drawing such that two vertices u and v are adjacent if and only if the disk having u and v as antipodal points does not contain any other vertex of the drawing. Trees that admit a Gabriel drawing are characterized in [3]. Lubiw and Sleumer [25] show that every maximal outerplanar graph is Gabriel drawable. In [19] the characterization is extended to all outerplanar graphs. The area required by Gabriel drawings is investigated in [24].

Delaunay triangulations are planar straight-line drawings with all internal faces triangles and such that three vertices form a face if and only of the the disk passing through them does not contain any other vertex of the triangulation. To our knowledge, no complete combinatorial characterization of Delaunay drawable triangulations has been given to date. Di Battista and Vismara [7] give a characterization based on a non-linear system of equations involving the angles in the triangulation. Dillencourt has shown that all Delaunay drawable triangulations are 1-tough, and have perfect matchings [11], and that all maximal outerplanar graphs are Delaunay drawable [10]. Dillencourt and Smith [12] show that any triangulation without chords or non facial triangles is Delaunay drawable. A survey on the problem of drawing a graph as a given type of geometric graph can be found in [5].

In this paper we consider a special type of straight-line drawing, a *minimum weight drawing*, which has applications in areas including computational geometry and numerical analysis. Let C be a class of graphs, let P be a set of points in the plane, and let G be a graph such that

1. G has vertex set P,
2. the edges of G are straight-line segments connecting pairs of points of P,
3. $G \in C$, and
4. the sum of the lengths of the edges of G is minimized over all graphs satisfying 1–3.

We call such a graph G a *minimum weight representative of* C. Given a graph $G \in C$, we say that G has a *minimum weight drawing for class* C if there exists a set P of points in the plane such that G is a minimum weight representative of C. For example, a *minimum spanning tree* of a set P of points is a connected, straight-line drawing that has P as vertex set and minimizes the total edge length. So, letting C be the class of all trees, a tree G has a minimum weight drawing if there exists a set P of points in the plane such that G is isomorphic to a minimum spanning tree of P. A *minimum weight triangulation* of a set P is a triangulation of P having minimum total edge length. Letting C be the class of all planar triangulations, a planar triangulation G has a minimum weight drawing if there exists a set P of points in the plane such that G is isomorphic to a minimum weight triangulation of P.

The problem of testing whether a tree admits a minimum weight drawing is essentially solved. Monma and Suri [27] proved that each tree with maximum vertex degree at most five can be drawn as a minimum spanning tree of some set of vertices by providing a linear time (real RAM) algorithm. In the same paper it is shown that no tree having at least one vertex with degree greater than six can be drawn as a minimum spanning tree. As for trees having maximum degree equal to six, Eades and Whitesides [13] showed that it is NP-hard to decide whether such trees can be drawn as minimum spanning trees. The representability of trees as minimum spanning trees in three-dimensional space was studied in [23].

Surprisingly, little seems to be known about the problem of constructing a minimum weight drawing of a planar triangulation. Moreover, it is still not known whether computing a minimum weight triangulation of a set of points in the plane is an NP-complete problem (see Garey and Johnson [14]). Several papers have been published on this last problem, either providing partial solutions, or giving efficient approximation heuristics. See, for example, the work by Meijer and Rappaport [26], Lingas [22], Keil [15], Dickerson et al. [8], Kirkpatrick [16], Aichholzer et al. [1], Cheng and Xu [4], and Dickerson and Montague [9], Levcopoulos and Krznaric [21, 20].

The problem examined in this paper is that of characterizing those triangulations that admit a minimum weight drawing. In [18, 17] it was shown that all maximal outerplanar triangulation are minimum weight drawable and a linear time (real RAM) drawing algorithm for computing a minimum weight drawing of these graphs was given.

Our contributions here are the following:

1. Exploit the relationship between locally minimum weight triangulations and minimum weight triangulations to construct classes of triangulations that are not minimum weight drawable.
2. Exhibit an infinite class of triangulations that can be drawn as minimum weight triangulations but not as Delaunay triangulations.
3. Establish the minimum weight drawability of several non-trivial classes of triangulations, including wheels, spined triangulations, and nested triangulations. (For definitions see Section 2)
4. Investigate the combinatorial structure of minimum weight triangulations by means of the notion of *skeleton* of a triangulation, that is the graph induced by the set of its internal vertices. We show that any forest can arise as the skeleton of a minimum weight drawable triangulation.
5. Present linear time, real RAM algorithms that accept as input a minimum weight drawable triangulation of one of the types mentioned above and produce as output a minimum weight drawing of that triangulation.

Our algorithmic techniques generalize drawing strategies that have been devised in recent years to compute proximity drawings of graphs (see, e.g., [2, 6, 19]).

For reasons of space, the proofs have been omitted in this extended abstract.

2 Preliminaries

We first discuss some terminology and define classes of triangulations that are of interest in this paper. We then recall some basic properties of minimum weight triangulations.

Many classes of graphs arise from using some geometric constraint to define edges on a set of points in the plane. Three such classes, Gabriel graphs, Delaunay triangulations and minimum weight triangulations, have already been mentioned. Given a certain class C of graphs defined by such a method, we can define a graph G as being C *drawable* if G is isomorphic to some member of C. Thus in the rest of the paper we will refer to certain triangulations as being, for example, *Delaunay drawable, minimum weight drawable, locally minimum weight drawable* (see below for a definition) and so forth.

We make particular use of some special types of triangulations. A *fan* consists of a cycle $a, v_1, v_2, \ldots, v_n$ along with edges from a to each vertex v_i. The vertex a is called the *apex* of the fan, and the vertices v_i are called the *neighbors* of a. The edges av_i are called the *radial* edges of the fan.

A *wheel graph* (or *wheel*, for brevity) is a triangulation consisting of a cycle and a single vertex c, called the *center* of the wheel, adjacent to all vertices on the cycle. The edges from c to the vertices of the cycle are called the *spokes* of the wheel.

The class of *k-nested triangulations* can be recursively defined as follows. A three-cycle is a 0-nested triangulation. For $k > 0$, a *k-nested triangulation* is defined to be one having a triangular outer face, the deletion of which results in a $k - 1$-nested triangulation. Figure 1 (b) shows a 3-nested triangulation.

Another family of triangulation we will study in this paper is the class of *k-spined triangulations*. Each element of the class is obtained by *k-spining* K_4. Let a_o, a_1, a_2 be the vertices on the outer face and let c be the interior vertex of K_4. *k-spining* K_4 consists of replacing edge ca_2 with a path of k vertices such that each new vertex is adjacent to both a_0 and a_1. The resulting triangulation is a *k-spined* triangulation. Figure 1 (c) and (d) show a 1-spined and a 2-spined triangulation, respectively. The Delaunay drawability of spined triangulations is studied in [10].

The *skeleton* of a triangulation is the graph induced by the set of its internal vertices. For example, the skeleton of a maximal outerplanar graph is the empty graph, the skeleton of a wheel graph consists of just one vertex (the center of the wheel). Figure 1 (a) shows a triangulation whose skeleton is a tree. In the figure, the skeleton is highlighted.

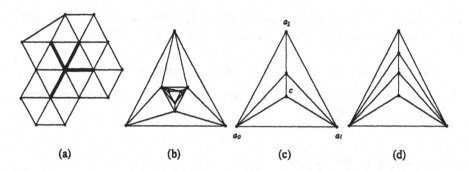

Fig. 1. (a) A Triangulation whose skeleton is a tree, (b) A 3-nested triangulation, (c) A 1-spined triangulation, and (d) A 2-spined triangulation.

We list some basic properties of minimum weight triangulations.

Property 2.1 *A minimum weight triangulation of a set S of points in the plane always includes the shortest edge between two points of S and the all edges of the convex hull of S.*

Property 2.2 *The minimum weight triangulation of a set S of points in the plane need not contain the minimum spanning tree of S as a subgraph.*

If T is a triangulation, and abd and bcd are two triangles of T which form a convex quadrilateral, then the operation of replacing edge bd with edge ac is called an *edge flip*. A triangulation T is *locally minimum weight* if no single edge flip reduces the weight of T.

Property 2.3 *If a triangulation cannot be drawn as a locally minimum weight triangulation, then it cannot be drawn as a minimum weight triangulation.*

In this paper we will explore the relationship between minimum weight drawable triangulations and Delaunay drawable triangulations. To our knowledge, no complete combinatorial characterization of Delaunay drawable triangulations has been to date given [10, 12, 11]. Di Battista and Vismara [7] give a characterization based on a non linear system of equations involving the angles in the triangulation. Dillencourt [10, 11] gives a set of necessary conditions for Delaunay drawability of a triangulation. In the following theorem, T is a triangulation, P a subset of the set S of the vertices of T, $|P|$ is the cardinality of P, and $T - P$ is the graph obtained by removing P (and all attached edges) from T.

Theorem 1. *[10, 11] If a triangulation T can be drawn as a Delaunay triangulation, then for any given $P \subseteq S$ the following two conditions hold.*

1. *T is 1-tough, that is , $T - P$ has at most $|P|$ components.*

2. *$T - P$ contains at most $|P| - 2$ components that do not contain any vertex of the outer face of T.*

3 Forbidden Triangulations

In Section 3.1 we show classes of *minimum weight forbidden* triangulations, i.e. triangulations that are not minimum weight drawable. Interestingly, the triangulations in these classes are also *Delaunay forbidden*, that is, they cannot be drawn as Delaunay triangulations. This observation leads us to compare minimum weight and Delaunay forbidden triangulations. In Section 3.2 we show an infinite family of triangulations that are minimum weight drawable and Delaunay forbidden.

3.1 Minimum weight forbidden triangulations

Our first lemma gives a set of necessary conditions that a triangulation must satisfy in order to be a locally minimum weight triangulation.

To any vertex v and incident face f of a triangulation T, we associate a variable $\alpha = \alpha(v, f)$ called the *angle of f at v*; collectively these variables are called the *angles of T*. Any straight-line drawing of T determines values for these variables, and thus certain relations among the angles of T must hold. If the triangulation is to be locally minimum weight, additional relations among the *magnitudes* of the angles must hold. We define the *magnitude of α* by $m(\alpha) = 1$ if α is obtuse and $m(\alpha) = 0$ otherwise.

Lemma 2. *Let T be a locally minimum weight triangulation. Then there exists an assignment of values to the angles of T such that the following condition holds:*

For each pair of faces f_1 and f_2 of T that share an edge e, the two angles α_1 in f_1 and α_2 in f_2 opposite e satisfy $m(\alpha_1) + m(\alpha_2) \leq 1$.

We can thus use Property 2.3 and Lemma 2 to identify forbidden triangulations.

Theorem 3. *Any triangulation containing either the graph of Figure 2 (a) or the graph of Figure 2 (b) as an induced subgraph is minimum weight forbidden.*

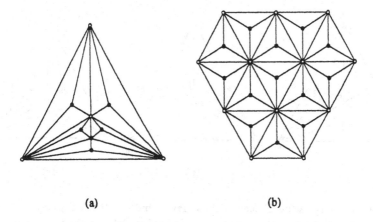

(a)   (b)

Fig. 2. Two examples of triangulations that cannot be drawn as minimum weight triangulations.

3.2   Minimum weight drawable and Delaunay forbidden triangulations

Lemma 4. *Every graph in the class described by Theorem 3 is also Delaunay forbidden.*

The above lemma motivates us to investigate the relationship between Delaunay forbidden and minimum weight forbidden triangulations.

Lemma 5. *The triangulation of Figure 3 is Delaunay forbidden and minimum weight drawable.*

The result of Lemma 5 can be generalized to an infinite class of triangulations that are Delaunay forbidden and minimum weight drawable.

Theorem 6. *There exists an infinite class of triangulations that are Delaunay forbidden and minimum weight drawable. The graphs of this class are obtained by k-spining some K_4 in the triangulation of Figure 3 (a) for $k > 0$.*

Figure 3 (c) shows an example of a triangulation that belongs to the class described in Theorem 6, obtained by 1-spining in Figure 3 (a) the K_4 induced the set of vertices $\{a_0, c, a_1, b\}$.

4   Classes of Minimum Weight Drawable Triangulations

In this section we construct several classes of minimum weight drawable triangulations. The first two subsections discuss triangulations whose skeleton is either

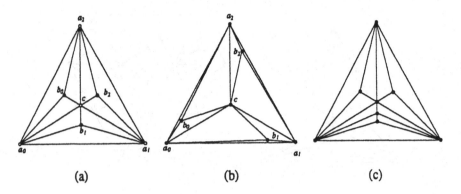

Fig. 3. (a) A triangulation T that is Delaunay forbidden, (b) A drawing of T that is a minimum weight triangulation, and (c) A forbidden Delaunay and minimum weight drawable triangulation obtained by 1-spining the subgraph of T induced by the vertices $\{a_0, c, a_1, b\}$.

a forest (Subsection 4.1) or a set of nested triangles (Subsection 4.2). Wheel graphs are a special case of the first class of triangulations. The last subsection (Subsection 4.3) is devoted to k-spined triangulations.

4.1 Wheel Graphs and Skeletons

In the previous section we established that a triangulation T may contain a subgraph which prevents T from being drawn as a minimum weight triangulation. This leads us to examine the structure of the skeleton of the triangulation. We show here that any forest can occur as the skeleton of a minimum weight triangulation. This is accomplished by describing methods for drawing any fan as a minimum weight triangulation, extending a minimum weight drawing of a fan by adding an additional fan, and repeating the previous step to build minimum weight drawings the skeletons of which are arbitrary forests.

The simplest fan of interest to us is a *kite*: a fan with apex a having exactly three neighbors b, c, d. Clearly such a graph is minimum weight drawable; in fact any fan is minimum weight drawable. We will be interested in a certain type of minimum weight drawing of a fan. Let $d(x, y)$ denote the Euclidean distance between points x and y in the plane.

Lemma 7. *Let $F = \{a, v_1, \ldots, v_n\}$ be a fan with apex a, Consider a drawing of F such that $v_1, \ldots, v_n$ are collinear and for all $1 \leq i < j < k \leq n$, $d(v_i, v_k) > d(v_j, a)$. Then there exists $r > 0$ such that if each v_i is moved away from a along the line $v_i a$ by a distance of at most r so as to form a convex drawing of F, the drawing so formed is a minimum weight drawing.*

Using this lemma, a minimum weight drawing of any wheel can be obtained as follows: Delete one of the exterior vertices of the wheel, draw the fan that remains as in the preceding lemma, then replace the deleted vertex suitably far away from the rest of the fan.

We now describe a method for joining a minimum weight drawing of a given triangulation to one of a fan; it will be of use in our later constructions. Let F be a fan with apex a having neighbors $\{v_1, \ldots, v_n\}$ which has been drawn according to Lemma 7. Fix a particular vertex $v_i, 1 < i < n$ and let $\theta, \tau > 0$ be any angles satisfying the following conditions:

1. $\theta + \tau$ is no larger than $\angle v_{i-1} v_i v_{i+1}$;
2. if $i \geq 3$ then $\theta + \angle v_{i-2} v_{i-1} v_i < \pi$; and
3. if $i \leq n - 2$ then $\tau + \angle v_{i+2} v_{i+1} v_i < \pi$

where all angles are internal angles of F. Let L_{i-1} be the line through v_{i-1} making counterclockwise angle θ with segment $v_{i-1} v_i$, and let L_{i+1} be the line through v_{i+1} making clockwise angle τ with segment $v_{i+1} v_i$. Let R be the quadrilateral region bounded by L_{i-1}, L_{i+1}, $v_{i-1} v_i$, and $v_{i+1} v_i$, and let $D_r(v_1)$ be the circle of radius r centered at v_1, where r satisfies

$$r < \frac{1}{2} \min\{d(v_i, v_k) - d(v_j, a) : 1 \leq i < j < k \leq n\}.$$

Finally, $S(r, \theta, \tau) = D_r(v_1) \cap R$; $S(r, \theta, \tau)$ is called a *safe region* of F at v_i.

Lemma 8. *Let $F = \{a, v_1, \ldots, v_n\}$ be a fan with apex a, let $S(r, \theta, \tau)$ be a safe region of F at v_i, $1 < i < n$, and let X be a finite subset of $S(r, \theta, \tau)$. Then a minimum weight triangulation of $P = \{a, v_1, \ldots, v_n\} \cup X$ can be obtained as the union of a minimum weight drawing of F, any minimum weight triangulation of $X \cup \{v_i\}$, and all edges from v_{i-1} and v_{i+1} to $X \cup \{v_i\}$ which cross no edge of the triangulation of $X \cup \{v_i\}$.*

Observe that the preceding lemma can be used on a fan consisting of a single kite; this gives a second method for constructing a wheel having $n > 3$ spokes: just attach a fan having $n - 3$ radial edges to a kite. Note also that given any fan $\overline{F}$, we can choose X in the preceding lemma such that the only minimum weight triangulation of $X \cup \{v_i\}$ is isomorphic to $\overline{F}$. Thus we can view Lemma 8 as providing a method for gluing any fan $\overline{F}$ with apex u to another fan F at (non-apex) vertex v by identifying u with v. The ideas contained in the proof of the lemma can also be used to design an efficient algorithm for producing a minimum weight drawing of a wheel. Since constructing a minimum weight drawing of a wheel with three spokes is trivial, it is straightforward to establish the following result.

Theorem 9. *Every wheel is minimum weight drawable, and its minimum weight drawing can be computed in linear time in the real RAM model of computation.*

In order to prove that every forest can occur as the skeleton of a minimum weight triangulation, we require a method of attaching several triangulations to an existing fan. The preceding lemma, along with the next definition, will give us one.

Definition 10. Let L be a set of line segments in the plane. Two regions A and B of the plane are *mutually invisible with respect to* L if for each pair of points $a \in A$ and $b \in B$, segment pq intersects some segment of L.

Lemma 11. *Let* $F = \{a, v_1, \ldots, v_n\}$ *be a fan with apex* a, *let* $F' \subset F$ *be a set of pairwise non-adjacent neighbors of* a *such that* $v_1, v_n \notin F'$, *and for each* $v_i \in F'$, *consider the kite* $\{v_{i-1}, v_i, v_{i+1}, a\}$ *in* F. *Choose* $r > 0$ *as in Lemma 8, and, for each* $v_i \in F'$, *choose* $\{\theta_i, \tau_i\}$ *so that the safe regions* $S(r, \theta_i, \tau_i)$ *are mutually invisible with respect to* F. *Finally, for each* $v_i \in F'$, *let* X_i *be a finite subset of* $S(r, \theta_i, \tau_i)$. *Then a minimum weight triangulation of* $P = \{a, v_1, \ldots, v_n\} \cup \{X_i : v_i \in F'\}$ *can be obtained as the union of a minimum weight drawing of* F, *any minimum weight triangulations of the sets* $X_i \cup v_i$, *for each* $v_i \in F'$, *and all edges from each* v_{i-1} *and* v_{i+1} *to* $X_i \cup v_i$ *which cross no edge of the triangulation of* $X_i \cup v_i$.

Theorem 12. *Any forest can be realized as the skeleton of some minimum weight triangulation.*

4.2 k-nested Triangulations

Lemma 13. *Let* T *be a triangulation with triangular outer face, and let* T' *be the triangulation obtained by deleting the vertices on the outer face of* T. *If* T' *has a triangular outer face and is minimum weight drawable, so is* T.

An application of Lemma 13 is the following.

Theorem 14. *Every* k-nested triangulation is minimum weight drawable, and its minimum weight drawing can be computed in linear time in the real RAM model of computation.

4.3 k-spined Triangulations

Lemma 15. *Let* T *be a triangulation with triangular outer face* f_0, *and let* T' *be obtained by adding a vertex* v *in the outer face of* T *and connecting* v *with all vertices of* f_0. *If* T *is minimum weight drawable, so is* T'.

An application of Lemma 15 is the following.

Theorem 16. *Every k-spined triangulation is minimum weight drawable, and its minimum weight drawing can be computed in linear time in the real RAM model of computation.*

5 Open Problems

Several problems remain open towards characterizing which graphs admit a minimum weight drawing; among the most relevant are: Does every triangulation whose skeleton is a forest admit a minimum weight drawing? Are there any Delaunay drawable but minimum weight forbidden triangulations? Are there other necessary conditions of the type expressed by Lemma 2.

References

1. O. Aichholzer, F. Aurenhammer, S.-W. Chen, N. Katoh, M. Taschwer, G. Rote, and Y.-F. Xu. Triangulations intersect nicely. *Discrete Comput. Geom.*, 16:339–359, 1996.
2. P. Bose, G. Di Battista, W. Lenhart, and G. Liotta. Proximity constraints and representable trees. In R. Tamassia and I. G. Tollis, editors, *Graph Drawing (Proc. GD '94)*, volume 894 of *Lecture Notes Comput. Sci.*, pages 340–351. Springer-Verlag, 1995.
3. P. Bose, W. Lenhart, and G. Liotta. Characterizing proximity trees. *Algorithmica*, 16:83–110, 1996. (special issue on Graph Drawing, edited by G. Di Battista and R. Tamassia).
4. S.-W. Cheng and Y.-F. Xu. Approaching the largest β-skeleton within a minimum weight triangulation. In *Proc. 12th Annu. ACM Sympos. Comput. Geom.*, pages 196–203, 1996.
5. G. Di Battista, W. Lenhart, and G. Liotta. Proximity drawability: a survey. In R. Tamassia and I. G. Tollis, editors, *Graph Drawing (Proc. GD '94)*, volume 894 of *Lecture Notes Comput. Sci.*, pages 328–339. Springer-Verlag, 1995.
6. G. Di Battista, G. Liotta, and S. H. Whitesides. The strength of weak proximity. In F. J. Brandenburg, editor, *Graph Drawing (Proc. GD '95)*, volume 1027 of *Lecture Notes Comput. Sci.*, pages 178–189. Springer-Verlag, 1996.
7. G. Di Battista and L. Vismara. Angles of planar triangular graphs. *SIAM J. Discrete Math.*, 9(3):349–359, 1996.
8. M. T. Dickerson, S. A. McElfresh, and M. H. Montague. New algorithms and empirical findings on minimum weight triangulation heuristics. In *Proc. 11th Annu. ACM Sympos. Comput. Geom.*, pages 238–247, 1995.
9. M. T. Dickerson and M. H. Montague. A (usually?) connected subgraph of the minimum weight triangulation. In *Proc. 12th Annu. ACM Sympos. Comput. Geom.*, pages 204–213, 1996.
10. M. B. Dillencourt. Realizability of Delaunay triangulations. *Inform. Process. Lett.*, 33:283–287, 1990.
11. M. B. Dillencourt. Toughness and Delaunay triangulations. *Discrete Comput. Geom.*, 5:575–601, 1990.

12. M. B. Dillencourt and W. D. Smith. Graph-theoretical conditions for inscribability and Delaunay realizability. In *Proc. 6th Canad. Conf. Comput. Geom.*, pages 287–292, 1994.

13. P. Eades and S. Whitesides. The realization problem for Euclidean minimum spanning trees is NP-hard. *Algorithmica*, 16:60–82, 1996. (special issue on Graph Drawing, edited by G. Di Battista and R. Tamassia).

14. M. R. Garey and D. S. Johnson. *Computers and Intractability: A Guide to the Theory of NP-Completeness*. W. H. Freeman, New York, NY, 1979.

15. M. Keil. Computing a subgraph of the minimum weight triangulation. *Comput. Geom. Theory Appl.*, 4:13–26, 1994.

16. D. G. Kirkpatrick. A note on Delaunay and optimal triangulations. *Inform. Process. Lett.*, 10:127–128, 1980.

17. W. Lenhart and G. Liotta. Drawing outerplanar minimum weight triangulations. *Inform. Process. Lett.*, 6(12):253–260, 1996.

18. W. Lenhart and G. Liotta. How to draw outerplanar minimum weight triangulations. In F. J. Brandenburg, editor, *Graph Drawing (Proc. GD '95)*, volume 1027 of *Lecture Notes Comput. Sci.*, pages 373–384. Springer-Verlag, 1996.

19. W. Lenhart and G. Liotta. Proximity drawings of outerplanar graphs. In S. North, editor, *Graph Drawing (Proc. GD '96)*, volume 1190 of *Lecture Notes Comput. Sci.*, pages 286–302. Springer-Verlag, 1997.

20. C. Levcopoulos and D. Krznaric. Quasi-greedy triangulations approximating the minimum weight triangulation. In *Proc. 7th ACM-SIAM Sympos. Discrete Algorithms*, pages 392–401, 1996.

21. C. Levcopoulos and D. Krznaric. Tight lower bounds for minimum weight triangulation heuristics. *Information Processing Letters*, (57):129–135, 1996.

22. A. Lingas. A new heuristic for minimum weight triangulation. *SIAM J. Algebraic Discrete Methods*, 8(4):646–658, 1987.

23. G. Liotta and G. Di Battista. Computing proximity drawings of trees in the 3-dimensional space. In *Proc. 4th Workshop Algorithms Data Struct.*, volume 955 of *Lecture Notes Comput. Sci.*, pages 239–250. Springer-Verlag, 1995.

24. G. Liotta, R. Tamassia, I. G. Tollis, and P. Vocca. Area requirement of Gabriel drawings. In *Algorithms and Complexity (Proc. CIAC' 97)*, volume 955 of *Lecture Notes Comput. Sci.*, pages 239–250. Springer-Verlag, 1995.

25. A. Lubiw and N. Sleumer. Maximal outerplanar graphs are relative neighborhood graphs. In *Proc. 5th Canad. Conf. Comput. Geom.*, pages 198–203, 1993.

26. H. Meijer and D. Rappaport. Computing the minimum weight triangulation of a set of linearly ordered points. *Information Processing Letters*, (42):35–38, 1992.

27. C. Monma and S. Suri. Transitions in geometric minimum spanning trees. *Discrete Comput. Geom.*, 8:265–293, 1992.

A Polyhedral Approach to the Multi-Layer Crossing Minimization Problem

(Extended Abstract)

Michael Jünger[1]*, Eva K. Lee[2]**, Petra Mutzel[3]*, and Thomas Odenthal[4]

[1] Institut für Informatik, Universität zu Köln, mjuenger@informatik.uni-koeln.de
[2] Ind. & Sys. Eng., Georgia Institute of Technology, evakylee@isye.gatech.edu
[3] Max-Planck-Institut für Informatik, Saarbrücken, mutzel@mpi-sb.mpg.de
[4] Ind. Eng. & Op. Res., Columbia University, odenthal@ieor.columbia.edu

Abstract. We study the multi-layer crossing minimization problem from a polyhedral point of view. After the introduction of an integer programming formulation of the multi-layer crossing minimization problem, we examine the 2-layer case and derive several classes of facets of the associated polytope. Preliminary computational results for 2- and 3-layer instances indicate, that the usage of the corresponding facet-defining inequalities in a branch-and-cut approach may only lead to a practically useful algorithm, if deeper polyhedral studies are conducted.

1 Introduction

The *crossing number* of a graph G is the minimum number of edge-crossings needed in any embedding of G in the plane. The *Crossing Number* (CN) problem, that is, to determine the crossing number of a given graph G, is well-known to be $\mathcal{NP}$-hard (Garey and Johnson, 1983). Thus, there is little hope to find a polynomial-time algorithm for (CN) on general graphs. Moreover, formulae for the crossing number are only known for restricted types of graphs. We refer the reader to Richter and Thomassen (1994) and Shahrokhi et al. (1995) for surveys.

A (proper) *k-layered* graph is a graph whose node-set is partitioned into k layers with the property that edges join only nodes in consecutive layers. Such graphs arise for example in workflow diagrams and are usually drawn such that all nodes in the same layer are placed on a horizontal line.

A natural combination of the two concepts crossing number and layered graphs is the *multi-layer crossing minimization problem*, which is to minimize the number of edge-crossings of a multi-layered graph. Harary (1969), Harary and Schwenk (1972), Watkins (1970) and Warfield (1977) give first structural results.

* Partially supported by DFG-Grant Ju204/7-1, Forschungsschwerpunkt "Effiziente Algorithmen für diskrete Probleme und ihre Anwendungen" and by ESPRIT LTR Project No. 20244 – ALCOM-IT.
** Supported in part by NSF grants CCR-9501584 and DMS-9527124.

From an algorithmic point of view, there have been extensive studies on crossing minimization for 2-layered graphs. In particular, during the 80's a lot of effort was focused on the design of heuristics for drawing such graphs with few crossings. Among others, we mention the barycenter heuristic (Sugiyama et al., 1981), the median heuristic (Eades and Wormald, 1994) and the greedy heuristics (Eades and Kelly, 1986). As reported in Jünger and Mutzel (1996), the barycenter heuristic yields the best results in terms of number of crossings and solution time.

Other approaches involve formulating the problems as integer programs. In the case of the crossing number problem, the geometric relationships between nodes and edges in the graph are difficult to be represented by an integer program. However, if the graph under consideration is "well-structured", this approach may be applicable and polyhedral theory and branch-and-cut algorithms may be able to provide an exact solution. This approach was first employed in Jünger and Mutzel (1996) for the one-sided *two-layer crossing minimization* problem. This problem is to find the minimum number of edge-crossings in a 2-layered graph, when one node-permutation of the two layers is fixed. Jünger and Mutzel (1996) showed that this problem can be transformed into a linear ordering problem which can then be solved by a branch-and-cut algorithm.

In this paper, we first address the (two-sided) two-layer crossing minimization problem. We first formulate this problem as an integer program, in which the optimal solution corresponds to a crossing-minimal representation of the graph. Next, we extend the integer programming formulation to a more general case – the proper *multi-layer crossing minimization* problem – which is to determine the minimum number of crossings in a (proper) multi-layered graph.

A branch-and-bound approach (Valls et al., 1996a, Jünger and Mutzel, 1996) and a tabu search (Valls et al., 1996b) were employed to determine the minimum crossing numbers in 2-layered graphs. In this paper, we will focus on polyhedral combinatorics to take advantage of the inherent structural properties of the problem. In Section 3, we define the polytope associated with the 2-layer crossing number problem as $\mathcal{P}_{CROSS}(G)$. Several classes of facet-defining inequalities for $\mathcal{P}_{CROSS}(G)$ will be reported. Most of the inequalities are derived from a combinatorial characterization of 2-layer planar graphs, i.e., graphs that can be drawn on two layers without edge-crossings. In Section 4, separation procedures for each class of facet-defining inequalities are described. Based on these facets, we implemented a branch-and-cut algorithm and compare its performance with a pure branch-and-bound algorithm.

2 The Integer Programming Formulations

Let $G = (V_1, V_2, E)$ be a bipartite graph with $|V_1| = n_1$ and $|V_2| = n_2$. The *Two-Layer Crossing Minimization* (TLCM) problem consists of determining the minimum number of crossings among the two layers of G such that nodes can be permuted in both layers and edges are drawn as straight lines. Any solution is uniquely determined by the permutations π_1 and π_2 of V_1 and V_2. We define

$x_{ij} = 1$ if $\pi_1(i) < \pi_1(j)$, 0 otherwise; and $y_{ij} = 1$ if $\pi_2(i) < \pi_2(j)$, 0 otherwise. For a given π_1 and π_2, the number of crossings can be expressed as

$$C(\pi_1, \pi_2) = \sum_{i=1}^{n_1-1} \sum_{k=i+1}^{n_1} \sum_{j \in N(i)} \sum_{l \in N(k)} x_{ik} \cdot y_{lj} + x_{ki} \cdot y_{jl}, \qquad (1)$$

where $N(v) = \{w \in V \mid e = (v, w) \in E\}$ denotes the set of neighbors of $v \in V = V_1 \cup V_2$ in G.

For a fixed permutation π_1 of V_1, this problem can be transformed to a linear ordering problem and solved with a branch-and-cut algorithm in short computation time (Jünger and Mutzel, 1996). The case of two freely permutable layers was handled by a branch-and-bound algorithm in which trivial lower bounds were employed for partial permutations of the smaller layer and the branch-and-cut algorithm for one-sided crossing minimization was applied for complete permutations of the smaller layer. In this paper, we describe an approach for solving (TLCM) directly.

We first present a nonlinear integer program for finding the minimum number of crossings for a 2-layered graph.

$$
\begin{aligned}
\min \quad & C(\pi_1, \pi_2) \\
\text{s.t.} \quad & x_{ij} + x_{jk} + x_{ki} \leq 2 & & 1 \leq i < j < k \leq n_1 \ (2) \\
& y_{ij} + y_{jk} + y_{ki} \leq 2 & & 1 \leq i < j < k \leq n_2 \ (3) \\
& x_{ij} + x_{ji} = 1 & & 1 \leq i < j \leq n_1 \quad\ (4) \\
& y_{ij} + y_{ji} = 1 & & 1 \leq i < j \leq n_2 \quad\ (5) \\
& x_{ij}, y_{ij} \in \{0, 1\}.
\end{aligned}
\qquad \text{(NTLCM)}
$$

The 3-cycle-constraints, (2) and (3), constraints (4) and (5) and the binary restriction on the variables are needed to ensure that the resulting vectors x and y are incidence vectors of feasible linear orderings of the nodes on both layers (Grötschel et al., 1985).

To formulate the problem as an integer linear program, we introduce the *crossing variables*, c_{ijkl}, to denote if edges (i, j) and (k, l) cross (where $i < j$, $k < l$, $i < k$, and $j \neq l$). This leads to the following integer linear programming formulation for (TLCM).

For notational convenience, we assume without loss of generality that nodes in the first layer have smaller indices than nodes in the second layer and that $i, k \in V_1$ and $j, l \in V_2$. Moreover, since two edges incident at a node cannot result in a crossing, we will employ the notation "for all $(i, j), (k, l) \in E$" to describe the situation in which $i < j$, $k < l$, $i < k$, and $j \neq l$.

$$
\begin{aligned}
\min \quad & \sum_{(i,j),(k,l) \in E} c_{ijkl} \\
\text{s.t.} \quad & x_{ik} + y_{lj} - c_{ijkl} \leq 1 & & (i,j),(k,l) \in E \ (6) \\
& x_{ki} + y_{jl} - c_{ijkl} \leq 1 & & (i,j),(k,l) \in E \ (7) \\
& (2),\ (3),\ (4)\ \text{and}\ (5) \\
& x_{ij}, y_{ij}, c_{ijkl} \in \{0, 1\}.
\end{aligned}
\qquad \text{(ITLCMa)}
$$

Theorem 1. *(NTLCM) and (ITLCMa) are equivalent formulations for (TLCM).*

Proof. Let (x, y, c) be a feasible solution to (ITLCMa). Clearly, (x, y) is also feasible for (NTLCM). From (6) and (7), if $x_{ik} = y_{lj} = 1$ or $x_{ki} = y_{jl} = 1$, then $c_{ijkl} = 1$. On the other hand, if either $x_{ik} = 0$ or $y_{lj} = 0$, then c_{ijkl} can be either 0 or 1. However, for the sake of minimization, we must have $c_{ijkl} = 1$ if and only if $x_{ik} = y_{lj} = 1$ or $x_{ki} = y_{jl} = 1$. In other words, $c_{ijkl} = 1$ if and only if $x_{ik} \cdot y_{lj} = 1$ or $x_{ki} \cdot y_{jl} = 1$.

To prove the other direction, let (x, y) be a feasible solution to (NTLCM). Define $c_{ijkl} = x_{ik} \cdot y_{lj} + x_{ki} \cdot y_{jl}$. Clearly, (x, y, c) is feasible for (ITLCMa). $\square$

Formulation (ITLCMa) can be reduced via variable substitutions from equations (4) & (5) to the following formulation (ITLCMb).

$$\min \sum_{(i,j),(k,l) \in E} c_{ijkl}$$

s.t.

$$
\begin{aligned}
-c_{ijkl} \le y_{jl} - x_{ik} \le c_{ijkl} \quad & (i,j),(k,l) \in E, j < l \ (8) \\
1 - c_{ijkl} \le y_{lj} + x_{ik} \le 1 + c_{ijkl} \quad & (i,j),(k,l) \in E, j > l \ (9) \\
0 \le x_{ij} + x_{jk} - x_{ik} \le 1 \quad & 1 \le i < j < k \le n_1 \quad (10) \\
0 \le y_{ij} + y_{jk} - y_{ik} \le 1 \quad & 1 \le i < j < k \le n_2 \quad (11) \\
x_{ij}, y_{ij}, c_{ijkl} \in \{0, 1\}. &
\end{aligned}
$$

A generalization of this approach to the (proper) *Multi-Layer Crossing Minimization* (MLCM) problem is straightforward. A (proper) *multi-layered* graph, $G = (V, E)$, is a graph in which the node-set V is partitioned into disjoints sets: $V = V_1 \cup V_2 \cup \ldots \cup V_p$ with $|V_i| = n_i$ such that for all edges $(u, v) \in E$, we must have $u \in V_i$, $v \in V_{i+1}$, for some $i = 1, \ldots, p - 1$. In other words, edges are only allowed between successive layers. Indeed, we can write $E = E_1 \cup E_2 \cup \ldots \cup E_{p-1}$ with E_i the edge-set between nodes from layer i and $i+1, i = 1, \ldots, p-1$. Using this property, we only need to couple successive layers with crossing variables. The corresponding integer program for (MLCM) can be written as:

$$\min \sum_{r=1}^{p-1} \sum_{(i,j),(k,l) \in E_r} c_{ijkl}^r$$

s.t.

$$
\begin{aligned}
-c_{ijkl}^r \le x_{jl}^{r+1} - x_{ik}^r \le c_{ijkl}^r \quad & (i,j),(k,l) \in E_r, j < l \ (12) \\
1 - c_{ijkl}^r \le x_{lj}^{r+1} + x_{ik}^r \le 1 + c_{ijkl}^r \quad & (i,j),(k,l) \in E_r, j > l \ (13) \\
0 \le x_{ij}^r + x_{jk}^r - x_{ik}^r \le 1 \quad & 1 \le i < j < k \le n_r \quad (14) \\
x_{ij}^r, y_{ij}^r, c_{ijkl}^r \in \{0, 1\}. &
\end{aligned}
$$

where $r = 1, \ldots, p - 1$.

In a general multi-layered graph the restriction on the edges is relaxed in such a way that edges are allowed to span more than two layers. A solution for (MLCM) can be adapted to a general multi-layered graph by introducing a dummy-node on a layer whenever an edge crosses this layer (see Eades and Wormald (1994) for the transformation). Applying this technique, our integer programming formulation can be employed within popular layout-algorithms (Sugiyama et al., 1981).

3 Polyhedral Study

Given a bipartite graph $G = (V_1, V_2, E)$ with $|V_1| = n_1$ and $|V_2| = n_2$, we define a *crossing configuration* to be a vector (x, y, c), where $x \in \mathbb{B}^{\binom{n_1}{2}}$ and $y \in \mathbb{B}^{\binom{n_2}{2}}$ are the linear ordering variables for the first and second layer, respectively, and $c \in \mathbb{B}^t$ is the vector of crossing variables, with t the total number of crossing variables for the graph. Let $s := \binom{n_1}{2} + \binom{n_2}{2} + t$. We define the polytope $\mathcal{P}_{CROSS}(G) \subseteq \mathbb{R}^s$ as

$$\mathcal{P}_{CROSS}(G) = conv\{(x, y, c) \in \mathbb{B}^s : (x, y, c) \text{ is a crossing configuration}\}.$$

In this section, we first show that $\mathcal{P}_{CROSS}(G)$ is full dimensional. We then present the various classes of facet-defining inequalities we obtained thus far. We denote a vector of all 1's by the vector e of appropriate dimension.

Theorem 2. *$\mathcal{P}_{CROSS}(G)$ is full dimensional.*

Sketch of Proof. It suffices to show that there exist $s + 1$ affinely independent crossing configurations. Let $\mathcal{P}_{LO1}$ ($\mathcal{P}_{LO2}$) be the linear ordering polytope associated with the x- (y-) variables. Since $\mathcal{P}_{LO1}$ is of dimension $\binom{n_1}{2}$, there exist $\binom{n_1}{2}$ affinely independent vectors, $x^1, \ldots, x^{\binom{n_1}{2}}$. Similarly, there exist $\binom{n_2}{2}$ affinely independent vectors for $\mathcal{P}_{LO2}$, $y^1, \ldots, y^{\binom{n_2}{2}}$. Thus the following $s + 1$ vectors are in $\mathcal{P}_{CROSS}(G)$ and are affinely independent:

  i. $(x^i, 0, e)$, $i = 1, \ldots, \binom{n_1}{2}$;

  ii. $(0, y^i, e)$, $i = 1, \ldots, \binom{n_2}{2}$;

  iii. $(\bar{x}^i, \bar{y}^i, e - e_i)$, $i = 1, \ldots, t$; where $\bar{x}^i, \bar{y}^i$ are chosen so that the vectors satisfy all the constraints for (ITLCMb) after setting exactly one crossing variable c_i to 0;

  iv. (e, e, e). □

Theorem 3. *The inequalities $c_{ijkl} \geq 0$ are redundant, while $c_{ijkl} \leq 1$ are facet-defining for $\mathcal{P}_{CROSS}(G)$.*

Proof. If we add the two parts of inequality (8) together, we clearly get $c_{ijkl} \geq 0$ for all edges $(i, j), (k, l) \in E, j < l$. Similarly, if we add the two parts of inequality (9) together, we get $c_{ijkl} \geq 0$ for all edges $(i, j), (k, l) \in E, l < j$. Hence $c_{ijkl} \geq 0$ is redundant.

Clearly, $c_{ijkl} \leq 1$ is valid for $\mathcal{P}_{CROSS}(G)$. To show the dimension, note that for each $ijkl$, all but one vector from Theorem 2 satisfy $c_{ijkl} = 1$, proving that it is facet-defining. □

In the next theorem we show that all facet-defining inequalities for $\mathcal{P}_{LO1}$ ($\mathcal{P}_{LO2}$) are facet-defining for $\mathcal{P}_{CROSS}(G)$.

Theorem 4. *Let $a^T x \leq a_0$ be a facet-defining inequality for $\mathcal{P}_{LO1}$ ($\mathcal{P}_{LO2}$). Then $a^T x \leq a_0$ is also facet-defining for $\mathcal{P}_{CROSS}(G)$.*

Sketch of Proof. Validity is trivial. Without loss of generality, we restrict ourselves to facet-defining inequalities corresponding to the x-variables.

Since $a^T x \leq a_0$ is a facet-defining inequality for $\mathcal{P}_{LO1}$, there exist $\binom{n_1}{2}$ affinely independent vectors satisfying $a^T x \leq a_0$ with equality, namely, $x^1, \ldots, x^{\binom{n_1}{2}}$. Now, order the y variables y_{ij} such that $i < j$. Let D_k be the set of the first k indices in vector y, and y^{D_k} be the corresponding characteristic vector. Then the following s vectors are affinely independent:

i. $(x^i, 0, e)$, $i = 1, \ldots, \binom{n_1}{2}$;
ii. (x^1, y^{D_k}, e), $k = 1, \ldots, \binom{n_2}{2}$;
iii. $(x^1, \bar{y}^i, e - e_i)$, $i = 1, \ldots, t$, where $\bar{y}^i$ is chosen such that all constraints in (ITLCMb) are satisfied. $\square$

Similar proof techniques are employed in all the theorems regarding facet-defining properties: proving validity for $\mathcal{P}_{CROSS}(G)$, and showing that the maximum number of affinely independent vectors satisfying the inequality at equality is s. For space reasons, we omit proofs for the remaining theorems.

Theorem 5. *Inequalities (8) and (9) are facet-defining for $\mathcal{P}_{CROSS}(G)$.*

Other classes of facet-defining inequalities can be derived from the following characterization of "2-layer planar" graphs (these are graphs that can be drawn on two layers without crossings) using forbidden subgraphs.

Theorem 6. [Harary and Schwenk, 1972, Tomii et al., 1977, Eades et al., 1986] *A 2-layer graph is 2-layer planar if and only if it contains no cycle and no 3-claw.*

In other words, whenever there is a cycle or a 3-claw (see Figure 1 below) in the graph, we have at least one crossing. Moreover, the exact 2-layer crossing number for cycle graphs is known.

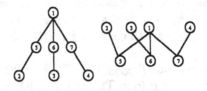

Fig. 1. 3-claw graph

Theorem 7. [**Harary and Schwenk, 1972, May and Szkatula, 1988**] *The 2-layer crossing number of a cycle C is $\frac{|C|-2}{2}$.*

In the next theorem, we state the valid inequalities arising from cycles as well as p-claws, i.e., claw graphs such as in Figure 1 with p legs.

Theorem 8. *The following inequalities are valid for $\mathcal{P}_{CROSS}(G)$.*

(i.) $\sum_{(i,j),(k,l)\in C} c_{ijkl} \geq \frac{|C|-2}{2}$ *for all cycles C in G.*

(ii.) $\sum_{(i,j),(k,l)\in W} c_{ijkl} \geq 1$ *for all 3-claws W in G.*

(iii.) $\sum_{(i,j),(k,l)\in W_p} c_{ijkl} \geq \begin{cases} \frac{p}{2}(\frac{p}{2}-1) & p \text{ even} \\ (\frac{p-1}{2})^2 & p \text{ odd} \end{cases}$ *for all p-claws W_p in G.*

Though these inequalities are not facet-defining, we can derive several classes of facets from them, which we will describe next. For notational convenience all integers representing nodes are to be considered modulo $|C|$.

Theorem 9. *Let $C = (0,\ldots,i-1,i,i+1,i+2,\ldots,|C-1|,0)$ be a cycle of length $|C|$ in the graph. For every edge $(i,i+1) \in C$, the following inequality is facet-defining for $\mathcal{P}_{CROSS}(G)$:*

$$\sum_{(k,l)\in C, k\neq i, l\neq i+1} c_{i,i+1,k,l} + c_{i,i-1,i+1,i+2} \geq 1.$$

Next, we consider any cycle $C = (0,\ldots,|C-1|,0)$ of length $|C|$ in the graph, where we fix an arbitrary ordering.

Theorem 10. *Let S be the set consisting of all pairs of edges $(i,j), (k,l) \in C$ for which edge (k,l) has an odd distance from edge (i,j) in the cycle, i.e., $k = i+2m$ or $i = k+2m$ for $m = 1, 2,\ldots,|C|/2$, respectively. Then the following inequality is facet-defining for $\mathcal{P}_{CROSS}(G)$:*

$$\sum_{(i,j),(k,l)\in S} c_{ijkl} \geq \frac{|C|-2}{2}.$$

We now turn to facet-defining inequalities based on 3-claws W.

Theorem 11. *Let T be the set consisting of all pairs of edges of W except those pairs of edges that are either both within the lower or the upper part of the 3-claw and are not adjacent to each other. Then the following inequality is facet-defining for $\mathcal{P}_{CROSS}(G)$:*

$$\sum_{(i,j),(k,l)\in T} c_{ijkl} \geq 1.$$

Another class of inequalities can be constructed from the *dome-path*. Each of these structures gives rise to two facet-defining inequalities.

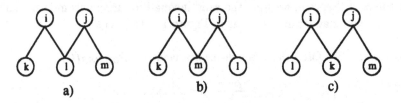

Fig. 2. Dome paths

Theorem 12. *For $k < l < m$, the following inequalities are facet-defining for $P_{CROSS}(G)$:*

$$x_{kl} - 2x_{km} + x_{lm} - c_{ikjl} - c_{iljm} \leq 0;$$
$$-x_{kl} + 2x_{km} - x_{lm} - c_{ikjl} - c_{iljm} \leq 0;$$
$$\qquad\qquad\text{Figure (2 a)}$$

$$2x_{kl} - x_{km} + x_{lm} - c_{ikjm} - c_{imjl} \leq 1;$$
$$-2x_{kl} + x_{km} - x_{lm} - c_{ikjm} - c_{imjl} \leq -1;$$
$$\qquad\qquad\text{Figure (2 b)}$$

$$x_{kl} - x_{km} + 2x_{lm} - c_{iljk} - c_{ikjm} \leq 1;$$
$$-x_{kl} + x_{km} - 2x_{lm} - c_{iljk} - c_{ikjm} \leq -1.$$
$$\qquad\qquad\text{Figure (2 c)}$$

Notice that in each of these inequalities, the linear-ordering variable associated with the two endnodes of the dome-path always has coefficient 2.

4 Separation Procedures and Computational Experience

In this section, we briefly highlight the separation procedures and their complexity for the classes of facets described in Section 3. We also report our computational experience with a branch-and-cut algorithm based on these facets.

Classes of inequalities from Theorem 9 can be separated in polynomial time. For every fixed edge (i, j), we set up an auxiliary graph $G' = (V', E')$ with $V' = V$ and $E' = E$ and edge-weights $w_{kl} = c_{ijkl}$ for every edge (k, l) not adjacent to i and j. Next we solve a series of shortest path problems, searching for the shortest path from node a to node b, where a and b are nodes adjacent to i and j, respectively. We perform this for all combinations of adjacent edges of node i and j. If the length of the obtained path plus the cost of the fixed edge is greater than 1, we have found a violated inequality. If we cannot find such a path, then no inequality of this class is violated. The running time of this separation procedure is $O(|E| \cdot S(|V|, |E|) \cdot 2d)$, where $S(|V|, |E|)$ denotes the time for solving a shortest path problem and d is the maximum node-degree in the graph.

In addition, inequalities derived from dome paths can be checked by enumeration, which can be done in polynomial time. Similarly, k-claws can be identified in polynomial time via enumeration, or by computing a matching between the nodes incident to a root-node and their adjacent edges (for a fixed k).

We perform a heuristic procedure for separating classes of facets from Theorem 10. We first choose a root node, and apply a breadth-first-search method to identify cycles in the graph. Once a cycle is found, the corresponding inequalities are formed and checked if they are violated by the current fractional solution.

Table 1 shows computational results for a set of 2-layered graphs which have appeared in the literature: [JM96a] and [JM96b] are the graphs of Figures 1 and 4 in Jünger and Mutzel (1996), respectively. [Mut97a] and [Mut97b] are the graphs of Figures 1 and 2 in Mutzel (1997). [KW95a] and [KW95b] are the two bipartite graphs of a 3-partite graph which arises from an illustration of dependencies in constraint negotiation (Kusiak and Wang, 1995). [Him97a] and [Him97b] are the two bipartite graphs of a 3-partite graph visualizing features of a graph-drawing-system. [MS88a] and [MS88b] are again the two bipartite parts of a 3-partite graph arising from computing invariants of matrices (May and Szkatula, 1988). [SM96a] and [SM96b] are two bipartite parts of a multi-layered social networks lattice (Shieh and McCreary, 1996). [Fuk96a] and [Fuk96b] are the two bipartite parts of a 3-partite graph which came up in the study of polyhedral face lattices (Fukuda, 1996). *Name, $|V_1|, |V_2|, |E|$, No. Var* denote the name of the problem, the number of nodes in each layer, the number of edges in the graph and the number of variables in the formulation, respectively. *LB* and *Opt. IP* denote the lower bound (the LP-value after adding cuts from Section 3 to the LP-relaxation and no subsequent branching) and the optimal objective value for the integer program, respectively.

For each instance in the table, the original lower bound before addition of cuts is 0. Comparing the values *LB* and *Opt. IP*, we observe that cutting planes helped to close the gap (calculated as (Opt. IP - LB) / Opt. IP) by 50% to 100%. In particular, out of 14 instances, the gap for 6 of them was completely closed after the addition of cuts.

In the rest of the table, computational results for 2-layered graphs for algorithm B-C, a branch-and-cut algorithm incorporating the facets from Section 3, and algorithm B-B, a pure branch-and-bound algorithm, are reported. We used a SUN ULTRA2 (167MHZ) workstation and CPLEX 4.0 as the branch-and-bound solver with strong branching selected as the branching variable selection (We tested various options in CPLEX and strong branching is far superior to other branching rules).

B-C cuts, B-C time, and *B-C nodes* denote the total number of cuts added, the total time required (with time limit set to 3600 seconds), and the total number of enumeration nodes needed by algorithm B-C. Finally, *B-B time* and *B-B nodes* denote respectively the time required to solve to optimality and the number of nodes searched by algorithm B-B for solving our formulation of (TLCM) when no cutting planes are included.

The analysis of the computational results has a mixed outcome. For small

Table 1. Computational results for 2-layered graphs

Name	$|V_1|$	$|V_2|$	$|E|$	No. Var.	LB	Opt. IP	B-C cuts	B-C time	B-C nodes	B-B time	B-B nodes
[JM96a]	8	8	19	193	10	19	45	5.18	9	2.68	14
[JM96b]	10	10	20	249	4	4	20	0.58	1	0.96	2
[Mut97a]	6	9	14	118	2	2	1	0.31	1	0.15	1
[Mut97b]	14	11	28	484	14	24	200	246.09	173	372.11	395
[KW95a]	8	7	15	127	9	10	21	0.42	3	0.41	6
[KW95b]	7	3	14	71	15	15	20	0.30	1	0.07	1
[Him97a]	8	16	26	409	10	19	24	8.87	7	8.17	8
[Him97b]	16	3	8	142	0	0	2	0.11	1	0.007	1
[MS88a]	10	15	39	791	35	35	70	10.83	3	31.77	17
[MS88b]	15	12	41	893	52	54	189	15.28	3	309.97	251
[SM96a]	26	18	29	484	0	0	0	0.21	1	0.012	1
[SM96b]	18	17	29	663	8	12	27	56.09	23	210.33	579
[Fuk96a]	8	12	24	334	18	35	134	3600 *	2372	71.03	153
[Fuk96b]	12	6	24	309	23	45	175	3600 +	1955	597.47	1636

*Not optimal with lower bound of 31 +Not optimal with lower bound of 39

problems (e.g., [JM96a]) algorithm B-B – although using more enumeration nodes than algorithm B-C – has a faster running time than algorithm B-C. Moreover, we mention that the branch-and-bound algorithm of Jünger and Mutzel (1996) was able to solve instances of (TLCM) to proven optimality when the smaller layer has 15 or fewer nodes in a fast computation time. Since cutting plane procedures can be rather expensive, on smaller problems, it is faster to simply apply a plain branch-and-bound approach.

For problems with a small gap (e.g., [MS88a] and [MS88b]), however, algorithm B-C performs best, both in terms of running time and number of enumeration nodes. On the other hand, we were not able to solve problems [Fuk96a] and [Fuk96b] to optimality with algorithm B-C within the time-limit. For those problems the gap is about 50% and this indicates that a deeper study of the facial structure of $P_{CROSS}(G)$ is needed in order to come up with a practical useful algorithm.

However, we caution that the success of a branch-and-cut approach depends on many different factors, to name a few, how often and how many cuts should be generated, how to integrate these cuts effectively within the tree search, and in what order should the cuts be generated. We anticipate further studies to determine the viability of this approach to our application.

We next focus on the (MLCM) problem and report our experience with algorithms B-C and B-B on some 3-partite graphs. These graphs are [KW95], [Him97], [MS88], and [SM96], which are the combined 3-partite graphs of the graphs described above. The contents of Table 2 is similar to that of Table 1.

All cutting planes arising from the two bipartite parts are valid cutting planes for the combined 3-partite graph. The analysis shows that for smaller problems ([KW95] and [Him97]) algorithm B-B has the best performance, whereas for the bigger problems ([MS88] and [SM96]), algorithm B-C is the winner. Both algorithms were not able to solve the combined 3-partite graph of [Fuk96a] and [Fuk96b] to optimality within the time-limit.

Table 2. Computational results for 3-layered graphs

Name	$|V_1|$	$|V_2|$	$|V_3|$	$|E_1|$	$|E_2|$	No. Var.	LB	Opt. IP	B-C cuts	B-C time	B-C nodes	B-B time	B-B nodes
[KW95]	8	7	3	15	14	198	24	27	153	2.57	15	0.76	9
[Him97]	8	16	3	26	8	551	10	21	82	21.61	15	8.55	7
[MS88]	10	15	12	39	41	1684	87	91	469	268.54	19	3553.85	908
[SM96]	26	18	17	29	29	1347	8	13	35	331.63	21	530.33	1579

References

Di Battista, G., Eades, P., Tamassia, R., Tollis, I.G.: Algorithms for drawing graphs: An annotated bibliography. Computational Geometry: Theory and Applications **4** (1994) 235–282.

Eades, P., Kelly, D.: Heuristics for Reducing Crossings in 2-Layered Networks. Ars Combinatoria **21-A** (1986) 89–98.

Eades, P., Wormald, N.C.: Edge Crossings in Drawings of Bipartite Graphs. Algorithmica **11** (1994) 379–403.

Eades, P., McKay, B.D., Wormald, N.C.: On an edge crossing problem. Proc. 9th Australian Computer Science Conference, Australian National University (1986) 327–334.

Fukuda, K.: Face Lattices. Personal Communication (1996).

Garey, M.R., Johnson, D.S.: Crossing Number is NP-Complete. SIAM Journal on Algebraic and Discrete Methods **4** (1983) 312–316.

Grötschel, M., Jünger, M., Reinelt, G.: Facets of the linear ordering polytope. Mathematical Programming **33** (1985) 43–60.

Harary, F.: Determinants, permanents and bipartite graphs. Mathematical Magazine **42** (1969) 146–148.

Harary, F., Schwenk, A.: A new crossing number for bipartite graphs. Utilitas Mathematica **1** (1972) 203–209.

Himsolt, M.: Personal Communication (1997).

Jünger, M., Mutzel, P.: 2-Layer Straightline Crossing Minimization: Performance of Exact and Heuristic Algorithms. Journal of Graph Algorithms and Applications (JGAA), (http://www.cs.brown.edu/publications/jgaa/), No. 1, Vol. 1, (1997) 1–25.

Kusiak, A., Wang, J.: Dependency Analysis in Constraint Negotiation. IEEE Trans. Sys. Man, Cybern. **25** (1995) 1301–1313.

May, M., Szkatula, K.: On the bipartite crossing number. Control and Cybernetics **17** No.1 (1988) 85–97.

Mutzel, P.: An Alternative Method for Crossing Minimization. Lecture Notes in Computer Science LNCS 1190 (1997) 318–333.

Richter, B.R., Thomassen, C.: A survey on crossing numbers. Manuscript, Carleton University and The Technical University of Denmark (1994).

Shahrokhi, F., Szelky, L.A., Vrtô, I.: Crossing Number of Graphs, Lower Bound Techniques and Algorithms: A Survey. Lecture Notes in Computer Science LNCS 894 (1995) 131–142.

Shieh, F., McCreary, C.,: Directed Graphs Drawing by Clan-Based Decomposition. Lecture Notes in Computer Science LNCS 1027 (1996) 472–482.

Sugiyama, K., Tagawa, S., Toda, M.: Methods for Visual Understanding of Hierarchical System Structures. IEEE Trans. Syst. Man, Cybern. SMC-11 (1981) 109–125.

Tomii, N., Kambayashi, Y., Shunzo, Y.: On Planarization Algorithms of 2-Level Graphs. Papers of tech. group on electronic computers, IECEJ, EC77-38 (1977) 1–12.

Valls, V., Marti, R., Lino, P.: A Branch and Bound Algorithm for Minimizing the Number of Crossing Arcs in Bipartite Graphs. Journal of Operational Research **90** (1996a) 303–319.

Valls, V., Marti, R., Lino, P.: A tabu thresholding algorithm for arc crossing minimization in bipartite graphs. Annals of Operations Research **63** (1996b) 223–251.

Warfield, J.N.: Crossing Theory and Hierarchy Mapping. IEEE Trans. Syst. Man, Cybern. SMC-7 (1977) 505–523.

Watkins, M.E.: A special crossing number for bipartite graphs: a research problem. Annals of New York Academy of Sciences **175** (1970) 405–410.

On Embedding an Outer-Planar Graph in a Point Set

Prosenjit Bose[1,2]

[1] School of Computer Science, Carleton University, 1125 Colonel By Drive, Ottawa, Ontario, Canada, K1S 5B6. E-mail: jit@scs.carleton.ca.
[2] Research supported by NSERC Grant OGP0183877 and a FIR Grant.

Abstract. Given an n-vertex outer-planar graph G and a set P of n points in the plane, we present an $O(n \log^3 n)$ time and $O(n)$ space algorithm to compute a straight-line embedding of G in P, improving upon the algorithm in [GMPP91, CU96] that requires $O(n^2)$ time. Our algorithm is near-optimal as there is an $\Omega(n \log n)$ lower bound for the problem [BMS95]. We present a simpler $O(nd)$ time and $O(n)$ space algorithm to compute a straight-line embedding of G in P where $\log n \leq d \leq 2n$ is the length of the longest vertex disjoint path in the dual of G. Therefore, the time complexity of the simpler algorithm varies between $O(n \log n)$ and $O(n^2)$ depending on the value of d. More efficient algorithms are presented for certain restricted cases. If the dual of G is a path, then an optimal $\Theta(n \log n)$ time algorithm is presented. If the given point set is in convex position then we show that $O(n)$ time suffices.

1 Introduction

The problem of deciding whether a certain combinatorial structure can be embedded in the plane, as well as computing an embedding of that structure has been a recurrent theme in many fields but particularly in graph drawing. From a graph drawing perspective (see [DETT94] for a survey of graph drawing), the traditional questions ask whether a graph can be embedded in the plane such that some criterion is satisfied e.g., that the area of the resulting embedding is small [CP95, KLTT93], that the symmetry present in the graph is revealed in the embedding [MA88], or that the graph is isomorphic to a proximity graph [EW94, MS91, BDLL95, BLL96] of the points in which the vertices are embedded.

The embedding problem that we address has a slightly different perspective: both the point set and the graph are given as input. We want to determine if the input graph can be straight-line embedded in the input point set. We say that an n-node graph $G = (V, E)$ can be *straight-line embedded* onto a set of n points P, if there exists a one-to-one mapping $\phi: V \to P$ from the nodes of G to the points of P such that edges of G intersect only at nodes. That is, edges $(\phi(u_1), \phi(v_1)) \cap (\phi(u_2), \phi(v_2)) = \emptyset$, for all $u_1 v_1 \neq u_2 v_2 \in E$.

The definition of a straight-line embedding implies that G must be a planar graph in order for a straight-line embedding of G onto P to exist. However, even if G is planar, there exist point sets that do not admit a straight-line embedding of G. See Figure 1 for such an example. This raises an interesting open question: Given a planar graph G and a point set P, can G be straight-line embedded in P? We believe that this problem is NP-complete. Although the question when G is a planar graph remains unanswered, progress has been made when G is restricted to a subclass of planar graphs.

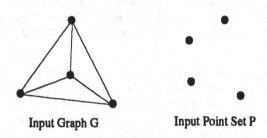

Input Graph G   Input Point Set P

Fig. 1. A graph that cannot be straight-line embedded in a point set.

When the class of graphs is restricted to trees, Perles at the 1990 DIMACS workshop on arrangements posed the following question: Given n points P in general position and an n-node tree T rooted at node ν, can T be straight-line embedded in P with ν at a specified point $p \in P$? Perles showed that this was always possible if p was on the *convex hull of P*, which is the smallest convex set containing the points P. Pach and Törőcsik [PT93] showed that it could if p was not the *deepest point* of P, obtained by repeatedly discarding points on the convex hull. Subsequently, Ikebe et. al [IPTT94] showed that there was always such an embedding using a quadratic time algorithm. In fact, all three algorithms use quadratic time. Finally, Bose et. al [BMS95] proved an $\Omega(n \log n)$ lower bound for the problem and provided a matching $O(n \log n)$ time embedding algorithm.

With the embedding problem being resolved when the input graphs are restricted to trees and unresolved when the input graphs are planar, a natural question to ask is what is the largest subclass of planar graphs that admits a straight-line embedding on any point set. Gritzmann et al. [GMPP91] first showed that the class of outer-planar graphs is the largest class of graphs that admits an embedding in any point set and provided an embedding algorithm that runs in $O(n^2)$ time (Castañeda and Urrutia [CU96] later rediscovered this theorem).

In this paper, we present an $O(n \log^3 n)$ time and $O(n)$ space algorithm to compute a straight-line embedding of an n-vertex outer-planar graph G in a set P of n points in the plane. Since a tree is an outer-planar graph, the $\Omega(n \log n)$ lower bound for trees [BMS95] also holds in this case, thereby implying that our algorithm is optimal to within a polylogarithmic factor. We present a simpler $O(nd)$ time and $O(n)$ space algorithm to compute a straight-line embedding of G in P where $\log n \leq d \leq 2n$ is the length of the longest vertex disjoint path in the dual of G. Therefore, the time complexity of the simpler algorithm varies between $O(n \log n)$ and $O(n^2)$. More efficient algorithms are presented for certain restricted cases. If the dual of G is a path, then an optimal $\Theta(n \log n)$ time algorithm is presented. If the given point set is in convex position then we show that $O(n)$ time suffices.

2 Notation and Preliminaries

We begin by defining some of the graph theoretic and geometric terminology used in this paper. For more details see [BM76] and [PS85].

A *graph* $G = (V, E)$ consists of a finite non empty set $V(G)$ of *vertices*, and a set $E(G)$ of unordered pairs of vertices known as *edges*. An edge $e \in E(G)$ consisting of vertices u and v is denoted by $e = uv$; u and v are called the *endpoints* of e and are said to be *adjacent* vertices or *neighbors*.

A *drawing* of a graph $G = (V, E)$ is a function which maps the vertices of G to points in the plane and edges of G to curves in the plane such that for each edge $e = uv$, the endpoints of the curve corresponding to e are the points in the plane corresponding to u and v. A drawing of G is called a *planar* drawing if no curve intersects itself or any other curve, except possibly at its endpoints. A graph is said to be planar if it admits a planar drawing. A *straight-line* drawing of a graph G is a drawing in which each edge corresponds to the line segment between its endpoints. All planar graphs admit straight-line planar drawings [Far48].

An outer-planar graph is a planar graph where every vertex is on the external face. A maximal outer-planar graph is an outer-planar graph that is no longer outer-planar with the addition of a single edge. Each internal face of a maximal outer-planar graph is a triangle. Note that an algorithm that can embed a maximal outer-planar graph can embed any outer-planar graph G simply by adding extra edges to G making it maximal, embedding the maximal graph and then removing the extra edges. Therefore, in the remainder of the paper, all outer-planar graphs are considered maximal.

Let G be a maximal outer-planar graph. Let $Ext(G)$ represent the external face of G. We adopt the convention that the vertices of a maximal outer-planar graph G are labelled $\{v_0, v_1, \ldots, v_{n-1}\}$ as they appear on $Ext(G)$ (i.e. v_i is adjacent to $v_{i+1}, i = 0, \ldots, n-1$ addition taken modulo n). An edge $e \in Ext(G)$ is an *external* edge of G.

The dual G^* of a maximal outer-planar graph $G = (V, E)$ is defined as follows. Each triangle or face (excluding the outer face) of G is a vertex of G^*. Two vertices of G^* are adjacent if the corresponding faces in G have an edge in common. Since G is maximal outer-planar, G^* is a tree with maximum vertex degree 3. See Figure 2 for an illustration.

All planar point sets are assumed to be in general position, i.e. no three points are collinear. Let P be a set of n points in the plane. Given $a, b \in P$, the open and closed line segments defined by a and b are denoted by (a, b) and $[a, b]$,

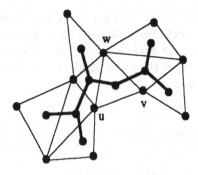

Fig. 2. A maximal outer-planar graph and its dual.

respectively. Given three points $a, b, c \in P$, by $\angle(a, b, c)$ we mean the clockwise angle between $[b, a]$ and $[b, c]$ (see Figure 3).

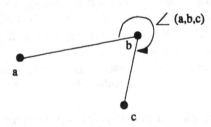

Fig. 3. Illustration of the angle (a, b, c).

3 Embedding Algorithm Outline

We begin by outlining a few of the ideas presented in [GMPP91] and [CU96]. A key concept in their embedding algorithms is the (r, s)-triangle. There are two types of (r, s)-triangles (defined below), one defined on a maximal outer-planar graph and the other defined on a point set.

In the discussion to follow, G is an n vertex maximal outer-planar graph and P is a set of n points in the plane.

Definition 1. Let u, v, w be three mutually adjacent vertices of G. Triangle $\triangle(u, v, w)$ is an (r, s)-triangle of G provided that uv is an external edge of G and the two components of $G \setminus \{u, v, w\}$ have r and s vertices, respectively, such that $r + s = n - 3$.

In Figure 2, $\Delta(u, v, w)$ is a $(5, 3)$-triangle of the graph.

Definition 2. Let r and s be two non-negative integers with $r + s = n - 3$. Let a and b be two consecutive vertices on the convex hull of P and $c \in P$. Triangle $\Delta(a, b, c)$ is an (r, s)-triangle of P provided the following holds:

1. No point of P lies in $\Delta(a, b, c)$.
2. There is a line l_c through c that intersects the interior of $\Delta(a, b, c)$ such that there are r points of $P \setminus \{a, b, c\}$ on one side of l_c and s points of $P \setminus \{a, b, c\}$ on the other side of l_c. These sets are denoted as P_r and P_s, respectively.

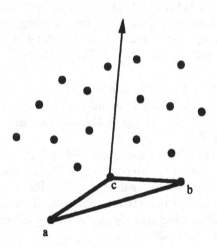

Fig. 4. An $(8, 7)$-triangle.

In Figure 4, $\Delta(a, b, c)$ is an $(8, 7)$-triangle of the point set.

The main idea behind the embedding algorithm is to find an (r, s)-triangle in G and map it to an (r, s)-triangle in P. The existence of an (r, s)-triangle in G follows from the fact that the dual of G is a binary tree. The proof of the existence of an (r, s)-triangle in P forms the basis of an embedding algorithm. A proof of the following lemma appears in [GMPP91] and [CU96]. We provide a similar but alternate proof in section 4.

Lemma 3. *[GMPP91] For any $r, s \geq 0$ such that $r + s = n - 3$ and any two consecutive vertices a, b on the convex hull of P, there always exists a point $c \in P$ such that $\Delta(a, b, c)$ is an (r, s)-triangle of P.*

Lemma 4. *[GMPP91] Let G be an n-node maximal outer-planar graph and P be a set of n points in the plane. Let a and b be two consecutive vertices on the convex hull of P. Let $e = v_i v_{i+1}$ be an external edge of G. There exists a straight-line embedding of G on P with the added constraint that v_i maps to a and v_{i+1} maps to b.*

Proof. We proceed by induction on the number of vertices of G. The result holds trivially if G has three vertices.

Inductive Hypothesis $(k < n, n > 3)$: Let G be a k-node maximal outer-planar graph and P be a set of k points in the plane. Let a and b be two consecutive vertices on the convex hull of P. Let $e = v_i v_{i+1}$ be an external edge of G. There exists a straight-line embedding of G on P with the added constraint that v_i maps to a and v_{i+1} maps to b.

Inductive Step $(k = n)$: Let G be an n-node maximal outer-planar graph and P be a set of n points in the plane. Let a and b be two consecutive vertices on the convex hull of P. Let $e = v_i v_{i+1}$ be an external edge of G.

Since G is maximal outer-planar, there is a unique vertex v_k adjacent to both v_i and v_{i+1}. The node in the dual G^* representing the triangle $\triangle(v_i, v_{i+1}, v_k)$ has degree at most two since edge $v_i v_{i+1}$ is an external edge. This implies that the removal of $\triangle(v_i, v_{i+1}, v_k)$ decomposes G into two components with cardinalities r and s respectively with $r + s = n - 3$. Therefore, $\triangle(v_i, v_{i+1}, v_k)$ is an (r, s)-triangle of G.

By lemma 3, there is a triangle $\triangle(a, b, c)$ that is an (r, s)-triangle of P. Let l_c be the line through c as defined in definition 2. Map v_i to a, v_{i+1} to b and v_k to c. By construction, the edge $[ac]$ is on the convex hull of P_r and the edge $[bc]$ is on the convex hull of P_s.

Let H_1 and H_2 be the subgraphs of G induced by $\{v_k, \ldots, v_i\}$ and $\{v_{i+1}, \ldots, v_k\}$. Both H_1 and H_2 are maximal outer-planar, and edge $v_i v_k$ is an external edge of H_1 and edge $v_{i+1} v_k$ is an external edge of H_2. Since both H_1 and H_2 have less than n vertices, by the inductive hypothesis, H_1 can be embedded in P_r with edge $v_i v_k$ mapping to $[ac]$ and H_2 can be embedded in P_s with edge $v_{i+1} v_k$ mapping to $[bc]$. The result follows.

As is often the case with inductive proofs, the proof of Lemma 4 directly implies an algorithm to embed an outer-planar graph on a point set. The main steps of the algorithm are outlined in Figure 5. The maximal outer-planar graph to be embedded is G. All index manipulation is done modulo n. The time taken by this algorithm depends on the time taken to perform steps 1-6. In essence, the time can be expressed recursively as $T(n) = T(n - k) + T(k) + \psi(n)$, with $1 \leq k \leq n - 1$ and where $\psi(n)$ represents the time taken to perform steps 1-6.

The adjacency information of the graph G can be stored in a standard data structure such as the doubly-connected edge list (DCEL) [PS85]. However, in the algorithm, there is no need to modify the adjacency information, but merely record the indicies of the vertices in the input graph in the recursive calls. All adjacency queries, such as those made in step 1 of the algorithm are made on the DCEL of G. Since each edge in G is adjacent to two triangles, the vertex v_k in step 1 can be found in constant time by identifying the unique vertex whose index k falls in the range delimited by I_s and I_e.

Step 2 can also be computed in constant time since the cardinalities of the two sets can be computed from indicies of the three vertices forming the triangle. Step 5 is a constant time operation. Finally, step 6 is also a constant time operation, given the indicies of the three vertices forming the triangle.

$Embed(I_s, I_e, v_i, v_{i+1}, P, a, b)$

I_s and I_e are the start and end indicies of the vertices on the external face of the graph.
The edge $v_i v_{i+1}$ is an external edge of the outer-planar graph.
P is a point set with points a, b on its convex hull.

1. Find the unique vertex v_k in G (where k lies in the interval defined by I_s and I_e) adjacent to v_i and v_{i+1}.
2. Since $\triangle(v_i, v_{i+1}, v_k)$ is an (r, s)-triangle of G, compute the cardinalities r and s.
3. Find $c \in P$, such that triangle $\triangle(a, b, c)$ that is an (r, s)-triangle of P.
4. Compute P_r and P_s, the sets on either side of the line l_c, respectively.
5. Map v_i to a, v_{i+1} to b, and v_k to c.
6. Let H_1 and H_2 be the subgraphs of G induced by $\{v_k, \ldots, v_i\}$ and $\{v_{i+1}, \ldots, v_k\}$. The start and end indices for H_1 are k and i, respectively and for H_2 are $i+1$ and k, respectively.
7. If the number of vertices in $H_1 \geq 3$ then $Embed(k, i, v_k, v_i, P_r, a, c)$.
8. If the number of vertices in $H_2 \geq 3$ then $Embed(i+1, k, v_{i+1}, v_k, P_s, b, c)$.

Fig. 5. Outline of algorithm to embed G in P.

Therefore, the main difficulty comes from steps 3 and 4: computing an (r, s)-triangle in a point set. The complexity of the whole algorithm depends on these two steps since the other four steps are constant time operations. In the next section, we present a method for computing an (r, s)-triangle in a point set in $O(n)$ time with no preprocessing which will form the basis of our embedding algorithms.

4 Simple Embedding Algorithm

In this section, we present a simple method for finding an (r, s)-triangle in a point set and show how it is used in the simple embedding algorithm.

Let P be an n point set with a and b two adjacent vertices on the convex hull of P. In the discussion to follow, for any line l through a and not b, the open half-plane containing b shall be referred to as the right half-plane of l; similarly, for any line l through b and not a, the open half-plane containing a shall be referred to as the left half-plane of l.

Lemma 5. *For any $r, s \geq 0$ such that $r + s = n - 3$ and any two consecutive vertices a, b on the convex hull of P, there always exists a point $c \in P$ such that $\triangle(a, b, c)$ is an (r, s)-triangle of P.*

Proof. Let l_1 be a line through a with $s + 1$ points of P (excluding b) in the right half-plane of l_1. Let $L(a)$ represent these $s + 1$ points. Let l_2 be a line through b and a point c from $L(a)$ such that the left half-plane of l_2 contains no points of $L(a)$ (refer to Figure 6). Let l_3 be the line through a and c. Triangle $\triangle(a, b, c)$ is an (r, s)-triangle of P. Since there are at least s points of P in the right half-plane of l_2 and at most s points in the right half-plane of l_3 there must be a line l_c through c intersecting the interior of $\triangle(a, b, c)$ with r points to one side and s points to the other (excluding a, b, c).

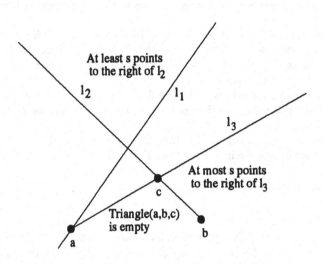

Fig. 6. Computing an (r, s)-triangle.

From the proof of lemma 5, the steps involved in finding point $c \in P$ and line l_c such that $\triangle(a, b, c)$ is an (r, s)-triangle of P are summerized below:

1. Find the line l_1 through a.
2. Compute the set $L(a)$.
3. Find the line l_2 through b, identifying point c.
4. Find the line l_c.

Recall that selecting the ith smallest element in an unsorted list of n elements can be achieved in $O(n)$ time (see [CLR90]). Let $x \in P$ be the point such that $\angle(x, a, b)$ is the $(s + 2)^{nd}$ smallest. The point x can be found in $O(n)$ time using linear selection. The line l_1 through a and x has $s + 1$ elements (excluding b) in the right half-plane. The set $L(a)$ can be constructed in $O(n)$ time once l_1 is found. Given $L(a)$, notice that l_2 is simply the through b and the point $y \in L(a)$ such that $\angle(a, b, y)$ is the smallest over all points in $L(a)$. Therefore, l_2 can be computed in $O(n)$ time. Finally, l_c can be found in $O(n)$ time by computing the point $z \in P$ such that $\angle(z, c, b)$ is the $(s + 1)^{st}$ among all points in P. Therefore, given a set of n points, an (r, s)-triangle can be computed in $O(n)$ time. This immediately gives an $O(n^2)$ time and $O(n)$ space algorithm for embedding an outer-planar graph in a point set since the recurrence for the algorithm *Embed(...)* becomes $T(n) = T(n - k) + T(k) + O(n)$ which solves to $O(n^2)$.

Upon further consideration of the recurrence, we notice that the complexity of the algorithm is actually dependant on the length of the longest path in the dual of G. The algorithm is initiated with an initial invocation of $Embed(0, n - 1, v_i, v_{i+1}, P, a, b)$. At each invocation, the algorithm embeds an (r, s)-triangle and makes at most two recursive calls with smaller problem instances. The calling relation forms a binary tree, which we refer to as the *recursion tree* for graph G, denoted as RT_G. An internal node of this tree has at least one child, and is an instance of $Embed(\ldots)$ where an (r, s)-triangle is embedded with at least one of r or s being non-zero initiating at least one recursive call. A leaf of the recursion tree is an instance of $Embed(\ldots)$ where the size of the graph to be embedded is 3. The root of the tree represents the initial call and the depth of a node in the tree represents its level of recursion. Let d be the length of longest path in the dual tree G^*. The depth of the recursion tree RT_G cannot exceed d since every root to leaf path in RT_G represents a path in G^*.

Lemma 6. *The depth of the recursion tree RT_G does not exceed d, where d is the length of the longest path in the dual tree G^*.*

Since at each level, the graph G is partitioned, the sum of the sizes of all the problems at a particular level of RT_G is $O(n)$. The amount of time spent in one invocation of $Embed(\ldots)$ excluding recursive calls is linear in the size of the graph. All of the steps of the algorithm (refer to Figure 5) are constant time except for the two steps involving the computation of an (r, s)-triangle, which we showed is linear in the size of the problem. Therefore, the amount of time spent by the algorithm is $O(n)$ per level of RT_G.

Theorem 7. *Given an n-vertex outer-planar graph G and a set P of n points in the plane, G can be straight-line embedded in P in $O(nd)$ time and $O(n)$ space where d is the length of the longest path in the dual of G.*

5   Near-optimal Embedding Algorithm

Our more efficient algorithm for embedding outer-planar graphs uses segments from the convex hull to avoid intersections between embedded edges. Consequently, we need efficient access to the convex hull of points. Moreover, we need the ability to insert and delete points from the convex hull as we embed (r, s)-triangles. Overmars and van Leeuwen's [OvL81] dynamic convex hull structure permits arbitrary insertion into and deletion from a set of points while maintaining the convex hull of the point set. Each update (insertion or deletion) costs at most $O(\log^2 n)$ time over a sequence of $O(n)$ updates.

If the points of P are placed in a dynamic convex hull maintenance structure that supports insertions and deletions in $O(\log^2 n)$ time then we can find an (r, s)-triangle without resorting to a linear time selection. We review the method for computing an (r, s)-triangle given the maintenance structure. CM will refer to the convex hull maintenance structure. We can insert and delete points from

CM in $O(\log^2 n)$ time. Given a point $x \in CM$, we can recover the point adjacent to x on the current convex hull in $O(\log n)$ time.

Without loss of generality, assume that $s \leq r$. Let $x \in P$ be the point such that $\angle(x, a, b)$ is the $(s + 2)^{nd}$ smallest. The point x can be found in $O(s \log^2 n)$ time by deleting $s + 2$ times the convex hull point adjacent to a starting with b. Store the deleted points in order of deletion into $L(a)$. The line l_1 through a and x has $s + 1$ elements (excluding b) in the right half-plane. Given $L(a)$, notice that l_2 is simply the line through b and the point $y \in L(a)$ such that $\angle(a, b, y)$ is the smallest over all points in $L(a)$. Therefore, l_2 can be computed in $O(s)$ time since $L(a)$ has $s + 2$ points. Finally, to compute l_c re-insert all the points of $L(a)$ into CM. Delete convex hull points adjacent to a starting with b until c is adjacent to a. If there are s points (excluding b) to the right of the line through a and c then l_c is this line. Otherwise, to find l_c continue deleting the convex hull points adjacent to c (different from a) until $s + 1$ points have been deleted in total from CM. Store all the deleted points in $L(a)$. Notice that CM is now a convex hull maintenance structure for P_r. There are $s + 1$ points in P_s. In $O(s \log^2 s)$, a CM structure can be built for P_s. Therefore, the revised complexity of the algorithm is $T(n) = T(n - k) + T(k) + O(\min(k, n - k) \log^2 n)$ where $1 \leq k \leq n - 1$. This recurrence solves to $T(n) = O(n \log^3 n)$. Building the initial CM for P cost $O(n \log^2 n)$, therefore, we have the following theorem.

Theorem 8. *Given an n-vertex outer-planar graph G and a set P of n points in the plane, G can be straight-line embedded in P in $O(n \log^3 n)$ time and $O(n)$ space.*

Our algorithm is optimal to within a polylogarithmic factor since an $\Omega(n \log n)$ lower bound for the problem was shown in [BMS95].

6 Restricted Case

If the dual of G is a tree, then notice that G can be embedded simply by computing $(r, 1)$-triangles. This immediately implies that our near-optimal algorithm will run in time $T(n) = T(n - 1) + O(\log^2 n)$ which solves to $T(n) = O(n \log^2 n)$. However, when computing $(r, 1)$-triangles, we do not need to re-insert points into the convex hull maintenance structure in order to compute l_c. Since, we do not need to insert points into the convex hull but simply delete them; we opt for a deletion-only convex hull maintenance structure [Cha85, HS92], which provides better amortized time complexities for point deletions than Overmars and van Leeuwen's method.

In [HS92], the point deletion operation removes a point from the convex hull maintenance structure in $O(\log n)$ amortized time (amortized over the sequence of n deletions). Consequently, by using a deletion-only convex hull structure, the running time of the algorithm is summerized by the recurrence $T(n) = T(n - 1) + O(\log n)$ which resolves to $T(n)$ is $O(n \log n)$. This is optimal since the lower bound proved in [BMS95] still holds in this restricted case.

Theorem 9. *If the dual of the input graph G is a path, G can be embedded into a point set P in optimal $\Theta(n \log n)$ time.*

If the input point set P is in convex position, then $O(n)$ time and space is sufficient. We assume that the input point set is given in an array A, ordered in clockwise fashion as the points appear on the convex hull of P. Given an (r, s)-triangle of G, finding an (r, s)-triangle in P can be achieved in $O(1)$ time by simply finding the index into array A which splits the array into two sub-arrays of size r and s respectively. Therefore, the recurrence for algorithm *Embed(...)* is $T(n) = T(n - k) + T(k) + O(1)$ which implies that $T(n)$ is $O(n)$.

Theorem 10. *If the input point set P is in convex position then $O(n)$ time and space suffice to straight-line embed G into P.*

7 Conclusions

We presented an $O(n \log^3 n)$ time and $O(n)$ space algorithm to compute a straight-line embedding of an n-vertex outer-planar graph G in a set P of n points in the plane. Since a tree is an outer-planar graph, the $\Omega(n \log n)$ lower bound for trees [BMS95] also holds in this case, thereby implying that our algorithm is optimal to within a polylogarithmic factor. We presented a simpler $O(nd)$ time and $O(n)$ space algorithm to compute a straight-line embedding of G in P where $\log n \leq d \leq 2n$ is the length of the longest vertex disjoint path in the dual of G. Finally, we showed that if the dual of G is a path, then $\Theta(n \log n)$ time and $O(n)$ space are sufficient and if the input point set is in convex position then $O(n)$ time and space suffice.

We conclude with two open problems:

1. Can a $\log^2 n$ factor be shaved off our embedding algorithm, i.e. is there an optimal $O(n \log n)$ time algorithm to embed an outer-planar graph on a point set?
2. Given a planar graph G and a point set P, what is the complexity of deciding if G can be embedded into P?

Acknowledgements: The author wishes to thank Kilani Ghoudi for helpful discussions on this topic, and Janos Pach for pointing out some missing references.

References

[BDLL95] P. Bose, G. Di Battista, W. Lenhart, and G. Liotta. Proximity constraints and representable trees. In R. Tamassia and I. G. Tollis, editors, *Graph Drawing (Proc. GD '94)*, volume 894 of *Lecture Notes in Computer Science*, pages 340–351. Springer-Verlag, 1995.

[BLL96] P. Bose, W. Lenhart, and G. Liotta. Characterizing proximity trees. *Algorithmica*, 16:83–110, 1996.

[BM76]   J.A. Bondy and U.S.R. Murty. *Graph Theory with Applications*. Elsevier Science, New York, NY, 1976.

[BMS95]   P. Bose, M. McAllister, and J. Snoeyink. Optimal algorithms to embed trees in a point set. *Journal of Graph Algorithms and Applications*, to appear. Also appears in Proceedings of Graph Drawing GD'95, LNCS 1027, pp. 64-75, 1995.

[Cha85]   B. Chazelle. On the convex layers of a planar set. *IEEE Trans. on Inf. Theory*, IT-31:509-517, 1985.

[CLR90]   T. Cormen, C. Leiserson, and R. Rivest. *Introduction to algorithms*. MIT Press, Cambridge, Mass., 1990.

[CP95]   P. Crescenzi and A. Piperno. Optimal-area upward drawings of AVL trees. In R. Tamassia and I. G. Tollis, editors, *Graph Drawing (Proc. GD '94)*, volume 894 of *Lecture Notes in Computer Science*, pages 307-317. Springer-Verlag, 1995.

[CU96]   N. Castaneda and J. Urrutia. Straight line embeddings of planar graphs on point sets. In *Proc. Eighth Canadian Conf. on Comp. Geom.*, pages 312-318, 1996.

[DETT94]   G. Di Battista, P. Eades, R. Tamassia, and I. G. Tollis. Algorithms for drawing graphs: an annotated bibliography. *Comput. Geom. Theory Appl.*, 4:235-282, 1994.

[EW94]   P. Eades and S. Whitesides. The realization problem for Euclidean minimum spanning trees is NP-hard. In *Proc. 10th Annu. ACM Sympos. Comput. Geom.*, pages 49-56, 1994.

[Far48]   I. Fary. On straight line representation of planar graphs. *Acta Sci. Math. Szeged*, 11:229-233, 1948.

[GMPP91]   P. Gritzmann, B. Mohar, J. Pach, and R. Pollack. Embedding a planar triangulation with vertices at specified points (solution to problem e3341. *American Mathematical Monthly*, 98:165-166, 1991.

[HS92]   John Hershberger and Subhash Suri. Applications of a semi-dynamic convex hull algorithm. *BIT*, 32:249-267, 1992.

[IPTT94]   Y. Ikebe, M. Perles, A. Tamura, and S. Tokunaga. The rooted tree embedding problem into points in the plane. *Discrete & Computational Geometry*, 11:51-63, 1994.

[KLTT93]   G. Kant, G. Liotta, R. Tamassia, and I. Tollis. Area requirement of visibility representations of trees. In *Proc. 5th Canad. Conf. Comput. Geom.*, pages 192-197, Waterloo, Canada, 1993.

[MA88]   J. Manning and M. J. Atallah. Fast detection and display of symmetry in trees. *Congressus Numerantium*, 64:159-169, 1988.

[MS91]   C. Monma and S. Suri. Transitions in geometric minimum spanning trees. In *Proc. 7th Annu. ACM Sympos. Comput. Geom.*, pages 239-249, 1991.

[OvL81]   M. Overmars and J. van Leeuwen. Maintenance of configurations in the plane. *Journal of Computer and System Sciences*, 23:166-204, 1981.

[PS85]   F. P. Preparata and M. I. Shamos. *Computational Geometry: an Introduction*. Springer-Verlag, New York, NY, 1985.

[PT93]   János Pach and Jenő Törőcsik. Layout of rooted trees. In W. T. Trotter, editor, *Planar Graphs*, volume 9 of *DIMACS Series*, pages 131-137. American Mathematical Society, 1993.

Bipartite Crossing Numbers of Meshes and Hypercubes

Farhad Shahrokhi[1,*], Ondrej Sýkora[2,**], László A. Székely[3,***], Imrich Vrťo[2]

[1] Department of Computer Science, University of North Texas
P.O.Box 13886, Denton, TX 76203-3886, USA
[2] Institute for Informatics, Slovak Academy of Sciences
P.O.Box 56, 840 00 Bratislava, Slovak Republic
[3] Department of Mathematics, University of South Carolina
Columbia, SC 29208, USA

Abstract. Let $G = (V_0, V_1, E)$ be a connected bipartite graph, where V_0, V_1 is the bipartition of the vertex set $V(G)$ into independent sets. A *bipartite drawing* of G consists of placing the vertices of V_0 and V_1 into distinct points on two parallel lines x_0, x_1, respectively, and then drawing each edge with one straight line segment which connects the points of x_0 and x_1 where the endvertices of the edge were placed. The *bipartite crossing number* of G, denoted by $bcr(G)$ is the minimum number of crossings of edges over all bipartite drawings of G. We develop a new lower bound method for estimating $bcr(G)$. It relates bipartite crossing numbers to edge isoperimetric inequalities and Laplacian eigenvalues of graphs. We apply the method, which is suitable for "well structured" graphs, to hypercubes and 2-dimensional meshes. E.g. for the $n-$dimensional hypercube graph we get $n4^{n-2} - O(4^n) \leq bcr(Q_n) \leq n4^{n-1}$. We also consider a more general setting of the method which uses eigenvalues, but as a trade-off for generality, often gives weaker results.

1 Introduction

Let $G = (V_0, V_1, E)$ be a connected bipartite graph, where V_0, V_1 is the bipartition of vertices into independent sets. A *bipartite drawing* of G consists of placing the vertices of V_0 and V_1 into distinct points on two parallel lines x_0, x_1, respectively, and then drawing each edge with one straight line segment which connects the points of x_0 and x_1 where the endvertices of the edge were placed. The *bipartite crossing number* of G, denoted by $bcr(G)$ is the minimum number

---

[*] The research of the first author was supported in part by the NSF grant CCR 9528228.

[**] The research of the 2nd and the 4th author was partially supported by the Alexander von Humboldt Foundation, by the Slovak Scientific Grant Agency grant No. 95/5305/277 and by grant of EU INCO-COP 96-0195.

[***] The research of the third author was supported in part by the NSF grant DMS 9701211 and the Hungarian NSF grants T 016 358 and T 019 367.

of crossings of edges over all bipartite drawings of G. The bipartite crossing number is one of the parameters which strongly influences the aesthetics of drawings of hierarchical graphs.

The problem of finding the bipartite crossing number is a variant of the standard crossing number problem, see e.g. [21], and was first introduced by Harary [13] and Harary and Schwenk [14]. In [14], they proved $bcr(G) = 0$ iff G is a caterpillar. Further, they obtained the exact values of the bipartite crossing numbers of subdivisions of complete and complete bipartite graphs. For even cycles they showed $bcr(C_{2n}) = n - 1$. The bipartite crossing number problem was also proposed by Watkins [26] independently. Some basic observations on $bcr(G)$ were made by May and Szkatula [19]. The bipartite crossing number problem is known to be NP-complete [12] but can be solved in polynomial time for bipartite permutation graphs [22], and trees [20]. A great deal of research has been devoted to the design of algorithms and heuristics for solving this problem [7, 11, 16, 23, 25]. Brandenburg, Jünger und Mutzel [6] have called for some entirely new approaches because the usual heuristics do not give good results for the bipartite crossing number. The latest progress in this area was made in [20] in which we have shown an intimate relationship between the bipartite crossing number problem and the optimal linear arrangement problem. These result have led to the first provably good approximation algorithms for computing $bcr(G)$. The restricted problem when the positions of the vertices of V_0 are given is also NP-complete [11] and frequently appears in drawing of hierarchical graphs [7, 25], see also the survey [10]. Eades and Wormald [11] designed a polynomial time algorithm which approximates the bipartite crossing number within a multiplicative factor of 3 in this restricted problem.

In this paper we develop a new lower bound argument which is suitable for "well structured graphs". The argument relates the bipartite crossing number of a graph to the edge isoperimetric inequality of the graph. We apply our new lower bound technique for two instances of important graphs: hypercubes and two-dimensional meshes. We obtain here substantially better lower bounds on the bipartite crossing number of theses graphs than the bounds that one can obtain by using the general results in [20]. For the meshes we get $\frac{3}{4}m^2n - \frac{1}{4}m^3 - \frac{3}{2}mn - \frac{3}{2}m^2 - 3m \leq bcr(M(m,n)) \leq \frac{3}{2}m^2n$, and for the n-dimensional hypercube graph we get $n4^{n-2} - O(4^n) \leq bcr(Q_n) \leq n4^{n-1}$. It is worth mentioning that especially for the $3 \times n$ mesh $M(3,n)$ we get an exact result $bcr(M(3,n)) = 5n - 6$.

We close the paper by describing a general framework to give lower bounds for bipartite crossing numbers using Laplacian eigenvalues and we develop our ideas on the two examples.

2   Notations and Definitions

Let δ and Δ denote minimum and maximum degrees of G, respectively. The degree of vertex v will be denoted by $d(v)$. For $X \subseteq V(G)$, let $vol(X)$ denote $\sum_{v \in X} d(v)$. For a bipartite drawing $D(G)$ of a graph G, let $bcr(D(G))$ denote the number of the crossings in $D(G)$ (i.e. the number of unordered pairs of crossing edges), and define the bipartite crossing number of G by $bcr(G) = \min_{D(G)} bcr(D(G))$.

Given an arbitrary graph $G = (V, E)$, and a bijection $F : V \to \{1, 2, 3, ..., |V|\}$, define the *length* of F, as $\sum_{uv \in E} |F(u) - F(v)|$. The *optimal linear arrangement problem* is to find a bijection F, of minimum length. This minimum value is denoted by $L(G)$, the optimal linear arrangement value of G.

For $X \subseteq V$ define

$$\partial(X) = \{uv \in E : u \in X, v \in V - X\}.$$

We call $\partial(X)$ the edge boundary of X. The general objective is to find a good approximating and easily described real function $f(x)$ such that $|\partial(X)| \geq f(|X|)$, for all $X \subseteq V(G)$. Such an inequality is called an *edge isoperimetric inequality* [2, 3]; edge isoperimetric inequalities have several applications in graph theory and computer science.

For $X, Y \subset V$ define $cut(X, Y)$ to be a smallest size set of edges in G which separates X from Y in G. Note that $|cut(X, Y)| \leq |\partial(X)|$.

For $m \leq n$, let $M(m, n)$ denote the 2-dimensional mesh i.e. the graph defined by the Cartesian product of an $m-$vertex path with an $n-$vertex path. Let Q_n denote the $n-$dimensional hypercube graph, i.e. the Cartesian product of n 2-vertex paths.

3 Meshes

In [20] we have shown that

$$\frac{\delta}{96} L(G) \leq bcr(G) \leq \Delta L(G),\tag{1}$$

for $\delta \geq 2$ (the lower bound holds for $|E(G)| \geq 1.05|V(G)|$ and in [20] the constant in the lower bound actually was 1/60). By composing our formula (1) with a result from [18]:

$$L(M(m, n)) = \left(m^2 n - \frac{\sqrt{2} - 1}{3} m^3 \right) + O(mn),$$

we get lower and upper bounds for $bcr(M(m, n))$. The issue here is to improve these bounds by constant multiplicative factors.

Now we will improve the upper bound by a "column after column" drawing and the lower bound by a new lower bound argument.

Theorem 3.1 *For a mesh $M(m, n), 4 \leq m \leq n$ it holds:*

$$\frac{3}{4} m^2 n - \frac{1}{4} m^3 - \frac{7}{2} mn - \frac{1}{2} m^2 + 6n + \frac{9}{2} m \leq bcr(M(m, n)) \leq \frac{3}{2} m^2 n.$$

Proof. Upper bound. First place the vertices of $M(m,n)$ on the line x_0 by the column after column manner. Then project the vertices of V_1 on x_1. The result of the theorem follows by counting the crossings between every two consecutive columns. Their number equals $(3m^2 - 7m + 4)/2$.

Lower bound. For the sake of simplicity assume that both m and n are even. Consider a bipartite drawing of $M(m,n)$. Then $|V_0| = mn/2$. For $k = 1, 2, ..., mn/2$, let A_k denote the set of the first k vertices on x_0 from the left. A variant of the Menger's theorem [24] says that the maximum number of edge disjoint paths between A_k and $V_0 - A_k$ equals $|cut(A_k, V_0 - A_k)|$.

Claim. Each of these paths, except for those ending in the $(k+1)$-st vertex v_{k+1} on x_0 from the left, must cross all but one edges adjacent to the $(k+1)$-st vertex v_{k+1}.

Define a function

$$f(x) = \begin{cases} 2\sqrt{x}, & \text{if } 0 \leq x \leq m^2/4, \\ m, & \text{if } m^2/4 \leq x \leq mn - m^2/4, \\ 2\sqrt{mn - x}, & \text{if } mn - m^2/4 \leq x \leq mn. \end{cases}$$

Now we use an edge isoperimetric inequality for meshes. It is known [1, 4] that for any $X \subset M(m,n)$, $|\partial(X)| \geq f(|X|)$ holds. The set $cut(A_k, V_0 - A_k)$ partitions $V(G)$ into X and $V(G) - X$ such that $A_k \subset X$ and $V_0 - A_k \subset V(G) - X$. (Actually $X = X(A_k)$ depends on A_k, but for the sake of simplicity we do not show this dependence in our notation.) Clearly $|cut(A_k, V_0 - A_k)| = |\partial(X)|$. As $A_k \subset X \subset V(G) - (V_0 - A_k)$, the concavity of f gives

$$|cut(A_k, V_0 - A_k)| \geq \min\{f(|A_k|), f(|V(G) - (V_0 - A_k)|)\} = \min\{f(k), f(\frac{mn}{2} + k)\}.$$

There are at most $m + n$ vertices in V_0 whose degree is less than 4. We are going to give a lower bound for the number of crossings that happen to edges adjacent to v_{k+1}. It is convenient not counting the contribution of vertices v_{k+1} whose degree is less than 4. Hence if k runs from 1 to $mn/2 - 1$, using only vertices v_{k+1} whose degree is 4, the Claim yields that all $(A_k, V_0 - A_k)$ paths but 4 intersect at least 3 of the edges adjacent to v_{k+1}. Below a sum with a prime indicates $m + n$ missing terms. We obtain

$$bcr(M(m,n)) \geq \frac{3}{2} \sum_{k=1}^{\frac{mn}{2}-1} {}' (|cut(A_k, V_0 - A_k)| - 4)$$

The denominator 2 occurs before the sum is because each crossing is counted at most twice. Further,

$$bcr(M(m,n)) \geq \frac{3}{2} \sum_{k=1}^{\frac{mn}{2}-1} {}' \min\{f(k), f(\frac{mn}{2} + k)\} - 3mn$$

$$= \frac{3}{2} \sum_{k=1}^{\frac{m^2}{4}} \min\{f(k), f(\frac{mn}{2} + k)\} + \frac{3}{2} \sum_{k=\frac{m^2}{4}+1}^{\frac{mn}{2}-\frac{m^2}{4}-1} {}' \min\{f(k), f(\frac{mn}{2} + k)\}$$

$$+\frac{3}{2}\sum_{k=\frac{mn}{2}-\frac{m^2}{4}}^{\frac{mn}{2}-1}\min\{f(k),f(\frac{mn}{2}+k)\}-3mn$$

$$\geq 6\sum_{k=1}^{\frac{m^2}{4}}\sqrt{k}+\frac{3}{4}(m^2n-m^3-4m)-3mn-(m+n)\frac{3m}{2}$$

$$\geq 6\int_0^{\frac{m^2}{4}}\sqrt{x}dx+\frac{3}{4}(m^2n-m^3-4m)-3mn-(m+n)\frac{3m}{2}$$

$$\geq \frac{3}{4}m^2n-\frac{1}{4}m^3-\frac{9}{2}mn-\frac{3}{2}m^2-3m.$$

We excluded from Theorem 3.1 the cases $m=2,3$. The result $bcr(M(2,n))=n-1$ can be easily deduced from the optimal bipartite drawing of the even cycle C_{2n}, [14].

Theorem 3.2 *For $n\geq 3$ it holds:*

$$bcr(M(3,n))=5n-6.$$

Proof. Upper bound. First place all vertices of $M(3,n)$ on x_0 in a column after column manner. Then project the vertices of V_1 on x_1. It is easy to see by induction on n that the resulting drawing has $5n-6$ crossings.

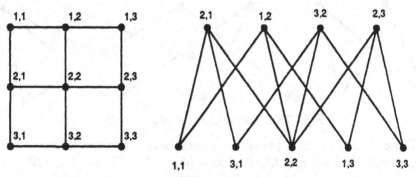

Fig. 1 : Mesh $M(3,3)$ and its optimal bipartite drawing

Lower bound. Imagine that $M(3,n)$ consists of 3 row and m column vertices. Let $M(3,3)$ denote the submesh induced by the last 3 column vertices. We proceed by induction on n. By a case analysis we can show that $bcr(M(3,3))=9$. Suppose that $bcr(M(3,n-1))\geq 5(n-1)-6$, for $n\geq 4$ and consider $M(3,n)$. Using a case analysis again one can show that the edges incident to the last column vertices in $M(3,n)$ contain at least 5 crossings. In fact this can be verified considering the submesh $M(3,3)$ only. Therefore

$$bcr(M(3,n))\geq bcr(M(3,n-1))+5\geq 5n-6.\qquad\qquad\square$$

4 Hypercubes

For hypercubes we could apply again our result (1) from [20] and compose it with a result of Harper who showed in [15] that: $L(Q_n) = 2^{n-1}(2^n - 1)$. This would give us lower and upper bounds on $bcr(Q_n)$. The issue here is to obtain better bounds than those. The upper bound is given by a recursive drawing. Then we adjust the lower bound argument from the previous section to hypercubes.

Theorem 4.1 *For $n \geq 3$ it holds:*

$$n4^{n-2} - O(4^n) < bcr(Q_n) \leq n4^{n-1}$$

Proof. Upper bound. Draw Q_n recursively. The drawing of Q_2 is unique. Assume that we have a drawing $D(Q_{n-1})$ with

$$bcr(D(Q_{n-1})) \leq (2n - 5)2^{2n-5} - ((n - 1)^2 - (n - 1) - 1)2^{n-3}.$$

Place a copy of the drawing $D(Q_{n-1})$ next to the drawing $D(Q_{n-1})$ and add 2^{n-1} new edges to complete the drawing of $D(Q_n)$. There are 2^{2n-4} crossings of new edges with new edges and $(n - 1)2^{n-1}(2^{n-2} - 1)$ crossings of new edges with old edges. See Fig. 2. We have:

$$bcr(D(Q_n)) \leq 2bcr(D(Q_{n-1})) + 2^{2n-4} + (n - 1)2^{n-1}(2^{n-2} - 1)$$
$$= (2n - 3)2^{2n-3} - (n^2 - n - 1)2^{n-2} < n4^{n-1}.$$

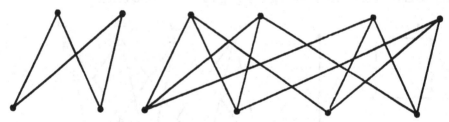

Fig. 2 : Bipartite drawings of Q_2 and Q_3.

Lower bound. We apply the same argument as for 2-dimensional meshes. Consider a bipartite drawing of Q_n. Note that $|V_0| = 2^{n-1}$. For $k = 1, 2, ..., 2^{n-1} - 1$, let $A_k \subset V_0$ denote the set of the first k vertices on x_0 from the left. Following Bollobás and Leader [5] define a function $f(x)$ as follows:

$$f(x) = \begin{cases} x(n - \log x), & \text{if } 1 \leq x \leq 2^{n-1}, \\ (2^n - x)(n - \log(2^n - x)), & \text{if } 2^{n-1} \leq x \leq 2^n. \end{cases}$$

(Here log denotes logarithm based 2.) An edge isoperimetric inequality for hypercubes (see e.g [9]) says that for any $X \subset Q_n$, the inequality $|\partial(X)| \geq f(|X|)$ holds. Following the reasoning applied for meshes (i.e. $|cut(A_k, V_0 - A_k)| \geq \min (f(k), f(2^{n-1} + k))$ for $1 \leq k \leq 2^{n-1}$) we show that

$$|cut(A_k, V_0 - A_k)| \geq \min\{k(n - \log k), (2^{n-1} - k)(n - \log(2^{n-1} - k))\}.$$

Hence if k runs from 1 to $2^{n-1} - 1$ we get

$$bcr(Q_n)$$

$$\geq \frac{n-1}{2} \sum_{k=1}^{2^{n-1}-1} (|cut(A_k, V_0 - A_k)| - n)$$

$$\geq \frac{n-1}{2} \sum_{k=1}^{2^{n-1}-1} \min\{k(n - \log k), (2^{n-1} - k)(n - \log(2^{n-1} - k))\} - n(n-1)2^{n-2}$$

$$\geq (n-1) \sum_{k=1}^{2^{n-2}-1} k(n - \log k) + (n-1)2^{n-2} - n(n-1)2^{n-2}$$

$$= n(n-1)2^{n-3}(2^{n-2} - 1) + (n-1)2^{n-2} - n(n-1)2^{n-2} - (n-1) \sum_{k=1}^{2^{n-2}-1} k \log k,$$

where we used that for $k \leq 2^{n-2}$ it holds $k(n - \log k) \leq (2^{n-1} - k)(n - \log(2^{n-1} - k))$. Observe that

$$\sum_{k=1}^{2^{n-2}-1} k \log k < \int_1^{2^{n-2}} x \log x \, dx = (n-2)2^{2n-5} - \frac{1}{\ln 2} 2^{2n-6} + \frac{1}{4 \ln 2}.$$

Substituting this into the previous inequality we get the result. □

5 A general lower bound method based on eigenvalues

Our basic reference to spectral graph theory is Fan Chung's recent book [8]. We use Laplacian eigenvalues of a graph like [8] and define λ_G as the smallest positive Laplacian eigenvalue of the graph G.

The connection between eigenvalues and isoperimetric inequalities has been subject of study since long. We recall the following theorem from Section 2.2 of [8]: For $X \subseteq V(G)$

$$|\partial X| \geq \frac{\lambda_G}{2} \min(vol(X), vol(V(G) - X)). \tag{2}$$

Assume now $G = (V_0, V_1, E)$ is a bipartite graph in an optimal bipartite drawing D. We apply to $bcr(G)$ the lower bound technique developed for hypercubes and meshes. Let v_i denote the i-th vertex in V_0 and A_i denotes the set of the first i vertices in V_0.

$$|cut(A_i, V_0 - A_i)| = |\partial X(A_i)| \geq \frac{\lambda_G}{2} \min(vol(X(A_i)), vol(V(G) - X(A_i))) \tag{3}$$

$$\geq \frac{\lambda_G}{2} \min(vol(A_i), vol(V_0 - A_i)). \tag{4}$$

Using (2) for estimating $|\partial X|$, instead of an explicit function $f(x)$ that is rarely known, we end up with the estimate

$$2bcr(G) \geq \sum_{i=1}^{|V_0|-1} (d(v_{i+1}) - 1) \left(\frac{\lambda_G}{2} \min(vol(A_i), vol(V_0 - A_i)) - d(v_{i+1}) \right). \quad (5)$$

Formula (5) applies very easily when all degrees in V_0 are the same. This is the case, for example, for hypercubes. In this case $\lambda_{Q_n} = 2/n$ (p. 6 in [8]) and we easily can derive the lower bound of our Theorem 4.1 with a slightly weaker (halved) multiplicative constant.

It is instructive to see how far can we get with eigenvalues if we try this approach to meshes. The first problem that we face is that the graph is not regular. It is unclear how to give good lower bounds for $\sum_i a_i b_i$ in the RHS of (5) in general. This problem can be overcome since almost all degrees are 4 if both n, m grow large.

The bigger problem is that $\lambda_{M(n,m)}$ is simply not large enough. The smallest positive Laplacian eigenvalue of an n-vertex path is $1 - \cos\frac{\pi}{n-1} = \Theta(\frac{1}{n^2})$ [8], p. 6. By the results of [8], p. 37, $\lambda_{M(n,m)} = \Theta(1 - \cos\frac{\pi}{n-1}) = \Theta(\frac{1}{n^2})$, since $n \geq m$. Therefore in this way we cannot get anything as good as our lower bound in Theorem 3.1.

Note that [17] has developed a lower bound for the linear arrangement value of a graph in terms of the smallest positive Laplacian eigenvalue. This lower bound can be combined with (1) to obtain a bound (5). However, due to the large constant in (1), formula (5) is expected to give tighter bounds.

6 Concluding Remarks

We introduced a new method for establishing lower bounds on bipartite crossing numbers. The method is based on edge isoperimetric inequalities. By means of the new method we essentially improved the previous lower bounds for meshes and hypercubes. The method is especially well suited for "well structured graphs", for which good edge isoperimetric inequalities and/or the smallest positive Laplacian eigenvalue are known [2]. Our upper and lower bounds still leave space for possible improvements. We believe that the upper bounds are closer to the optimal values than the lower bounds.

Acknowledgment. The research was partly done while the second and fourth authors were visiting Department of Mathematics and Informatics of University in Passau. They thank Prof. F.-J. Brandenburg for invitation and for the inspiring atmosphere in his section.

The figures were produced by system INCA of Olivier Bouden, LaBRI, Bordeaux I.

References

1. Ahlswede, R., Bezrukov, S. L., Edge isoperimetric theorems for integer point arrays, *Appl. Math. Lett.* **8** (1995), 75-80.

2. Bezrukov, S. L., Edge isoperimetric problems on graphs, *Technical Report*, Department of Computer Science, University of Paderborn, 1997.

3. Bollobás, B., Combinatorics, Chapter 16, Cambridge Uni. Press, 1986.

4. Bollobás, B., Leader, I., Edge-isoperimetric inequalities in the grid, *Combinatorica* 11 (1991), 299-314.

5. Bollobás, B., Leader, I., Matchings and paths in cubes, *SIAM J. Discrete Mathematics*, to appear.

6. Brandenburg, F. J., Jünger, M., Mutzel, P., Algorithms for automatic graph drawing, *Technical Report*, Max Planck Institute, MPI-I-97-1-007, Saarbrücken, March 1997, (in German).

7. Catarci, T., The assignment heuristics for crossing reduction, *IEEE Transactions on Systems, Man and Cybernetics* 25 (1995), 515-521.

8. Chung, F. R. K., *Spectral Graph Theory*, Regional Conference Series in Mathematics Number 92, American Mathematical Society, Providence, RI, 1997.

9. Chung, F. R. K., Füredi, Z., Graham, R. L., Seymour, P. D., On induced subgraphs of the cube, *J. Combinatorial Theory* (A) 49 (1988), 180-187.

10. Di Battista, J., Eades, P., Tamassia, R., Tollis, I.G., Algorithms for drawing graphs: an annotated bibliography, *Computational Geometry* 4 (1994), 235-282.

11. Eades, P., Wormald, N., Edge crossings in drawings of bipartite graphs, *Algorithmica* 11 (1994), 379-403.

12. Garey, M.R., Johnson, D.S., Crossing number is NP-complete, *SIAM J. Algebraic and Discrete Methods* 4 (1983), 312-316.

13. Harary, F., Determinants, permanents and bipartite graphs, *Mathematical Magazine* 42 (1969), 146-148.

14. Harary, F., Schwenk, A., A new crossing number for bipartite graphs, *Utilitas Mathematica* 1 (1972), 203-209.

15. Harper, L. H., Optimal assignements of numbers to vertices, *SIAM J. Applied Mathematics* 12 (1964), 131-135.

16. Jünger, M., Mutzel, P., Exact and heuristic algorithm for 2-layer straightline crossing number, in: *Proc. Graph Drawing'95*, Lecture Notes in Computer Science 1027, Springer Verlag, Berlin, 1996, 337-348.

17. Juvan, M., Mohar, B., Optimal linear labelings and eigenvalues of graphs, *Discrete Applied Mathematics* 36 (1992), 153-168.

18. Muradyan, D.O., Piliposian, T.E., Minimal numberings of vertices of a rectangular lattice, Akad. Nauk Armjan. SSR Doklady 70 (1980), 21-27, (in Russian).

19. May, M., Szkatula, K., On the bipartite crossing number, *Control and Cybernetics* 17 (1988), 85-98.

20. Shahrokhi, F., Sýkora, O., Székely, L. A., Vrťo, On bipartite crossings, biplanar subgraphs, and the linear arrangement problem, in: Proc. *5th Workshop Algorithms and Data Structures, (WADS'97), August 6-8, 1997 Halifax, Nova Scotia, Canada*, Lecture Notes in Computer Science Vol. 1272, Springer-Verlag, 55-68.

21. F. Shahrokhi, L. A. Székely, I. Vrťo, Crossing numbers of graphs, lower bound techniques and algorithms: a survey, in: Proc. *DIMACS Workshop on Graph Drawing'94*, Lecture Notes Computer Science 894, Springer Verlag, Berlin, 1995, 131-142.

22. Spinrad, J., Brandstädt, A., Stewart, L., Bipartite permutation graphs, *Discrete Applied Mathematics* 19, 1987, 279-292.

23. Sugiyama, K., Tagawa, S., Toda, M., Methods for visual understanding of hierarchical systems structures, *IEEE Transactions on Systems, Man and Cybernetics* 11 (1981), 109-125.

24. Tutte, W., T., Graph Theory, Addison-Wesley Publishing Company, Reading, 1984.
25. Warfield, J., Crossing theory and hierarchy mapping, *IEEE Transactions on Systems, Man and Cybernetics* 7 (1977), 502-523.
26. Watkins, M.E., A special crossing number for bipartite graphs: a research problem, *Annals of New York Academy Sciences* 175 (1970), 405-410.

Three-Dimensional Grid Drawings of Graphs

János Pach*, Torsten Thiele** and Géza Tóth***

Courant Institute, New York University

Abstract. A three-dimensional *grid drawing* of a graph G is a placement of the vertices at distinct integer points so that the straight-line segments representing the edges of G are pairwise non-crossing. It is shown that for any fixed $r \geq 2$, every r-colorable graph of n vertices has a three-dimensional grid drawing that fits into a box of volume $O(n^2)$. The order of magnitude of this bound cannot be improved.

1 Introduction

In a *grid drawing* of a graph, the vertices are represented by distinct points with integer coordinates and the edges are represented by straight-line segments connecting the corresponding pairs of points. Grid drawings in the plane have a vast literature [BE]. In particular, it is known that every planar graph of n vertices has a two-dimensional grid drawing that fits into a rectangle of area $O(n^2)$, and this bound is asymptotically tight [FP],[S].

The possibility of three-dimensional representations of graphs was suggested by software engineers [MR]. The analysis of the volume requirement of such representations was initiated in [CE], where the following statement was proved. Every graph of n vertices has a three-dimensional grid drawing in a rectangular box of volume $O(n^3)$, and this bound cannot be improved. To establish the first half of the statement, it is sufficient to consider representations of *complete graphs*. Cohen et al. used a generalization of a well-known construction of Erdős showing that the vertices of a complete graph K_n can be placed at the points $(i, i^2 \bmod p, i^3 \bmod p)$, $1 \leq i \leq n$, where p is a prime between n and $2n$. Since no four of these points lie in the same plane, the resulting straight-line drawing of K_n has no crossing edges.

A complete r-partite graph is called *balanced*, if any two of its classes have the same number of points, or their sizes differ by one. Let $K_r(n)$ denote a balanced complete r-partite graph with $n \geq r$ vertices. That is, the vertex set of $K_r(n)$ splits into r disjoint classes, $V_1, V_2, \ldots, V_r$, such that $|V_i| = \lfloor n/r \rfloor$ or $\lceil n/r \rceil$, and two vertices are connected by an edge if and only if they belong to different V_i's.

It was pointed out by Cohen et al. [CE] that $K_2(n)$ has a three-dimensional grid drawing within a box of volume $O(n^2)$, and they asked whether this bound is optimal. T. Calamoneri and A. Sterbini [CS] proved that any such drawing

* Supported by NSF grant CCR-94-24398 and PSC-CUNY Research Award 667339
** Supported by the Deutsche Forschungsgemeinschaft
*** Supported by OTKA-T-020914, OTKA-F-22234 and the Alfred Sloan Foundation

requires a box whose volume is at least $\Omega(n^{3/2})$. Furthermore, they have shown that $K_3(n)$ and $K_4(n)$ also permit three-dimensional grid drawings of volume $O(n^2)$, and conjectured that the same is true for $K_r(n)$, for any fixed $r > 4$.

The aim of the present note is to answer the question of Cohen et al. and to verify the conjecture of Calamoneri and Sterbini. A graph is called r-*colorable* if its vertices can be colored by r colors so that no two adjacent vertices receive the same color. Equivalently, G is r-colorable if it is a subgraph of a complete r-partite graph.

Theorem. *For every $r \geq 2$ fixed, any r-colorable graph of n vertices has a three-dimensional grid drawing that fits into a rectangular box of volume $O(n^2)$. The order of magnitude of this bound cannot be improved.*

2 Proof of the Theorem

To prove the second assertion, it is enough to establish an $\Omega(n^2)$ lower bound on b, the number of integer points in a box B accommodating a three-dimensional grid drawing of the balanced complete bipartite graph $K_2(n)$ with vertex classes V_1, V_2. Clearly, $K_2(n)$ is r-colorable for any $r \geq 2$. Fix a grid drawing and consider the set of all vectors pointing from a vertex of V_1 to a vertex of V_2. The number of such vectors is $\lceil \frac{n}{2} \rceil \cdot \lfloor \frac{n}{2} \rfloor \geq \frac{n^2-1}{4}$, and no two of them can be identical, otherwise the corresponding four points would induce a parallelogram whose diagonals cross each other. (In fact, no two such vectors can point in the same direction.) On the other hand, the total number of vectors determined by two gridpoints in B is smaller than $8b$. Thus, $8b > \frac{n^2-1}{4}$, as required.

The proof of the upper bound is based on the following.

Lemma. *For any $r \geq 2$ and for any n divisible by r, the balanced complete r-partite graph $K_r(n)$ has a three-dimensional grid drawing that fits into a rectangular box of size $r \times 4n \times 4rn$.*

Proof. Let p be the smallest prime with $p \geq 2r - 1$ and set $N := p \cdot \frac{n}{r}$. Note that $p < 4r$ and thus $N < 4n$. For any $0 \leq i \leq r - 1$, let

$$V_i := \{(i, t, it) : 0 \leq t < N, t \equiv i^2 \pmod{p}\}.$$

These sets are pairwise disjoint, and each of them has precisely $\frac{N}{p} = \frac{n}{r}$ elements. Connect any two points belonging to different V_i's by a straight-line segment. The resulting drawing of $K_r(n)$ fits into a rectangular box of size $r \times 4n \times 4rn$, as desired.

It remains to show that no two edges of this drawing cross each other. Suppose, for contradiction, that there are two crossing edges, e and e'. We distinguish three different cases, according to the number of distinct classes V_i, the endpoints of e and e' belong to.

<u>Case 1</u>: The endpoints of e and e' are from four distinct classes, $V_{i_1}, V_{i_2}, V_{i_3},$ and V_{i_4}.

Let $(i_\alpha, t_\alpha, i_\alpha t_\alpha), 1 \le \alpha \le 4$, be the corresponding endpoints, so that $t_\alpha \equiv i_\alpha^2$ (mod p). Then these points lie in a plane, and the determinant

$$D = \begin{vmatrix} 1 & i_1 & t_1 & i_1 t_1 \\ 1 & i_2 & t_2 & i_2 t_2 \\ 1 & i_3 & t_3 & i_3 t_3 \\ 1 & i_4 & t_4 & i_4 t_4 \end{vmatrix}$$

vanishes. Therefore, it must also vanish modulo p. However, modulo p this determinant reduces to the Vandermonde determinant. Thus,

$$D \equiv \begin{vmatrix} 1 & i_1 & i_1^2 & i_1^3 \\ 1 & i_2 & i_2^2 & i_2^3 \\ 1 & i_3 & i_3^2 & i_3^3 \\ 1 & i_4 & i_4^2 & i_4^3 \end{vmatrix} = \prod_{1 \le \beta < \alpha \le 4} (i_\alpha - i_\beta),$$

which is non-zero modulo p, a contradiction.

<u>Case 2</u>: The endpoints of e and e' are from three different classes, V_i, V_j, V_k.

Assume without loss of generality that two of these points, (i, t_1, it_1) and (i, t_2, it_2), belong to V_i, and the other two are $(j, s, js) \in V_j$ and $(k, u, ku) \in V_k$. These four points cannot be coplanar (hence e and e' cannot cross each other), because the corresponding test determinant

$$\begin{vmatrix} 1 & i & t_1 & it_1 \\ 1 & i & t_2 & it_2 \\ 1 & j & s & js \\ 1 & k & u & ku \end{vmatrix} = (j - i)(k - i)(t_2 - t_1)(s - u)$$

is non-zero. To see this, observe that $s - u \equiv j^2 - k^2 \equiv (j-k)(j+k) \not\equiv 0 \pmod p$. This is the point where we use the assumption that $0 \le j, k \le r - 1 \le (p-1)/2$.

<u>Case 3</u>: The endpoints of e and e' are from two different classes, V_i and V_j.

Let these points be $(i, t_1, it_1), (i, t_2, it_2) \in V_i$ and $(j, s_1, js_1), (j, s_2, js_2) \in V_j$. Now the corresponding test determinant

$$\begin{vmatrix} 1 & i & t_1 & it_1 \\ 1 & i & t_2 & it_2 \\ 1 & j & s_1 & js_1 \\ 1 & j & s_2 & js_2 \end{vmatrix} = (j - i)^2 (t_2 - t_1)(s_1 - s_2)$$

does not vanish, therefore e and e' cannot cross each other.

This contradiction completes the proof of the Lemma. □

Now we return to the proof of the upper bound of the theorem. In fact, we can deduce a more precise statement.

Corollary. *There exists a $c > 0$ such that for any $r \geq 2$, any r-colorable graph of n vertices has a three-dimensional grid drawing that fits into a rectangular box of volume $cr^2 n^2$.*

Proof. Fix an r-colorable graph G with n vertices. We split every color class into smaller parts such that all but one of them have *exactly* and the last one *at most* $\lceil \frac{n}{r} \rceil$ points. This defines a decomposition of the vertex set of G into at most $2r - 1$ classes, whose sizes do not exceed $\lceil \frac{n}{r} \rceil$, and no two points belonging to the same class are connected by an edge. In other words, G is a subgraph of a balanced complete $(2r - 1)$-partite graph K with $(2r - 1)\lceil \frac{n}{r} \rceil < 2n + 2r$ vertices. Applying the Lemma to K, the Corollary follows. □

3 Remarks and open problems

A The rectangular box used in the proof of the Theorem has two sides of size $O(n)$. We can use rectangular boxes of different shapes to represent $K_2(n)$.

Proposition. *There is a three dimensional grid drawing of $K_2(n)$ which fits into a rectangular box of size $O(n) \times O(\sqrt{n}) \times O(\sqrt{n})$.*

Proof. Let V_1 and V_2 be the vertex classes of $K_2(n)$ and let

$$V_1 = \{(i, 0, 0) \mid 0 \leq i \leq \lceil n/2 \rceil - 1\},$$

$$V_2 = \{(0, a, b) \mid 0 \leq a, b \leq k, (a, b) = 1\}$$

where $k \approx \pi\sqrt{n/6}$.
Then

$$|V_1| = \lceil n/2 \rceil,$$

$$|V_2| = \sum_{i=1}^{k} \phi(i) \approx \frac{3k^2}{\pi^2} \approx n/2$$

(see [HW]). To see that there is no crossing, we observe that the points of V_2 lie on a horizontal plane and there are no two of these points on a line through $(0, 0, 0)$ (see Figure). □

B. It would be interesting to determine $S_r = S_r(n)$, the set of triples (s_1, s_2, s_3) for which every r-colorable graph of n vertices has a grid drawing that fits into a box of size $s_1 \times s_2 \times s_3$. In particular, what is the smallest $s = s(r)$ with $(s, s, s) \in S_r$? It is not hard to see that $s_i s_j \geq \lfloor n/2 \rfloor$ and $s_i > \frac{r}{5}$ holds for every $(s_1, s_2, s_3) \in S_r$ ($r \geq 2$).

C. Since any graph of n vertices with fixed maximum degree $r - 1$ is r-colorable, it follows from the Theorem that any such graph permits a grid drawing in a box of volume $O(n^2)$. It seems likely that for every fixed r, this bound can be substantially improved. We cannot even decide if every graph with maximum degree 3 has a grid drawing of volume $O(n)$.

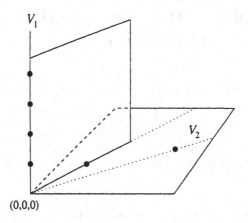

V_1

V_2

(0,0,0)

Figure

References

[BE] G. Di Battista, P. Eades, R. Tamassia, and I.G. Tollis, Algorithms for drawing graphs: an annotated bibliography, *Computational Geometry: Theory and Applications* 4 (1994), 235–282.

[CE] R.F. Cohen, P. Eades, T. Lin, and F. Ruskey, Three-dimensional graph drawing, in: *Graph Drawing (GD '94, R. Tamassia and I.G. Tollis, eds.)*, *Lecture Notes in Computer Science* 1027, Springer-Verlag, Berlin, 1995, 1–11.

[CS] T. Calamoneri and A. Sterbini, Drawing 2-, 3-, and 4-colorable graphs in $O(n^2)$ volume, in: *Graph Drawing (GD '96, S. North, ed.)*, *Lecture Notes in Computer Science* 1190, Springer-Verlag, Berlin, 1997, 53-62.

[FP] H. de Fraysseix, J. Pach, and R. Pollack, How to draw a planar graph on a grid, *Combinatorica* 10 (1990), 41–51.

[HW] G. H. Hardy, E. M. Wright, *An Introduction to the Theory of Numbers*, University Press, Oxford, 1954.

[MR] J. Mackinley, G. Robertson, and S. Card, Cone trees: Animated 3d visualizations of hierarchical information, in: *Proceedings of SIGCHI Conference on Human Factors in Computing*, 1991, 189–194.

[S] W. Schnyder, Embedding planar graphs on the grid, in: *Proceedings of the 1st Annual ACM-SIAM Symposium on Discrete Algorithms*, 1990, 138-148.

Incremental Orthogonal Graph Drawing in Three Dimensions

Achilleas Papakostas   and   Ioannis G. Tollis*

Dept. of Computer Science
The University of Texas at Dallas
Richardson, TX 75083-0688
email: **papakost@utdallas.edu, tollis@utdallas.edu**

Abstract. We present two algorithms for orthogonal graph drawing in three dimensional space. For graphs of maximum degree six, the 3-D drawing is produced in linear time, has volume at most $4.66n^3$ and each edge has at most three bends. If the degree of the graph is arbitrary, the vertices are represented by solid 3-D boxes whose surface is proportional to their degree. The produced drawing has two bends per edge. Both algorithms guarantee no crossings and can be used under an interactive setting (i.e., vertices arrive and enter the drawing on-line), as well.

1 Introduction

Graph drawing addresses the problem of automatically generating geometric representations of abstract graphs or networks. For a survey of graph drawing algorithms and other related results see the annotated bibliography of Di Battista, Eades, Tamassia and Tollis [6]. Various algorithms have been introduced to produce orthogonal drawings of planar [1, 10, 20] or general [1, 10, 13, 15, 19] graphs of maximum degree three or four. For drawings of general graphs, the required area can be as little as $0.76n^2$ [13, 15], the total number of bends is no more than $2n + 2$, and there are at most two bends any edge [1, 13, 15]. There has been a recent trend in Graph Drawing to visualize graphs in the three dimensional space [2, 3, 5, 7, 8, 9, 17, 18]. Although the number of applications that require such a representation for graphs is still limited [2, 9, 12, 18], there is no doubt that 3-D Graph Drawing will find many applications in the future.

In [4] it is shown that any n-vertex graph has a 3-D drawing in a $n \times 2n \times 2n$ grid, so that all vertices are located on grid points, and no two edges cross. In the same paper, a technique to convert an orthogonal 2-D drawing of area $H \times V$ to a 3-D straight-line drawing of volume $\lceil \sqrt{H} \rceil \times \lceil \sqrt{H} \rceil \times V$ is also presented. Graph drawing in three dimensional space Naturally, orthogonal drawing in three dimensional space has also received attention recently [7, 8, 9, 17]. A 3-D orthogonal drawing typically has the following properties:

– vertices are points with integer coordinates in three dimensional space,

---

* Research supported in part by NIST, Advanced Technology Program grant number 70NANB5H1162.

- each edge is a polyline sequence of consecutive straight line segments; each one of these line segments is parallel either to the x-axis, y-axis, or z-axis,
- the meeting point of two consecutive straight line segments of the same edge is a bend and has integer coordinates, and
- line segments coming from routes of two different edges are not allowed to overlap.

A very interesting upper bound on the volume for 3-D orthogonal drawings of graphs of maximum degree six is shown in [8]. More specifically, the volume for such drawings is at most $O(\sqrt{n}) \times O(\sqrt{n}) \times O(\sqrt{n})$, while each edge has at most seven bends, and no two edges cross. This improves the result in [7], where the volume upper bound was the same but the drawings allowed up to 16 bends per edge. If we require that each edge have at most three bends, then another algorithm is presented in [8] that requires volume exactly $27n^3$ (the produced drawings have no crossings). Both algorithms run in $O(n^{\frac{3}{2}})$ time. Note that Kolmogorov and Bardzin [11] show an existential lower bound of $\Omega(n^{\frac{3}{2}})$ on the volume occupied by 3-D orthogonal drawings.

In this paper we present one algorithm for producing 3-D orthogonal drawings of graphs of maximum degree six, and a second algorithm that produces 3-D orthogonal drawings of graphs of arbitrary degree. Note that there has not been any previous work that dealt with the theory of 3-D orthogonal drawing of graphs of arbitrary degree. Both algorithms are based on the "Relative-Coordinates" paradigm for vertex insertion [14, 16]. As such, both algorithms support interactive environments where vertices arrive and enter the drawing on-line. An important feature of this work is that both algorithms guarantee no edge crossings.

Given an n-vertex graph G of maximum degree six, our first algorithm produces a 3-D orthogonal drawing of G whose volume is at most $4.66n^3$, in linear time. Moreover, each edge of the drawing has at most three bends. Hence, our algorithm outperforms the algorithm of [8] in terms of both running time and volume of the drawing. Our second algorithm uses solid three dimensional boxes to represent vertices. The surface of each such box is proportional to the degree of the represented vertex. The produced 3-D orthogonal drawings have at most two bends per edge, and volume $O((\frac{m}{3} + O(n))^3)$, where m is the number of edges of the drawing.

2   Preliminaries

Clearly, for each graph of maximum degree six, there is a 3-D orthogonal drawing. The system of coordinates typically used in three dimensional space is based on three axes x, y, z so that each one of them is perpendicular to the other two (see Fig. 1a). Three different planes are formed by the three possible ways we can pair these axes: The xz-plane is defined by the x, z-axes, the yz-plane is defined by the y, z-axes, and the xy-plane is defined by the x, y-axes. Each one of these planes is called a *base* plane; each base plane is perpendicular to the other two.

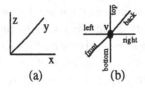

Fig. 1. (a) Coordinates system for 3-D drawing, (b) possible directions from where an edge can enter v.

Each vertex of a 3-D drawing has six possible directions around it from where incident edges may enter the vertex. The two directions parallel to the z-axis are *top* (extending towards the positive part of the z-axis) and *bottom* (extending towards the negative part of the z-axis). *Front* and *back* directions are parallel to the y-axis and they extend towards the negative and positive parts of the y-axis, respectively. The remaining two directions are parallel to the x-axis and are called *left* (extending towards the negative part of the x-axis) and *right*. See also Fig. 1b.

Two directions parallel to the same axis are *opposite* directions. Two directions parallel to two different axes are *orthogonal* directions. If there is no edge entering a vertex v from a specific direction of v, this direction is called *free direction* of v. A *plane* free direction is a left, right, front, or back free direction. Consider vertices $v_1, v_2, \cdots v_r$, where $r \geq 2$, having plane free directions $fd_1, fd_2, \cdots fd_r$ which extend towards the same direction (e.g., they are all left free directions). The set of the fd_i's forms a *beam*. If the fd_i's are left (resp. right, front, back) free directions, then their beam is a left (resp. right, front, back) beam. Vertices $v_1, v_2, \cdots v_r$ are the *origins* of the beam. Two beams are opposite (resp. orthogonal) if the free direction of one beam is opposite (resp. orthogonal) to the free direction of the other.

The volume of a 3-D drawing is the volume of the smallest rectangular parallelepiped that encloses the 3-D drawing. Also recall [14, 16] the following terminology: The *current* drawing is the drawing before the insertion of a new vertex v; the number of vertices of the current drawing that will be connected with v through new edges, is v's *local* degree. We call these vertices *adjacent* vertices of v.

3 Drawing Graphs with Maximum Degree Six

In this section we present our incremental algorithm for producing orthogonal drawings of graphs of maximum degree six in the three dimensional space. The incremental nature of our algorithm comes from the fact that a user is allowed to insert vertices (along with edges to other existing vertices) into the current drawing in any order. The algorithm supports such vertex insertions at any moment t, as long as each request observes the following rules:

- we start the drawing from scratch, that is the very first current drawing is the empty graph,
- the degree of any vertex of the current drawing at any time t is at most six, and
- the graph represented by the current drawing is always connected.

Our technique follows the Relative-Coordinates scenario [14, 16]. This means that the decision about where to place a new vertex and how its incident edges will be routed depends entirely on the free directions around the adjacent vertices. The properties of the Relative-Coordinates scenario [14, 16] are also properties of the 3-D drawings produced by our algorithm and guarantee a "smooth" transition from the current drawing to the next. The notation $u \to p \to p'$ means that from vertex u we draw a straight line segment that intersects plane p perpendicularly, and from the intersection point we draw another segment to plane p' that intersects p' perpendicularly as well. We use the notation $p_{a,v}$, where $a = x, y$, or z and v some vertex, to denote the plane which is perpendicular to the a-axis and contains vertex v. As we will see later, our 3-D orthogonal drawing is built in an upward fashion (i.e., it grows along the positive z-axis). For this reason, we always keep the following basic rule during the interactive drawing process:

Basic Rule: No vertex has a bottom free direction in any current drawing.

Most of the edges we route follow one of five fundamental routes, described below, depending on their available free directions. Assume that w and w' are two vertices of the current drawing and w has higher z-coordinate than w'. In the first three fundamental routes, edge (w', w) always enters w' from its left (or other plane) free direction. In the remaining two fundamental routes, edge (w', w) enters w' from its top free direction. Note that these fundamental routes can generalize to other situations (besides the examples shown in Fig. 2).

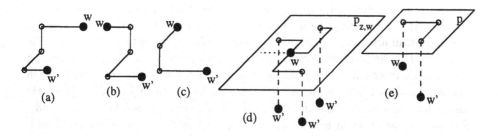

Fig. 2. (a) First, (b) Second, (c) Third Fundamental Routes, (d) Same-Plane, (e) Over-The-Top Routes.

- **First Fundamental Route:** Edge (w', w) enters w from its left free direction. We open up a new plane p to the left of the leftmost plane of the current drawing. Edge (w', w) is routed with three bends as follows: $w' \rightarrow p \rightarrow p_{y,w} \rightarrow p_{z,w} \rightarrow w$. This is shown in Fig. 2a. The small empty circles of this figure denote the three bends of the route.

- **Second Fundamental Route:** Vertex w has lower x-coordinate than w', and edge (w', w) enters w from its right free direction. We open up a new plane p parallel to the yz-plane and one unit to the right of w. Edge (w', w) is routed with three bends (see Fig. 2b), as follows: $w' \rightarrow p \rightarrow p_{y,w} \rightarrow p_{z,w} \rightarrow w$.

- **Third Fundamental Route:** Vertex w has lower x-coordinate and higher y-coordinate than w', and edge (w', w) enters w from its front free direction. No new plane is opened up and we route edge (w', w) with two bends (see Fig. 2c) as follows: $w' \rightarrow p_{x,w} \rightarrow p_{z,w} \rightarrow w$.

- **Same-Plane Route:** Edge (w', w) may enter w from any one of its plane free directions. We draw a straight line segment from w' intersecting plane $p_{z,w}$ perpendicularly. The remaining portion of edge (w', w) is routed exclusively in $p_{z,w}$, and may enter w from any one of its plane free directions with at most two bends (if two bends are required, then a new plane parallel either to the xz or yz-plane has to be inserted). This means that the whole route has at most three bends. In Fig. 2d we show three examples of the portions of three routes in plane $p_{z,w}$.

- **Over-The-Top Route:** Edge (w', w) enters w from its top free direction. A new plane p parallel to the xy-plane is inserted in the drawing, one unit above w. Edge (w', w) is routed with three bends (see Fig. 2e) as follows: $w' \rightarrow p \rightarrow p_{x,w} \rightarrow p_{y,w} \rightarrow w$. In other words, we draw a straight line segment intersecting p perpendicularly, route the edge in p bringing it directly on top of w with one bend, and then just draw the line segment from that point to w.

3.1 Overview of the Algorithm - Preprocessing

Assume that we start with an empty graph. The following gives an overview of the algorithm for placing the next vertex v in the current drawing. The steps of this algorithm are analyzed in this and the following subsections. Let v_1 be the first vertex to be inserted. Vertex v_1 has local degree zero. If v_2 is the second vertex to be inserted, then v_2 has local degree one and is connected with v_1. In Fig. 3a, we show the first two vertices inserted in an empty drawing. There are three observations to make about Fig. 3a. First, edge (v_1, v_2) has three bends. Second, a total of seven new planes are inserted in the empty drawing. Third, neither v_1 nor v_2 has a bottom free direction.

1. IF v is the first or second vertex to be inserted, THEN place them as discussed above.
2. ELSE
   (a) Find v's adjacent vertices $u_1, \cdots u_l$ in the current drawing.
   (b) FOR each adjacent vertex determine its connector (use the procedure described below).

(c) Find which Routing Case (see next section) v's insertion falls into.

(d) WITHIN a Routing Case:

    i. Determine the anchor vertex u_a and the cover vertex u_c.

    ii. Place v.

    iii. Route edge (u_a, v).

    iv. Route the remaining edges (u_i, v) except (u_c, v), using the three Fundamental Routes and/or the Same-Plane Route.

    v. Route edge (u_c, v).

Let v be the next vertex to be inserted in the current drawing and l $(1 \le l \le 6)$ be v's local degree. We find the l adjacent vertices $u_1, u_2, \cdots u_l$ of v. According to the Basic Rule, v must not have a bottom free direction after v is placed and all its l incident edges are routed. This means that exactly one of these edges must enter v from the bottom. The vertex which is the other endpoint of this edge is called *anchor* vertex, and is denoted by u_a. If $l = 6$, then the last one of v's incident edges to be routed enters v from its top free direction. The other endpoint of this edge is called *cover* vertex, and is denoted by u_c.

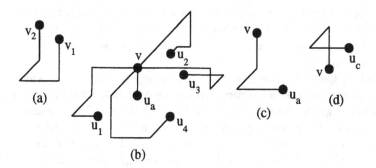

Fig. 3. (a) Inserting the first two vertices, (b) a Routing Case 1 example, (c) (u_a, v) of Routing Case 1 when u_a does not have top connector, (d) (u_c, v) of Routing Case 2 when u_c does not have top connector.

For each adjacent vertex u_i, we must pick one of its free directions which will be used for routing edge (u_i, v). The free direction picked for each u_i is called u_i's *connector*. Once a connector for an adjacent vertex u_i is determined, it remains the same throughout the whole process of placing v and routing its incident edges. If a connector of some u_i is a right (left, front, back, top) free direction, then it is called *right (left, front, back, top) connector*. Opposite, orthogonal, and plane connectors are defined in the same way as for free directions. Also, a beam of connectors is defined similarly to the beam of free directions. Let c_i be u_i's connector. We run the following procedure to determine the connector of each u_i.

1. Choose a free direction fd_i for each u_i so that:
   (a) The number of pairs $< fd_i, fd_j > (i \neq j$ and $1 \leq i, j \leq l)$ where fd_i and fd_j are opposite, is the smallest possible.
   (b) fd_i is top free direction, only if u_i has only this free direction left.
2. IF there are no two opposite beams among the fd_i's, THEN
   (a) FOR each u_i:
      i. $c_i := fd_i$.
3. ELSE IF there are two opposite beams B_1 and B_2, THEN
   (a) Consider the beam with the smallest cardinality; say B_1.
   (b) FOR each origin u_i of B_1:
      i. IF u_i's top free direction is available, THEN $c_i :=$ top connector.
      ii. ELSE $c_i := fd_i$.
   (c) FOR each u_i that is NOT an origin of B_1:
      i. $c_i := fd_i$.

3.2   Vertex Placement and Edge Routing

As we will see in this subsection, many of v's incident edges are routed using the fundamental routes. When this is the case, vertex v corresponds to w, and the adjacent vertex u_i, which is the other end of the route, corresponds to w'. Depending on the types of connectors that v's adjacent vertices have, we distinguish three Routing Cases, which are briefly discussed below (see [17] for a detailed description):

Routing Case 1: There is no beam among connectors c_i. Anchor u_a is selected among the adjacent vertices with top connectors. Edge (u_a, v) is a simple straight line segment from u_a to v (see Fig. 3b). If there is no adjacent vertex having top connector, any adjacent vertex can be the anchor vertex u_a. Then, edge (u_a, v) is routed with two bends as shown in Fig. 3c. This generalizes to cases where u_a has a different plane connector, through a rotation. The remaining edges (u_i, v) where u_i is not the anchor, are routed using the fundamental routes.

Routing Case 2: There is at least one beam among connectors c_i and there are no two opposite beams. If there is at least one adjacent vertex with top connector and $l = 6$, then this vertex is the cover u_c. If $l = 6$ and there is no adjacent vertex with top connector, then cover u_c is the adjacent vertex with highest z-coordinate which belongs to a beam. Then, we find the beam B_{max} having the highest cardinality without counting u_c. Assume that B_{max} is a left beam (the following discussion generalizes through rotation). Anchor u_a is always one of B_{max}'s origins. More specifically, it is the vertex whose y-coordinate is the median of the y-coordinates of all B_{max}'s origins. Edge (u_a, v) is routed with two bends (see Fig. 3c).

Vertex v is connected with the rest of its adjacent vertices using the fundamental routes. If cover u_c does not have top connector, then, by the way it was chosen, it has the following properties: (a) u_c has higher z-coordinate than v (see [17]), and (b) u_c has either left (if it is an origin of B_{max}), or back (if it is an origin of the other beam different from B_{max}) connector. If it has left connector, it is routed with two bends as shown in Fig. 3d (routing is similar when u_c has back connector).

Routing Case 3: There are two opposite beams among the u_i's. Clearly, in this routing case, we have that $l \geq 4$. B_{max} is the beam with the highest cardinality. Let us assume that B_{max} is a left beam (the discussion generalizes through rotation). Let B be the beam which is opposite to B_{max}. Anchor u_a is the median of the origins of beam B_{max} with respect to their y-coordinates. Edges connecting v with the origins of the two beams are basically routed using the fundamental routes. There are two exceptions to that:

The first exception deals with the situation where exactly one edge (u_B, v) (where u_B is an origin of B) is routed on top of the current drawing, all the way from the rightmost to the leftmost side of the drawing (see Fig. 4a). The second exception deals with the situation where $l = 6$ and there is no adjacent vertex with top connector. Vertex u_c is an origin of B, and edge (u_c, v) is routed with three bends as shown in Fig. 4b. Note that in this situation, vertex v is placed in plane p_{y,u_c}.

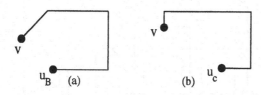

Fig. 4. Routing Case 3: (a) edge routing on top of current drawing, (b) edge (u_c, v) when u_c has no top connector.

In Fig. 5 we show the 3-D orthogonal drawing of K_7 produced by our algorithm. The numbers in the vertices denote the order in which the vertices were inserted. The volume of this drawing is $8 \times 8 \times 8 = 512 \leq 1.5n^3$, where $n = 7$. Observe that out of K_7's 21 edges, there is one edge with no bends, 12 edges require two bends each, and the remaining eight edges are routed with three bends each. Due to space limitations, we only present our two main theorems for orthogonal graph drawing in three dimensions of graphs of maximum degree six. For more details, see the full paper [17].

Theorem 1. *There is a 3-D orthogonal graph drawing algorithm for graphs of maximum degree six that allows on-line vertex insertion so that the following hold at any time t:*

- *after each vertex insertion, the coordinates of any vertex or bend of the current drawing shift by a small constant amount of units along the x, y, z-axes, effectively maintaining the general shape of the drawing,*
- *there are at most three bends along any edge,*
- *no two edges cross,*

- *the volume of the drawing is at most $4.66n^3(t)$, where $n(t)$ is the number of vertices in the drawing at time t, and*
- *vertex insertion takes constant time (if the screen needs to be refreshed after each vertex insertion, then it takes linear time).*

Our incremental algorithm can be used to produce a 3-D orthogonal drawing of a graph by first numbering its vertices and then inserting each vertex one at a time. By maintaining the drawing implicitly each insertion takes constant time. Hence we have the following:

Theorem 2. *Let G be an n-vertex connected graph of maximum degree six. There is a linear-time algorithm that produces a 3-D orthogonal drawing of G, so that each edge has at most three bends, no two edges cross, and the volume of the drawing is at most $4.66n^3$.*

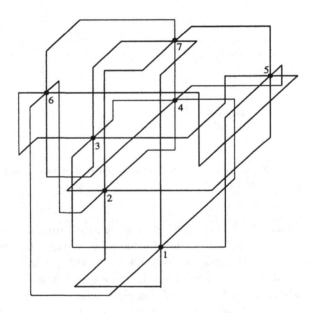

Fig. 5. 3-D orthogonal drawing of K_7 produced by our algorithm.

4 Drawing High Degree Graphs

In this section we give a very brief description of our model to support high degree three dimensional orthogonal graph drawing based on the Relative-Coordinates scenario. Our model allows vertices to arrive on-line and the degree of the vertices to increase arbitrarily. We represent vertices using three dimensional boxes.

When a vertex is inserted into the drawing, it is represented by a cubic box of size depending on the degree of the vertex. Edges that are adjacent to a vertex are attached to the surface of its box. The points on the box surface where edges can be attached are called *connectors*, and they have integer coordinates. As a result of edge routing, edges may require to attach to specific sides of incident boxes. If there are no available connectors on that side, we need to grow the box creating new connectors on that side. Our model for representing vertices in three dimensional space supports box growing. Figure 6a illustrates how box growing works. Finally, note that although the box of every vertex starts out having a cubic configuration, it may grow its size in various different ways in the course of the drawing process.

Our algorithm produces a 3-D drawing considering and placing one vertex at a time. Assume that v is the next vertex to be inserted in the current 3-D drawing. Let k be v's local degree, and let $u_1, u_2, \cdots u_k$ be v's adjacent vertices. For each vertex u_i we find the sides of box u_i that have available connectors. Then we find the side of the adjacent boxes on which most of these boxes have at least one available connector. This is the side where the edges connecting v with each u_i will be attached.

The next step is to create the box representing v and place it in the current 3-D drawing. Box v is a cube. Each newly inserted box is placed in such a way so that none of its connectors has the same x, y or z-coordinate as any other connector of any box of the current drawing. In [17] we describe in detail how v's position is computed. After v is inserted, the edges that connect v with its adjacent vertices are routed. We show that each edge can be routed with two bends without crossing other edges [17].

Theorem 3. *There is an algorithm to produce 3-D orthogonal drawings of graphs (not necessarily connected) which allows vertices to arrive on-line. The drawings have the following properties at any time t:*

- *vertices are represented by boxes and the surface of each box is at most six times the current degree of the vertex,*
- *each edge has two bends,*
- *no two edges cross,*
- *the volume is $O((\frac{m(t)}{3} + n(t))^3)$, where $m(t)$ and $n(t)$ are the number of edges and vertices at time t, and*
- *vertex insertion takes constant time.*

In practice, we expect the volume to be smaller than the upper bound given in the above theorem. This is because our analysis assumes that for each vertex insertion the boxes of all the adjacent vertices need to grow. We expect a box to grow very infrequently, since each box has several connectors on its sides. If the each inserted vertex is adjacent to no more than 16 vertices then the volume is bounded by $(\frac{m(t)}{3} + 2n(t))^3$. Figure 6b shows the 3-D orthogonal drawing of K_5 as produced by our algorithm. The box numbers denote the vertex insertion order.

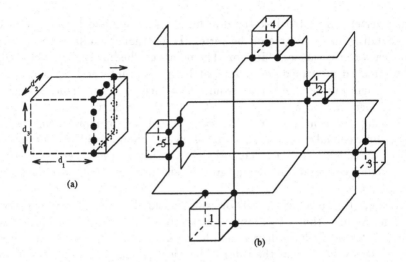

Fig. 6. (a) Box growing creates new connectors (denoted by solid and dotted circles), (b) 3-D orthogonal drawing of K_5, using boxes to represent vertices.

5   Conclusions and Open Problems

We presented incremental algorithms for producing orthogonal graph drawings in three dimensional space. The first algorithm deals with graphs of maximum degree six, and the produced drawings have volume at most $4.66n^3$. This improves the best known [8] volume requirement of exactly $27n^3$, while maintaining the same upper bound for the number of bends per edge. Our algorithm runs in linear time.

The second algorithm introduces 3-D orthogonal drawing for graphs of degree higher than six. Vertices are represented by solid three dimensional boxes whose surface is proportional to the degree of the vertex. The volume of the drawings is bounded by $O((\frac{m}{3} + n)^3)$ and each edge has only two bends.

Improving the upper bounds on the volume while keeping the number of bends per edge to three or less, is an interesting open problem.

References

1. T. Biedl and G. Kant, *A Better Heuristic for Orthogonal Graph Drawings*, Proc. 2nd Ann. European Symposium on Algorithms (ESA '94), Lecture Notes in Computer Science, vol. 855, pp. 24-35, Springer-Verlag, 1994.
2. M. Brown and M. Najork, *Algorithm animation using 3D interactive graphics*, Proc. ACM Symp. on User Interface Software and Technology, 1993, pp. 93-100.
3. I. Bruss and A. Frick, *Fast Interactive 3-D Visualization*, Proc. of Workshop GD '95, Lecture Notes in Comp. Sci. 1027, Springer-Verlag, 1995, pp. 99-110.

4. R. Cohen, P. Eades, T. Lin, F. Ruskey, *Three Dimensional Graph Drawing*, Proc. of DIMACS Workshop GD '94, Lecture Notes in Comp. Sci. 894, Springer-Verlag, 1994, pp. 1-11.

5. I. Cruz and J. Twarog, *3D Graph Drawing with Simulated Annealing*, Proc. of Workshop GD '95, Lecture Notes in Comp. Sci. 1027, Springer-Verlag, 1995, pp. 162-165.

6. G. Di Battista, P. Eades, R. Tamassia and I. Tollis, *Algorithms for Drawing Graphs: An Annotated Bibliography*, Computational Geometry: Theory and Applications, vol. 4, no 5, 1994, pp. 235-282. Also available via anonymous **ftp** from **ftp.cs.brown.edu**, **gdbiblio.tex.Z** and **gdbiblio.ps.Z** in **/pub/papers/compgeo**.

7. P. Eades, C. Stirk, S. Whitesides, *The Techniques of Kolmogorov and Bardzin for Three Dimensional Orthogonal Graph Drawings*, TR 95-07, Dept. of Computer Science, University of Newcastle, Australia, 1995. Also to appear in Information Processing Letters.

8. P. Eades, A. Symvonis, S. Whitesides, *Two Algorithms for Three Dimensional Orthogonal Graph Drawing*, Proc. of Workshop GD '96, Lecture Notes in Comp. Sci. 1190, Springer-Verlag, 1996, pp. 139-154.

9. A. Garg and R. Tamassia, *GIOTTO3D: A System for Visualizing Hierarchical Structures in 3D*, Proc. of Workshop GD '96, Lecture Notes in Comp. Sci. 1190, Springer-Verlag, 1996, pp. 193-200.

10. Goos Kant, *Drawing Planar Graphs Using the Canonical Ordering*, Algorithmica, vol. 16, no. 1, 1996, pp. 4-32.

11. A. N. Kolmogorov and Y. M. Bardzin, *About Realization of Sets in 3-dimensional Space*, Problems in Cybernetics, 1967, pp. 261-268.

12. J. MacKinley, G. Robertson, S. Card, *Cone Trees: Animated 3d visualizations of hierarchical information*, In Proc. of SIGCHI Conf. on Human Factors in Computing, pp. 189-194, 1991.

13. A. Papakostas and I. G. Tollis, *Algorithms for Area-Efficient Orthogonal Drawings*, Technical Report UTDCS-06-95, The University of Texas at Dallas, 1995.

14. A. Papakostas and I. G. Tollis, *Issues in Interactive Orthogonal Graph Drawing*, Proc. of Workshop GD '95, Lecture Notes in Comp. Sci, 1027, Springer-Verlag, 1995, pp. 419-430.

15. A. Papakostas and I. G. Tollis, *A Pairing Technique for Area-Efficient Orthogonal Drawings*, Proc. of Workshop GD '96, Lecture Notes in Comp. Sci. 1190, Springer-Verlag, 1996, pp. 355-370.

16. A. Papakostas, J. Six and I. G. Tollis, *Experimental and Theoretical Results in Interactive Graph Drawing*, Proc. of Workshop GD '96, Lecture Notes in Comp. Sci. 1190, Springer-Verlag, 1996, pp. 371-386.

17. A. Papakostas and I. G. Tollis, *Incremental Orthogonal Graph Drawing in Three Dimensions*, Technical Report UTDCS-02-97, The University of Texas at Dallas, 1997. (available through www.utdallas.edu/~tollis)

18. S. Reiss, *An engine for the 3D visualization of program information*, J. Visual Languages and Computing, vol. 6, no. 3, 1995.

19. Markus Schäffter, *Drawing Graphs on Rectangular Grids*, Discr. Appl. Math. 63 (1995) pp. 75-89.

20. R. Tamassia and I. Tollis, *Planar Grid Embeddings in Linear Time*, IEEE Trans. on Circuits and Systems CAS-36 (1989), pp. 1230-1234.

On Three-Dimensional Layout
of Interconnection Networks
(Extended Abstract) *

Tiziana Calamoneri[1] and Annalisa Massini[2]

[1] Dipartimento di Matematica and Dipartimento di Scienze dell'Informazione,
Università di Roma "La Sapienza", Italy - calamo@dsi.uniroma1.it.
[2] Dipartimento di Scienze dell'Informazione, Università di Roma "La Sapienza",
Italy - massini@dsi.uniroma1.it.

Abstract. In this paper we deal with the layout of interconnection networks on three-dimensional grids. In particular, in the first part we prove a general formula for calculating an exact value for the lower bound on the volume. Then we introduce the new notion of k-3D double channel routing and we use it to exhibit an optimal three-dimensional layout for butterfly networks. Finally, we show a method to lay out multigrid and X-tree networks in optimal volume.

1 Introduction and Preliminaries

Recent hardware advances have allowed three-dimensional circuits to have a cost low enough to make them commonly available. For this reason three-dimensional layouts of graphs on rectilinear grids are becoming of wide interest both in the study of the VLSI layout problem for integrated circuits and in the study of algorithms for drawing graphs. Indeed, the tie between VLSI layout studies and theoretical graph drawing is very strong since to lay out a network on a grid is equivalent to orthogonally draw the underlying graph.

To the best of our knowledge, not many papers have been written about three-dimensional grid drawing of graphs [2, 3, 4, 5, 6, 10] and all of them show results that are valid for very general graphs and therefore they do not work efficiently for structured and regular graphs such as the most commonly used interconnection networks. On the other hand, the importance of representing interconnection networks in three dimensions has already been stated in the 80's by Rosenberg [12]: the most relevant aims are to shorten wires and to save in material.

By virtue of the equivalence between layout of networks and drawing of graphs in the following we will prefer the network terminology instead of the graph theory one; therefore we will use the word 'node' instead of 'vertex' and 'layout' instead of 'drawing', while we will interchangeably use the terms 'graph' and 'network', 'edge' and 'wire'.

In this paper we focus our attention on *three-dimensional grid layout* of an interconnection network $\mathcal{G}$, that is a mapping of $\mathcal{G}$ in the three-dimensional grid

---

* The first author has been supported by Italian National Research Council.

such that nodes are mapped in grid-nodes and edges are mapped in independent grid-paths satisfying the following conditions:

- distinct grid-paths are edge-disjoint (then at most three paths can cross at a grid-node);
- grid-paths that share an intermediate grid-node must cross at that node (that is 'knock-knee' paths [9] are not allowed);
- a grid-path may touch no mapped node, except at its endpoints.

If the layout of a graph $\mathcal{G}$ can be enclosed in a $h \times w \times l$ three-dimensional grid, we say *layout volume* of $\mathcal{G}$ the product $h \times w \times l$.

In this work we give some results about lower and upper bounds on the layout volume of some interconnection networks. Namely, in the first part we prove a general formula for calculating an exact value for the lower bound on the three-dimensional layout volume. Then we introduce the new notion of k-3D double channel routing and we use it to exhibit an optimal three-dimensional layout for butterfly networks. Finally, we show a method to lay out multi-grid and X-tree networks in optimal volume.

2 Lower bound

In this section we prove a general formula giving an exact value for the lower bound on the layout volume of interconnection networks.

We obtain our result by generalizing to three dimensions the classical lower bound strategy for two dimensions invented in [13], and modified in [1]. In [12] the order of magnitude of the result obtained in Lemma 4 is given for reticulated graphs and extended to more general graphs. Before proving the general formula for the lower bound, we give some definitions and prove some preliminary results.

Definition 1. An *embedding of graph* $\mathcal{G}$ *into graph* $\mathcal{H}$ (which has at least as many nodes as $\mathcal{G}$) comprises a *one-to-one association* α of the nodes of $\mathcal{G}$ with nodes of $\mathcal{H}$, plus a *routing* ρ which associates each edge $\{u, v\}$ of $\mathcal{G}$ with a path in $\mathcal{H}$ that connects nodes $\alpha(u)$ and $\alpha(v)$. The *congestion* of embedding $\langle \alpha, \rho \rangle$ is the maximum, over all edges e in $\mathcal{H}$, of the number of edges in $\mathcal{G}$ whose ρ-routing paths contain edge e.

Definition 2. Let $\mathcal{G}$ be a graph having a designated set of $2c > 0$ nodes, called *special nodes*. The *minimum special bisection width* of a graph $\mathcal{G}$, MSBW($\mathcal{G}$), is the smallest number of edges whose removal partitions $\mathcal{G}$ into two disjoint subgraphs, each containing half of $\mathcal{G}$'s special nodes.

Lemma 3. *[8] Let ϵ be an embedding of graph $\mathcal{G}$ into graph $\mathcal{H}$ that has congestion C, then $MSBW(\mathcal{H}) \geq \frac{1}{C} MSBW(\mathcal{G})$.*

Now we prove a general formula to get a lower bound on the layout volume of a network, given its MSBW.

Lemma 4. *For any graph $\mathcal{H}$, the volume of the smallest three-dimensional layout of $\mathcal{H}$ is at least $\left(\frac{MSBW(\mathcal{H}) - 1}{2} \right)^{3/2}$.*

Sketch of proof. We consider an arbitrary layout of $\mathcal{H}$ in the grid of dimension $h \times w \times l$. A surface S with a single jog J (see Fig. 1) can be positioned on the grid in such a way that it cuts the layout of $\mathcal{H}$ into two subgraphs, each containing half of $\mathcal{H}$'s special nodes.

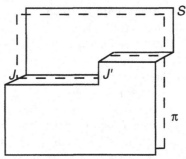

Fig. 1. Surface S with the jog J.

Removing the grid-edges crossed by S yields a bisection of $\mathcal{H}$. By definition, at least $MSBW(\mathcal{H})$ edges of $\mathcal{H}$ must cross surface S. By construction, at most $hl + l + 1 \leq 2hl + 1$ edges of the grid cross surface S. It follows that $2hl + 1 \geq MSBW(\mathcal{H})$. On the other hand, without loss of generality, we can choose $w \geq h \geq l$. Hence $h \times w \times l \geq \left(\frac{MSBW(\mathcal{H})-1}{2} \right)^{3/2}$.

As a consequence of Lemmas 4 and 3, if $MSBW(\mathcal{H})$ is not known, a lower bound on the layout volume of a network $\mathcal{H}$ can be computed through an embedding ϵ into $\mathcal{H}$ of a graph $\mathcal{G}$ if $MSBW(\mathcal{G})$ and the congestion C of ϵ are known. In this way, we have that a lower bound on the layout volume of $\mathcal{H}$ is no less than $\left(\frac{MSBW(\mathcal{H})-1}{2} \right)^{3/2} \geq \left(\frac{\frac{1}{C}MSBW(\mathcal{G})-1}{2} \right)^{3/2}$.

Since another lower bound on the layout volume of a graph is trivially given by the number of nodes of the graph, the following theorem derives:

Theorem 5. *Given a graph $\mathcal{H}$ with n nodes, a lower bound on its layout volume is given by* $\max \left\{ n, \left(\frac{MSBW(\mathcal{H})-1}{2} \right)^{3/2} \right\}$. *Alternatively, when an embedding of congestion C for an auxiliary graph $\mathcal{G}$ into $\mathcal{H}$ and $MSBW(\mathcal{G})$ are known, a lower bound on the layout volume of $\mathcal{G}$ is* $\max \left\{ n, \left(\frac{\frac{1}{C}MSBW(\mathcal{G})-1}{2} \right)^{3/2} \right\}$.

3 Upper bound of some Interconnection Networks

In this section we first give the definitions of all networks we are going to manage, then we exhibit a method to lay out each of them in a three-dimensional grid.

Definition 6. The *butterfly network* having N inputs $\mathcal{B}_N$, where $N = 2^n$, has nodes corresponding to pairs $\langle w, l \rangle$ where l is the level ($1 \leq l \leq \log N + 1$) and w is a $\log N$-bit binary number that denotes the column of the node. Two nodes

$\langle w, l \rangle$ and $\langle w', l' \rangle$ are linked by an edge if and only if $l' = l + 1$ and either:
1. w and w' are identical (straight-edge), or
2. w and w' differ in precisely the l-th bit (cross-edge).

Lemma 7. *[7] The subgraph of $\mathcal{B}_N$ induced by the nodes of levels $1, \ldots, h$ is the disjoint sum of $2^{\log N - h + 1}$ copies of $\mathcal{B}_{2^{h-1}}$ and the subgraph of $\mathcal{B}_N$ induced by the nodes of levels $h, \ldots, \log N + 1$ is the disjoint sum of 2^{h-1} copies of $\mathcal{B}_{2^{\log N - h + 1}}$.*

Definition 8. The $N \times N$ *multigrid network* $\mathcal{M}_N$, where $N = 2^n$, consists of $\log N + 1$ bidimensional arrays, each one of size $N/2^k \times N/2^k$ for $0 \le k \le \log N$. The arrays are interconnected so that node (i, j) on the $2^k \times 2^k$ array is connected to node $(2i, 2j)$ on the $2^{k+1} \times 2^{k+1}$ array for $1 \le i, j \le 2^k$, and $0 \le k < \log N$.

Definition 9. The N-leaf *X-tree* $\mathcal{T}_N$, where $N = 2^n$, is a complete N-leaf binary tree with edges added to connect consecutive nodes on the same level of the tree.

We can utilize Theorem 5 to compute, in particular, a lower bound on the layout volume of the interconnection networks just defined:

- a lower bound on the layout volume of a butterfly network $\mathcal{B}_N$ is $\left(\frac{N-1}{2}\right)^{3/2}$ and can be obtained by considering the embedding described in [1].
- the number of nodes of a multigrid $\mathcal{M}_N$ constitutes a lower bound on its layout volume, that is $\frac{4N^2-1}{3}$. Indeed, the formula involving $MSBW(\mathcal{M}_N) = \Theta(N)$ produces a worse value.
- similar considerations hold for an N-leaf X-tree $\mathcal{T}_N$, whose $MSBW$ is $\Theta(\log N)$, and therefore a lower bound on its layout volume is $2N - 1$.

For what concerns the upper bound on the layout volume we divide the next part into three subsections, one for each network.

3.1 Butterfly Network

It is easy to obtain an optimal three-dimensional layout of a butterfly network by using the forerunner intuition of Wise [14] used to better visualize a butterfly network in the space. This idea is based on opportunely putting and connecting in the space $O(\sqrt{N})$ copies of any bidimensional optimal layout of a butterfly with $O(\sqrt{N})$ inputs (possible in view of Lemma 7). A drawback of such a nice layout is that the maximum wire length is $O(\sqrt{N})$, and most of the wires reach this upper bound.

In the following we will describe a method to lay out a $\mathcal{B}_N$ in the three-dimensional grid so that all its wires have maximum length $O(N^{1/4})$ but one (additive) edge-level characterized by having maximum wire length $O(\sqrt{N})$.

From now on, we will assume that $\log N$ is even; when $\log N$ is odd it is easy to adjust the details, that we omit for the sake of brevity.

In view of Lemma 7 we can 'cut' $\mathcal{B}_N$ along its median node-level and get $\sqrt{N}$ copies of $\mathcal{B}_{\sqrt{N}}$ (O-group) whose output nodes must be re-connected to the input nodes of other $\sqrt{N}$ copies of $\mathcal{B}_{\sqrt{N}}$ (I-group) through an additive edge-level.

Hence, our layout consists of two main operations:

– three-dimensional layout of each copy of $\mathcal{B}_{\sqrt{N}}$;
– re-connection of the two groups of $\sqrt{N}$ copies of $\mathcal{B}_{\sqrt{N}}$ through an additive edge-level.

Three-dimensional layout of each copy of $\mathcal{B}_{\sqrt{N}}$

In order to explain how to manage this operation, we need to mark the following observation:

Observation 1 *An N-input butterfly network $\mathcal{B}_N$ can be covered by N edge-disjoint complete binary trees as follows:*
– *for any $i = 2, \ldots, \log N$, there are $2^{\log N - i}$ trees T_i having i levels, sharing their leaves with some tree $T_j, j > i$, and their internal nodes with some $T_k, k < i$;*
– *there are two trees $T_{\log N + 1}$ having $(\log N + 1)$ levels, sharing their leaves each other, and their internal nodes with some $T_k, k < \log N + 1$.*

An example of this covering for $\mathcal{B}_{16}$ is depicted in Fig. 2.

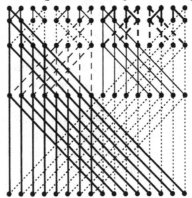

Fig. 2. Tree-covering of $\mathcal{B}_{16}$ (different trees are represented by different line types).

Consider an H-tree representation of $T_{\log \sqrt{N} + 1}$, call it $H_{\log \sqrt{N} + 1}$. Call H_i a plane representation of T_i obtained from $T_{\log \sqrt{N} + 1}$ by eliminating superfluous $\log \sqrt{N} + 1 - i$ levels. Then T_i is represented according to an H-tree scheme wasting some area. Observe that if the leaves of a tree T_j coincide with some internal nodes of a tree $T_i, i > j$, it is possible to lay out T_i and T_j in the three-dimensional grid by considering H_i and H_j on two parallel planes, such that the orthogonal projection of H_j on the plane containing H_i coincides, level by level, with H_i itself. To correctly connect H_i and H_j we have to connect duplicate nodes by a segment orthogonal to both planes and to eliminate the leaves of H_j, substituting them with bends (see Fig. 3).

In view of Observation 1, it remains to detail in which order the planes containing the $\sqrt{N}$ binary trees must be arranged. The following recursive pseudo-code allows one to assign a z-coordinate to each plane containing T_j ($z \leftarrow T_j$ for short). The first call of the procedure is $\text{PUT}(T_{\log \sqrt{N} + 1}, 0)$.

```
PROCEDURE PUT(T_j, VAR z);
BEGIN
    z ← T_j;
    z + 1 ← (T_2 sharing its leaves with level 2 of T_j);
    i := 3;
    WHILE (i < j) DO
    BEGIN
        PUT(T_i sharing its leaves with level i of T_j, z + 2);
        i:=i+1;
    END;
END.
```

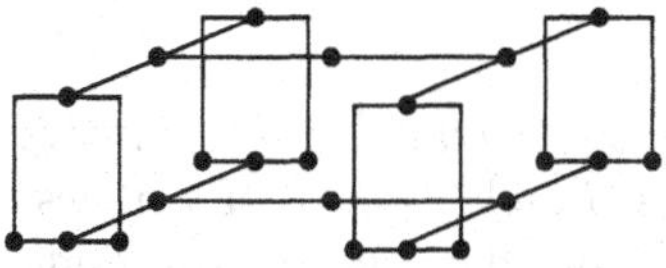

Fig. 3. Layout of two trees sharing some nodes.

After the procedure is terminated, half of $B_{\sqrt{N}}$ has been lain out. The remaining part can be symmetrically laid out in such a way that the planes containing trees $T_{\log \sqrt{N}+1}$ are consecutive.

As far as the procedure is concerned, vertical lines are guaranteed:

— not to cross tree-nodes of intermediate planes; indeed, the procedure puts the trees connected to a certain tree T_j such that as smaller they are as closer to T_j they are positioned;

— not to coincide with other vertical lines; indeed, no more than two trees can share the same nodes.

In view of the construction of the three-dimensional layout of $B_{\sqrt{N}}$, of Observation 1 and of the area of an H-tree, each butterfly $B_{\sqrt{N}}$ belonging both to the O-group and to the I-group take a $(2N^{1/4} - 1) \times (2N^{1/4} - 1) \times (N^{1/2})$ volume.

Re-connection between the two groups of $\sqrt{N}$ copies of $B_{\sqrt{N}}$

Let us consider the two groups of $\sqrt{N}$ copies of $B_{\sqrt{N}}$. Each group is positioned in the space to form a square with $N^{1/4}$ copies on each side, such that the correspondent trees of each copy lie on the same plane. The two groups are then positioned one in front of the other. Now we have to connect the duplicated nodes through an additive edge-level.

Before detailing this operation, we need to remind some known results. A *k-channel routing* involves a bidimensional grid and two sets S and S' each consisting of k nodes to be connected by a 1-1 function. S and S' are arranged on two opposite sides of the grid.

Lemma 10. *[11] The grid involved in any k-channel routing is not greater than $(k + 1) \times (\frac{3}{2}k + 2)$ and S and S' lie on the shorter sides.*

Coming back to the butterfly problem, observe that all the output nodes of the O-group and all the input nodes of the I-group can be provided of an outgoing link towards the opposite group and their extremes can be leaded to two parallel planes, having empty intersection with the layouts of each copy. If we number in the same way –from left to right, row by row– both the output nodes of any butterfly of the O-group and the input nodes of any butterfly of the I-group and the butterflies themselves of O- and I-groups, then each edge must connect the i-th output node of the j-th butterfly in the O-group to the j-th input node of the i-th butterfly in the I-group. Furthermore, it is easy to see that each row of output nodes in the O-group is routed to a row of input nodes in the I-group.

In order to solve this problem we define a new three-dimensional constrained routing, called k-3D double channel routing, to which we reduce the previous problem.

Definition 11. A *k-3D double channel routing* involves a three-dimensional grid (the channel) and two sets S and S', both of k nodes, to be connected by a 1-1 function f. S and S' are arranged on two opposite sides of the three-dimensional grid, on the nodes of a $\sqrt{k} \times \sqrt{k}$ grid. Function f associates to a node (x, y) of S a node (x', y') of S' such that $x' = g(x)$ and $y' = h(y)$, where functions g and h are two-dimensional $\sqrt{k}$-channel routings.

Theorem 12. *A three-dimensional grid of size* $(\sqrt{k} + 1) \times (\sqrt{k} + 1) \times (\frac{3}{2}\sqrt{k} + 2)$ *is enough to realize a k-3D double channel routing.*

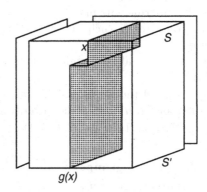

Fig. 4. Three-dimensional double channel routing.

Proof. Project the three-dimensional grid of the k-3D double channel routing on plane xz. It is easy to see that function g mapping rows of S in rows of S' can be considered as a two-dimensional channel routing on plane xz. Therefore, a $(\sqrt{k} + 1) \times (\frac{3}{2}\sqrt{k} + 2)$ two-dimensional grid is enough to realize such a channel routing (Lemma 10). When coming back to three dimensions, lines laid out to represent function g become (bent) planes. Each of such planes has on opposite horizontal sides a row x of S and its corresponding row $g(x)$ of S' and it is at least $\frac{3}{2}\sqrt{k} + 2$ long (see Fig. 4). Therefore, on each plane we can realize a two-dimensional channel routing given by function h, simply by adding an extra-plane, parallel to plane xz.

We use this theorem to lay out the additive edge-level between the O-group and the I-group in at most $\frac{3}{2}\sqrt{N}+2$ height.

Recombining all the arguments about the volume needed by the two operations of laying out each copy of $\mathcal{B}_{\sqrt{N}}$ and re-connecting the two groups of $\sqrt{N}$ copies of $\mathcal{B}_{\sqrt{N}}$, we can state the following theorem:

Theorem 13. *There exists a three-dimensional grid layout of a butterfly network with N inputs and N outputs $\mathcal{B}_N$ with volume $(2N^{1/2} - N^{1/4} + 1) \times (2N^{1/2} - N^{1/4}+1) \times (\frac{7}{2}N^{1/2}+2)$ and all edges have maximum wire length $O(N^{1/4})$, except N edges having maximum wire length $O(N^{1/2})$.*

3.2 Multi-Grid Network

In this subsection we will show how to lay out an $N \times N$ multigrid $\mathcal{M}_N$ in a three-dimensional grid of size $O(N) \times O(N) \times O(1)$ and maximum edge length $O(N)$. It remains an open problem to find an equal sided three-dimensional layout such that the maximum wire length is shortened.

First, we describe how to lay out all the bidimensional arrays (shortly arrays, where no confusion arises), then we show how to connect adjacent arrays.

All nodes and edges of all the arrays can be positioned on a unique plane π in the following way (see Fig. 5):

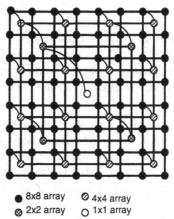

● 8x8 array   ⊘ 4x4 array
⊗ 2x2 array   ○ 1x1 array

Fig. 5. How to lay out all the bidimensional arrays in a $\mathcal{M}_8$.

- put all nodes of the $N \times N$ array at even coordinates, and connect them in the oblivious way;
- let v_k be the generic node on the $N/2^k \times N/2^k$ array. Put it at coordinates $(x+2^{k-1}, y+2^{k-1})$, where (x,y) are the coordinates of node v_{k-1} to which v_k is connected. Finally, lay out the edges of the current array in the oblivious way.

Edges connecting different arrays can be lain out as follows:

- from any node $v_k, 0 \leq k \leq \log N - 1$ that is endpoint of an edge towards a v_{k+1}, draw a unit length segment orthogonal to π going to an upper plane π' (u-lines);
- from any node $v_k, 1 \leq k \leq \log N$ that is endpoint of an edge towards a v_{k-1}, draw a broken line composed by: a unit length segment orthogonal to π going to a lower plane π'', a unit length segment along y coordinate on π'', and a segment orthogonal to π, going from π'' to π' (l-lines);
- on π', connect the endpoints of the u- and l-lines corresponding to the same edge by means of an L-like line.

Observe that, in view of the position of the nodes on π, both all these edges never cross any node and no collisions arise on π''.

It is easy to see that the area occupied on π by all the arrays is $(2N - 1) \times (2N-1)$ and that the addition of π' and π'' is enough to lay out all the remaining edges. Furthermore, the longest wires on π are N long (they belong to the 2×2 array); the longest edge connecting adjacent arrays connects the 2×2 and the 1×1 arrays and is $N + 4$ long. All these considerations lead to the following result:

Theorem 14. *There exists a three-dimensional grid layout of an $N \times N$ multi-grid $\mathcal{M}_N$ with volume $(2N - 1) \times (2N - 1) \times 3$ and all edges have maximum wire length $O(N)$.*

3.3 X-tree Network

In this subsection we will show how to lay out an N leaf X-tree $\mathcal{T}_N$ in a three-dimensional grid having $O(\sqrt{N}) \times O(\sqrt{N}) \times O(1)$ volume, that is optimum. The authors are going to prove that it is possible to lay out an N leaf X-tree in an equal sided three-dimensional grid, such that the maximum wire length is $N^{1/3}$ instead of $\sqrt{N}$.

From the definition itself of X-tree, we can distinguish in a $\mathcal{T}_N$ an N leaf complete binary tree and a set of $2N - 2 - \log N$ horizontal non-tree edges. It is easy to lay out the binary tree, as an H-tree on a bidimensional $O(\sqrt{N}) \times O(\sqrt{N})$ grid. From now on we will call π the plane where the H-tree lies.

It is also easy to lay out a part of the set of non-tree edges in view of the following observation:

Observation 2 *Consider the set of $N - 1$ non-tree edges lying alternately on each level. Each of them can be visualized on an N leaf complete binary tree as a couple of edges connecting two siblings, eliminating their father. See Fig. 6.*

It is possible to lay out all such $N - 1$ non-tree edges on a new plane π'; to this end, lead a unit length connection orthogonal to π towards π' from the extremes of such edges and lay out on π' the required connections. Then, on π' there is a kind of H-tree, whose nodes are substituted by knock-knees. We can eliminate them by using two parallel planes, π' and π'', instead of one.

To manage the set of the remaining non-tree edges, we use an inductive method.

Our claim is that given any $\mathcal{T}_k, k \geq 4$ and k power of two, its $2k - 2 - \log k$ non-tree edges can be positioned on the three-dimensional grid in the following way:

Fig. 6. Non-tree edges visualized as couples of tree edges.

a. $k/2$ non-tree edges lie on π;
b. $k - 1$ non-tree edges lie on π' and π'' (on a unique plane in Fig. 7);
c. the remaining $k/2 - \log k - 1$ non-tree edges lie on a further plane π'''.

The basis of the induction is represented by the three-dimensional layouts of $\mathcal{T}_4, \mathcal{T}_8$ and $\mathcal{T}_{16}$, all depicted in Fig. 7. $\mathcal{T}_4, \mathcal{T}_8$ are initial cases, while $\mathcal{T}_{16}$ is the first X-tree following our claim.

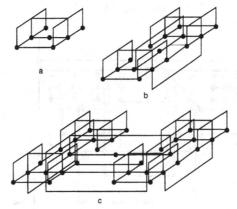

Fig. 7. Three-dimensional layout of $\mathcal{T}_4, \mathcal{T}_8$ and $\mathcal{T}_{16}$.

The inductive step consists in considering that each $\mathcal{T}_N$ is constituted by two copies of $\mathcal{T}_{N/2}$ connected by a newly introduced root and $\log N$ new non-tree horizontal edges (see Fig. 8). Our inductive hypothesis is that $N/4$ edges lie on π, $N/2 - 1$ lie on π' and π'' and the remaining $N/4 - \log N/2 - 1$ lie on a further plane π'''. The N leaf complete binary tree inside $\mathcal{T}_N$ can be laid out on π as union of the two $N/2$ leaf binary trees inside the two copies of $\mathcal{T}_{N/2}$ and of the new root.

Let us prove that our claim remains true for $\mathcal{T}_N$ if it is true for $\mathcal{T}_{N/2}$:

a. the $N/4 + N/4$ non-tree edges of $\mathcal{T}_{N/2}$ lying on π constitute all non-tree edges of $\mathcal{T}_N$ that must lie on π;

b. the non-tree edge connecting the two children of the root of $\mathcal{T}_N$ takes part in the special H-tree of planes π' and π''; therefore, non-tree edges we put on such planes are $(N/2 - 1) + (N/2 - 1)$ from the two $\mathcal{T}_{N/2}$ plus one, that is $N - 1$;

c. on π''' lie all non-tree edges of the two $\mathcal{T}_{N/2}$ lying on it plus all $\log N - 1$ non-tree edges connecting the two copies of $\mathcal{T}_{N/2}$ and not laid yet, that is $2(N/4 - \log N/2 - 1) + \log N - 1 = N/2 - \log N - 1$.

Fig. 8. A $\mathcal{T}_N$ as union of two $\mathcal{T}_{N/2}$ and non-tree edges.

It remains to detail how non-tree edges on π''' are settled. Observe that non-tree edges lying on π''' we add in the inductive phase connect the right-most nodes of a $\mathcal{T}_{N/2}$ to the left-most nodes of the other $\mathcal{T}_{N/2}$. As far as the H-tree is concerned, we can lay out on π''' directly only $\lceil \frac{\log N - 1}{2} \rceil$ of such edges; for the remaining non-tree edges we need $\lfloor \frac{\log N - 1}{2} \rfloor$ extra-lines on π''' with respect to the area occupied by the H-tree on π (see Fig. 9).

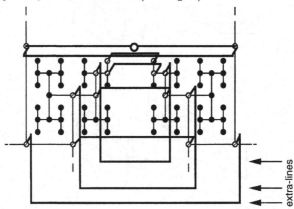

Fig. 9. Edges laid out on π and π''' during the inductive step.

Actually, at each inductive step, it is not necessary to add $\lfloor \frac{\log N - 1}{2} \rfloor$ extra-lines but only one, since we can use the extra-lines introduced in the previous steps. Possible knock-knees on π''' can again be avoided by means of a further parallel plane.

By following the previous construction, it is possible to express the layout volume of a $\mathcal{T}_N$ by means of a recursive formula, whose solution is:

$-\ 5 \times \left(\frac{11}{4}\sqrt{N} - 3\right) \times \left(\frac{19}{16}\sqrt{N} - 3\right)$ when $\log N$ is even;

$-\ 5 \times \left(\frac{23}{16}\sqrt{N/2} - 3\right) \times \left(\frac{35}{16}\sqrt{N/2} - 3\right)$ when $\log N$ is odd.

All the previous arguments lead to the following result:

Theorem 15. *There exists a three-dimensional grid layout of an N leaf X-tree $\mathcal{T}_N$ with volume $O(\sqrt{N}) \times O(\sqrt{N}) \times O(1)$ and all edges have maximum wire length $O(N^{1/2})$.*

Unfortunately, we did not succeed in applying our inductive method to the three-dimensional version of the H-tree introduced in [12], without increasing the volume of a non-constant factor. It would have implied an optimal layout in an equal sided volume with optimal wire length, that is $O(N^{1/3})$.

References

1. AVIOR, A. – CALAMONERI, T. – EVEN, S. – LITMAN, A. – ROSENBERG, A.L.: A Tight Layout of the Butterfly Network. *Proc. ACM SPAA '96*, ACM Press Ed., 1996, pp 170–175.
2. BIEDL, T.: New Lower Bounds for Orthogonal Graph Drawings. *Proc. GD '95*, LNCS 1027, Springer-Verlag, 1995, pp 28–39.
3. CALAMONERI, T. – STERBINI, A.: Drawing 2-, 3- and 4-colorable Graphs in $O(n^2)$ volume. *Proc. GD '96*, LNCS 1190, Springer-Verlag, 1996, pp 53–62.
4. COHEN, R.F. – EADES, P. – LIN, T. – RUSKEY, F.: Three-dimensional graph drawing. *Proc. GD '94*, LNCS 894, Springer-Verlag, 1994, pp 1-11. Also in *Algorithmica* 17(2), pp 199–208, 1997.
5. EADES, P. – FENG, Q.W.: Multilevel Visualization of Clustered Graphs. *Proc. GD '96*, LNCS 1190, Springer-Verlag, 1996, pp 101-112.
6. EADES, P. – SYMVONIS, A. – WHITESIDES, S.: Two Algorithms for Three Dimensional Orthogonal Graph Drawing. *Proc. GD '96*, LNCS 1190, Springer-Verlag, 1996, pp 139-154.
7. EVEN, S. – LITMAN, A.: Layered Cross Product – A Technique to Construct Interconnection Networks. *ACM SPAA '92*, ACM Press Ed., 60-69, 1992.
8. LEIGHTON, F.T.: *Complexity Issues in VLSI: Optimal Layouts for the Shuffle-Exchange Graph and Other Networks.* MIT Press, Cambridge, Mass, 1983.
9. MEHLORN, K. – PREPARATA, F.P. – SARRAFZADEH, M.: Channel routing in knock-knee mode: simplified algorithms and proofs. *Algorithmica* 1, 213-221, 1986.
10. PACH, J. – TÓTH, G.: Three-dimensional grid drawings of graphs, These Proceedings, 1997.
11. PINTER, R.Y.: On routing two-point nets across a channel. *19th ACM-IEEE Design Automation Conf.*, 894-902, 1982.
12. ROSENBERG, A.L.: Three-Dimensional VLSI: A Case Study. *Journal of the ACM*, 30(3), 1983, pp 397–416.
13. THOMPSON, C.D.: *A complexity theory for VLSI*. Ph.D. thesis, Carnegie-Mellon Univ. Pittsburgh, 1980.
14. WISE, D.S.: Compact layouts of banyan/FFT networks. *VLSI Systems and Computations* (H.T. Kung, B. Sproull, G. Steele, eds.) Computer Science Press, Rockville, Md., 1981, pp 186-195.

Orthogonal 3-D Graph Drawing *

T. Biedl[1] and T. Shermer[2] and S. Whitesides[1] and S. Wismath[3]

[1] School of Computer Science, McGill University, Montreal, PQ H3A2A7, Canada
[2] School of Computing Science, Simon Fraser University, Burnaby, BC V5A1A6, Canada
[3] Department of Mathematics and Computer Science, University of Lethbridge, Lethbridge, AB T1K3M4, Canada

Abstract. This paper studies 3-D orthogonal grid drawings for graphs of arbitrary degree, K_n in particular, with vertices drawn as boxes. It establishes an asymptotic lower bound for the volume of the bounding box of such drawings and exhibits a construction that achieves this bound. No edge route in this unconstrained construction bends more than three times.

For drawings constrained to have at most k bends on any edge route, simple constructions are given for $k = 1$ and $k = 2$. The unconstrained construction handles the $k \geq 3$ cases, while for $k = 0$ (no bends), it is proved here that not all graphs can be drawn.

1   Introduction

This paper offers methods for constructing 3-D orthogonal grid drawings for graphs of *arbitrary* degree. It also contributes a lower bound result for the volumes of such drawings, establishing that one of our constructions is in some sense optimal. To state the main results clearly, we explain, following some terminology, the drawing conventions and volume measure used.

A *grid point* is a point in R^3 whose coordinates are all integers. A *grid box* is the set of all points (x, y, z) in R^3 satisfying $x_0 \leq x \leq x_1$, $y_0 \leq y \leq y_1$ and $z_0 \leq z \leq z_1$ for some integers $x_0, x_1, y_0, y_1, z_0, z_1$. A *port* of a box is any grid point of the box that is extremal in at least one direction. A grid box is said to have *dimensions* $a \times b \times c$ whenever $x_1 = x_0 + a - 1$, $y_1 = y_0 + b - 1$, and $z_1 = z_0 + c - 1$. The *volume* of such a box is defined to be the number of grid points it contains, namely abc. For example, a single grid point is a $1 \times 1 \times 1$ box of volume 1. The *volume* of a drawing is the volume of its *bounding box*, which is the smallest volume grid box containing the drawing. Often we refer to the bounding box as an $X \times Y \times Z$-grid.

Throughout this paper, a *3-D orthogonal grid drawing* of a graph $G = (V, E)$ is a drawing that satisfies the following. Distinct vertices of V are represented

---

* The authors gratefully thank N.S.E.R.C. for financial assistance.

by disjoint grid boxes[4]. An edge $e = (v_1, v_2)$ of E is drawn as a simple path that follows grid lines, possibly turning ("bending") at grid points; the endpoints of the path for e are ports on the boxes representing v_1 and v_2. The intermediate points along the path for an edge do not belong to any vertex box, nor do they belong to any other edge path. See Figure 1. In what follows, graph theoretic terms such as *vertex* are typically used to refer both to the graph theoretic object and to its representation in a drawing.

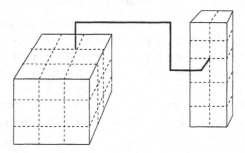

Fig. 1. Two boxes joined by a 4-bend edge.

For graphs drawn orthogonally in the 2-D grid, early research mainly considered graphs of maximum degree 4 and represented vertices as single grid points. More recently, 2-D orthogonal grid drawings of higher degree graphs have been investigated, where vertices have been drawn as rectangular boxes. See for example [FK96], [PT96], [BMT97].

At present, there are few results on 3-D orthogonal grid drawings. Rosenberg showed that any graph of maximum degree 6 can be embedded in a 3-D grid of volume $\mathcal{O}(n^{3/2})$, and that this is asymptotically optimal [Ros83]. No bounds on the number of bends were given. Recently, Eades, Symvonis and Whitesides gave a method for drawing graphs of maximum degree 6 in a grid of side-length $4\sqrt{n}$, with vertices represented by single grid points and each edge having at most 7 bends [ESW97]. They also gave a simple method for drawing such graphs in a grid of side-length $3n$, creating at most 3 bends on each edge. Papakostas and Tollis have proposed a more elaborate method that produces a drawing of volume at most $4.66n^3$ [PT97].

The focus of this paper is on 3-D orthogonal grid drawings of *complete* graphs. Since any simple graph G on n vertices is a subgraph of the complete graph K_n, a drawing of K_n immediately provides a drawing for G, since irrelevant edges may be deleted from the drawing. Complete graphs are also critical for many lower bound arguments.

---

[4] This paper allows vertices to be represented by degenerate boxes, i.e., by boxes that have dimension 1 with respect to one or more coordinate directions. Such degeneracies can be removed by adding additional grid lines, which increases the volume of the drawing by a multiplicative constant.

convention: From now on, the terms *drawing* and 3-D *orthogonal grid drawing* are used interchangably.

In informal language, the main results of this paper are as follows.

- For all sufficiently large n, K_n has no bend-free drawing.
- Any drawing of K_n has volume $\Omega(n^{2.5})$.
- K_n can be drawn in $O(n^3)$ volume with at most $k = 1$ bend per edge.
- K_n can be drawn in $O(n^3)$ volume with at most $k = 2$ bends per edge.
- K_n can be drawn in $O(n^{2.5})$ volume with at most $k = 3$ bends per edge.

Note that for $k \geq 3$, the upper and lower bounds on the volume match (within a constant factor) when a maximum of k bends per edge is allowed. The constructions of this paper have reasonably small constant factors for the volume. Only for the $k = 1$ and $k = 2$ cases do the bounds not match; in each of these cases we give an $O(n^3)$ volume drawing of K_n and leave as an open problem whether this drawing indeed has asymptotically optimum volume.

The results can be restated more precisely with the following terminology.

definition: Let $vol(n)$ denote the minimum possible volume of any drawing of K_n, and let $vol_k(n)$ denote the minimum possible volume for drawings of K_n that have k or fewer bends on any edge.

In these terms, the main results are that $vol_0(n)$ is undefined for large n, $vol(n)$ is in $\Omega(n^{2.5})$, $vol_1(n)$ and $vol_2(n)$ are in $O(n^3)$, and $vol_k(n)$ is in $\Theta(n^{2.5})$ for $k \geq 3$.

2 No Bends

The main result of this section is that there exist graphs that have no 0-bend 3-D orthogonal drawing. If no bends are permitted in the drawing, then the edges correspond to axis-parallel visibility lines between pairs of boxes. Such visibility representations have been studied in 2-D by Wismath [Wis85] and by Tamassia and Tollis [TT86], and in 3-D with 2-D objects [BEFLMRSW93], [FHW96]. A 3-D orthogonal drawing of a graph with no bends splits the edges into three classes, depending on the direction of visibility. Each class of edges forms a graph that has a visibility representation using only one direction of visibility. Our lower bound result depends on the fact that K_{56} has no such visibility representation, as shown by [FHW96].

The *3-Ramsey number* $R(r, b, g)$ is the smallest number such that any arbitrary colouring of the edges of $K_{R(r,b,g)}$ with colours red, blue and green induces either a red K_r, or a blue K_b, or a green K_g as a subgraph. This number exists and is finite; see for example [GRS80].

Theorem 1. *For all sufficiently large n (e.g., $n \geq R(56, 56, 56)$), K_n has no bend-free 3-D orthogonal grid drawing.*

One consequence of the previous theorem is that $\Omega(n^2)$ bends are required in any 3-D orthogonal grid drawing of K_n. Details are omitted.

3 A Lower Bound on the Volume

Recall that $vol(n)$ is the minimum possible volume for a drawing of K_n. This definition is valid since, as later sections show, every K_n has a drawing if edges are allowed to bend.

A *z-line* is a line that is parallel to the z-axis; *y-lines* and *x-lines* are defined analogously. A *z-plane* is a plane that is orthogonal to the z-axis; *x-planes* and *y-planes* are defined analogously.

Theorem 2. $vol(n) \in \Omega(n^{2.5})$.

Proof. The constants that appear below were chosen for convenience and have no special significance other than that they give a simple proof. We make no attempt here to produce a large constant multiplier for the $n^{2.5}$.

Consider a drawing of K_n in a grid of dimensions $X \times Y \times Z$.

Case 1: A line intersects many vertices

Assume there exists a z-line intersecting at least t vertices, where t is even and $t \geq \frac{1}{16}n$. Let $v_1, \ldots, v_t$ be any t of the vertices intersected by the z-line, listed in order of occurrence along the line, and let P_z be a z-plane (not necessarily with integer z-coordinate) that intersects none of these t vertices and that separates the first half of them from the second half.

Since the $\frac{1}{4}t^2$ edges connecting these two groups must cross the plane P_z, this plane must contain at least $\frac{1}{4}t^2$ points having integer x- and y-coordinates. Hence $XY \geq \frac{1}{4}t^2 > \frac{1}{1024}n^2$. Also, $Z \geq \frac{1}{16}n$ since the z-line intersects at least $\frac{1}{16}n$ vertices. Thus $vol(n) > \frac{1}{16384}n^3$.

Case 2: A plane intersects many vertices

Assume now that no x-line, y-line or z-line intersects as many as $\frac{1}{16}n$ vertices, but that there exists a z-plane P_z intersecting at least $\frac{1}{4}n$ vertices.

A vertex is *left* of an x-plane P_x if all the points in its grid box have x-coordinates less than x. The notion of *right* of P_x is analogous. As P_x is swept from smaller to larger values of x, the y-line determined by its intersection with P_z intersects fewer than $\frac{1}{16}n$ vertices, by assumption. As x increases, an integer $x = x_0$ is encountered where, for the last time, there are fewer than $\frac{1}{16}n$ vertices left of P_x and intersecting P_z.

The number of vertices that intersect P_z and that lie left of P_{x_0+1} is at least $\frac{1}{16}n$ but at most $\frac{2}{16}n - 2$. Thus at least $\frac{1}{16}n$ vertices intersect P_z and lie right of P_{x_0+1}. There are at least $\frac{1}{256}n^2$ edges between the vertices on the left and the vertices on the right, so $YZ \geq \frac{1}{256}n^2$. Apply exactly the same argument in the y-direction to obtain $XZ \geq \frac{1}{256}n^2$. Finally, note that $XY \geq \frac{1}{4}n$, since P_z intersects $\frac{1}{4}n$ vertices. Consequently, $XYZ = \sqrt{YZ \cdot XZ \cdot XY} \geq \frac{1}{512} n^{5/2}$.

Case 3: No plane intersects many vertices

Assume now that no plane intersects as many as $\frac{1}{4}n$ vertices. Consider P_x planes in order of increasing x value. By an argument analogous to the one in Case 2, a P_x will be encountered for which at least $\frac{1}{4}n$ vertices lie left of P_x, and at least $\frac{1}{4}n$ vertices lie right of P_x. Consequently, P_x contains at least $\frac{1}{16}n^2$ points with integer y- and z-coordinates, and $YZ \geq \frac{1}{16}n^2$. Since the same argument holds for the other two directions, $XYZ \geq (\frac{1}{16}n^2)^{3/2} = \frac{1}{64}n^3$.

For all sufficiently large n, the bound given by Case 2 is the smallest of the three; hence $vol(n) \in \Omega(n^{5/2})$.

4 Constructions

The lower bound of the previous section provides a volumetric goal for layout strategies. This section presents a construction that achieves this lower bound with a small constant factor. For the $k = 1$ case, two strategies are described and then modified to give a drawing for the $k = 2$ case. A simple construction that realizes the $\Omega(n^{2.5})$ lower bound for volume is described in subsection 4.3. The construction generates at most 3 bends on any edge and hence is valid for each $k \geq 3$. Whether the lower bound is attainable when $k = 1$ or 2 remains an open problem.

In each of the constructions, vertices are first placed as points in a 2-D x, y-plane. Next, all the edges are routed in the same xy-plane, with overlap and crossings of edges temporarily permitted. Then a number Z of z-planes is introduced, and edges are assigned to these planes so that no edges overlap or cross. The vertices are stretched into segments of z-lines.

4.1 Drawings of $O(n^3)$ volume for $k = 1$

In this section, we describe two strategies to draw K_n with at most $k = 1$ bends on any edge. The first layout scheme draws K_n in an $n \times n \times n$-grid. The second scheme then makes two drawings of $K_{n/2}$ (without recursion) using the first scheme; then it positions these drawings in an $\frac{n}{2} \times n \times \frac{n}{2}$-grid and supplies the edges between the two parts. For simplicity, assume below that n is divisible by 4.

Drawing K_n in an $n \times n \times n$-grid for $k = 1$ Enumerate the vertices as $v_1, \ldots, v_n$. Place vertex v_i at (i, i). Route edge $e = (v_i, v_j)$, where $i < j$, with one bend via $(i, i), (i, j), (j, j)$. Note that no vertex or part of an edge is placed at a point (x, y) with $y < x$.

Now partition the edges of K_n into edge sets $E_i^a, E_i^b, i = 1, \ldots, \frac{n}{2}$, defined as $E_i^a = \{(v_{i-l+1}, v_{i+l}) | l = 1, \ldots, \frac{n}{2}\}$ and $E_i^b = \{(v_{i-l}, v_{i+l}) | l = 1, \ldots, \frac{n}{2} - 1\}$ (all additions are modulo n). It is easy to check that these sets indeed partition the edges of K_n, and that no crossings or overlaps occur among edges in E_i^a nor among edges in E_i^b. Hence only n z-planes are needed. See Fig. 2. This gives the following lemma.

Lemma 3. *There exists a drawing of K_n in an $n \times n \times n$-grid with one bend per edge such that the points $\{(x, y, z) : y < x\}$ are unused.*

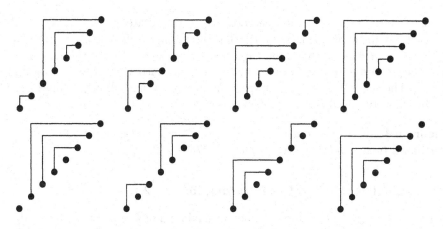

Fig. 2. The sets $E_b^1, \ldots, E_b^4$ for K_8.

Remark: Note that E_i^a and E_i^b can be drawn in the same plane by reflecting the edges of E_i^a with respect to the diagonal line through the vertices. This yields a drawing of K_n in an $n \times n \times \frac{n}{2}$-grid. This strategy is closely related to the *pagenumber* of a graph and in fact, may prove a useful idea for drawing sparse graphs. This idea yields, for example, a method for drawing planar graphs in $O(n^2)$ volume in an $n \times n \times 4$-grid, since it is known that planar graphs have pagenumber equal to 4 (see [Yan89]).

Drawing K_n in an $\frac{n}{2} \times n \times \frac{n}{2}$-grid for $k = 1$ Let K^1 and K^2 denote two drawings of $K_{n/2}$ as described in the previous lemma. Thus each drawing has an $\frac{n}{2} \times \frac{n}{2} \times \frac{n}{2}$ bounding box. Reflect the points in the box for K^2 through the $(y = 0)$-plane, so that all points in the reflected K^2 have negative y-coordinate. Then rotate this reflected K^2 so that vertex v_j of the rotated, reflected K^2 overlaps the points $(x, -j, j)$, where $1 \le x \le \frac{n}{2}$. See Fig. 3.

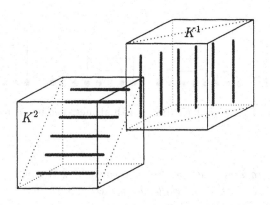

Fig. 3. K_1 and K_2

Each vertex v_i in K^1 sees each vertex v_j in the rotated, reflected K^2 along the y-line segment $[(i,i,j),(i,-j,j)]$. Therefore, these edges can be drawn as straight line segments, thus producing a drawing of K_n. Delete the unused y-plane of y-coordinate 0 to obtain a drawing with dimensions $X = Z = \frac{n}{2}$ and $Y = n$. There are $n^2/4$ edges drawn without a bend, and all other edges have one bend, so the total number of bends is $n^2/4 - n/2$.

Theorem 4. K_n *can be drawn in a* $\frac{n}{2} \times n \times \frac{n}{2}$-grid *with at most one bend per edge and a total number of bends equal to* $n^2/4 - n/2$.

4.2 A smaller $O(n^3)$ volume drawing for $k = 2$

A similar strategy can be applied when a maximum of $k = 2$ bends on an edge is allowed. In this section, K_n is drawn with at most two bends per edge by first making two copies of a drawing for $K_{\frac{n}{2}}$ and then placing them in a grid of side-length $\frac{n}{2}$ and supplying the edges connecting the two parts.

Drawing in an $n \times \frac{n}{2} \times n$-**grid** Enumerate the vertices as $\{v_1, \ldots, v_n\}$ and place v_i at $(x,y) = (i,1)$ in a 2-D (x,y)-plane. To route edge $e = (v_i, v_j)$, where $i < j$, let $y = \lceil \frac{j-i}{2} \rceil$ and route e via the points $(i,1), (i,y), (j,y), (j,1)$, creating two bends if $y > 1$ and no bends if $y = 1$.

Define the edge sets E_i^a and E_i^b as above. Again there are no crossings nor overlaps among edges in the same set and so n z-planes suffice. Since the largest y-coordinate is $\lceil \frac{n-1}{2} \rceil$, the bounding box has dimensions $n \times \frac{n}{2} \times n$. The edges (v_i, v_{i+1}) for $i = 1, \ldots, n-1$ are drawn straight; all other edges have two bends, so the total number of bends is $n^2 - 3n + 2$.

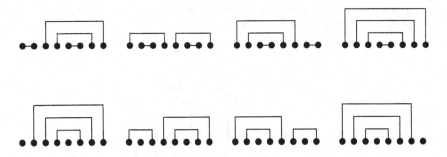

Fig. 4. The edge sets of K_8 drawn with at most two bends per edge.

Lemma 5. *The graph* K_n *can be drawn in an* $n \times \frac{n}{2} \times n$-grid, *with a total of* $n^2 - 3n + 2$ *bends and at most two bends per edge, such that vertex* v_i *overlaps the points* $(i, 1, z)$, *where* $1 \le z \le n$.

Drawing in an $\frac{n}{2} \times \frac{n}{2} \times \frac{n}{2}$-grid Let K^1 and K^2 denote two $K_{\frac{n}{2}}$'s drawn as described above. Thus each drawing has a bounding box of dimensions $\frac{n}{2} \times \frac{n}{4} \times \frac{n}{2}$. Reflect K^2 through the $(y = 0)$-plane, so that all points in the reflected K^2 have negative y-coordinate. Then rotate the reflected K^2 so that vertex v_j of the rotated, reflected K^2 now overlaps the points $(x, -1, j)$, where $1 \leq x \leq \frac{n}{2}$.

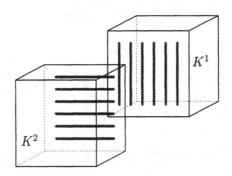

Fig. 5. Two $K_{\frac{n}{2}}$'s, with K^2 reflected and rotated

Each vertex v_i in K^1 sees each vertex v_j in the rotated, reflected K^2 along the y-line segment $[(i, 1, j), (i, -1, j)]$. Therefore, these edges can be drawn as straight lines, thus producing a drawing of K_n. Removing the unused y-plane of y-coordinate 0 yields a drawing of dimensions $X = Y = Z = \frac{n}{2}$. The total number of bends is $2(n^2/4 - \frac{3}{2}n + 2) = n^2/2 - 3n + 4$.

Theorem 6. K_n can be drawn in an $\frac{n}{2} \times \frac{n}{2} \times \frac{n}{2}$-grid with a total of $n^2/2 - 3n + 4$ bends and at most two bends per edge.

4.3 An $O(n^{2.5})$ Volume drawing for $k = 3$

In this section, we draw K_n with at most $k = 3$ bends on any edge and with volume $\mathcal{O}(n^{2.5})$. Case 2 of the lower bound proof suggests what general form such a drawing might take. For simplicity, assume below that $n = N^2$ for some integer N. Enumerate the vertices as ordered pairs (i, j), where $1 \leq i \leq N$, $1 \leq j \leq N$, and place vertex (i, j) at $(2i, 2j)$ in the 2-D x, y-plane. Suppose edge e joins vertex (i_1, j_1) and vertex (i_2, j_2). After possible renaming, we may assume that $i_1 \leq i_2$, and that if $i_1 = i_2$, then $j_1 > j_2$. Call e an *L-edge* if $j_1 > j_2$ and a *Γ-edge* otherwise.

Initially route each L-edge via the points $(2i_1, 2j_1), (2i_1+1, 2j_1), (2i_1+1, 2j_2+1), (2i_2, 2j_2 + 1), (2i_2, 2j_2)$, thus with three bends. Route each Γ-edge via points $(2i_1, 2j_1), (2i_1 + 1, 2j_1), (2i_1 + 1, 2j_2 - 1), (2i_2, 2j_2 - 1), (2i_2, 2j_2)$.

Split the L-edges into $N(N - 1)$ groups E_{d_x, d_y}, with $0 \leq d_x \leq N - 1$ and $1 \leq d_y \leq N - 1$. Each group E_{d_x, d_y} consists of those edges $((i_1, j_1), (i_2, j_2))$ for which $i_2 = i_1 + d_x$ and $j_2 = j_1 - d_y$. These groups cover all L-edges since $i_1 \leq i_2$ and $j_1 > j_2$ for any L-edge.

Now split each group E_{d_x,d_y} into at most $d_x + d_y$ sets of edges as follows. For $p = 0, \ldots, d_x + d_y - 1$, let $E^p_{d_x,d_y}$ be the edges in E_{d_x,d_y} for which $j_2 - i_1 = p$ modulo $(d_x + d_y)$. In other words, the lower left "corners" of the L-edges in $E^p_{d_x,d_y}$ lie on diagonals that intersect the y-axis at the value $2p$ modulo $(2d_x + 2d_y)$. See Fig. 6. It is easy to check that no two edges in $E^p_{d_x,d_y}$ overlap or intersect since the corners of the L's are placed on a sequence of diagonals having a vertical spacing of $2(d_x + d_y)$ between adjacent diagonals. Also, note that $E^p_{d_x,d_y}$ is non-empty only if $p \leq 2N - d_x - d_y$.[5]

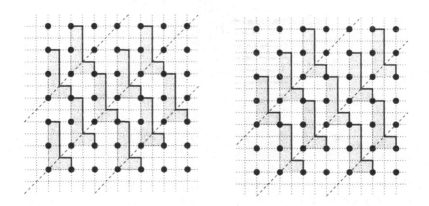

Fig. 6. The edge sets $E^0_{1,2}$ and $E^2_{1,2}$.

Assign a z-plane to each set $E^p_{d_x,d_y}$ to obtain a legal drawing of the L-edges. Route the Γ edges in an analogous fashion. This doubles the number of z-planes, yielding a drawing of K_n in a grid with $X = Y = 2N = 2\sqrt{n}$. The Z dimension is given by

$$2 \sum_{d_x=0}^{N-1} \sum_{d_y=1}^{N-1} \min\{d_x + d_y, 2N - d_x - d_y\}.$$

Some analysis shows that this sum is at most

$$2 \left[\sum_{k=1}^{N-1} k(2k-1) + (N-1)N \right] = \frac{2(N-1)N(2N-1)}{3} - (N-1)N + 2(N-1)N$$

which is less than $\frac{4}{3}N^3$. Every edge has three bends. However, the $2N(N-1) = 2n - 2\sqrt{n}$ edges where $d_x = 0$ and $d_y = 1$, or $d_x = 1$ and $d_y = 0$ can be drawn without a bend. So the total number of bends is $3(n^2/2 - n/2) - 3(2n - 2\sqrt{n}) = \frac{3}{2}n^2 - \frac{15}{2}n + 6\sqrt{n}$.

---

[5] A java applet demonstrating the sets and their routings for K_{100} can be found at http://www.cs.uleth.ca/~wismath/ortho.html.

Theorem 7. *If* $n = N^2$ *is a square, then* K_n *can be drawn in a* $2N \times 2N \times \frac{4}{3}N^3$-*grid (so volume* $\frac{16}{3}n^{2.5}$*) with* $\frac{3}{2}n^2 - \frac{15}{2}n + 6\sqrt{n}$ *bends and at most three bends per edge.*

5   Conclusions

This paper is one of the first to address volume and bend considerations for 3-D orthogonal grid drawings of graphs. The focus has been on K_n, since it is the most difficult graph on n vertices to draw in small volume or with restrictions on bends. In particular, we have

- provided a method for drawing K_n with volume that is provably within a constant factor (same constant for all n) of best possible in the case that at most k bends per edge are allowed, where $k \geq 3$;

- proved the non-existence of drawings of K_n for large n in the $k = 0$ case, where no bends are permitted;

- proved a lower bound of $\Omega(n^{2.5})$ and an upper bound of $O(n^3)$ on the volume of drawings of K_n when $k = 1$ and $k = 2$.

An open problem is to close the gap between the upper and lower bounds in the $k = 1$ and $k = 2$ cases, where at most 1 and at most 2 bends on each edge are permitted, respectively. The ideas and methods presented here may serve as a useful starting point for constructing drawings with good constant factors for volume and bends.

6   Acknowledgments

Thanks to Michael Kaufmann for discussions on orthogonal drawings. The joint results of this paper have also appeared as part of the PhD thesis of T. Biedl at Rutgers University.

References

[BEFLMRSW93] P. Bose, H. Everett, S. Fekete, A. Lubiw, H. Meijer, K. Romanik, T. Shermer and S. Whitesides. On a visibility representation for graphs in three dimensions. *Proc. Graph Drawing '93*, Paris, 1993, 38-39.

[BMT97] T. Biedl, B. Madden, and I. Tollis. The three-phase method: A unified approach to orthogonal graph drawing. In these proceedings.

[Bra96] F. Brandenburg, ed. *Symposium on Graph Drawing '95, Passau, Germany*, Springer-Verlag LNCS **1027**, 1996.

[ESW97] P. Eades, A. Symvonis and S. Whitesides. Two algorithms for three dimensional orthogonal graph drawing. In S. North, ed. *Symposium on Graph Drawing '96, Berkeley, California*, Springer Verlag LNCS **1190**, 1997, 139-154.

[FHW96] S. Fekete, M. Houle and S. Whitesides. New results on a visibility representation of graphs in 3d. In [Bra96], 234-241.

[FK96] U. Fößmeier and M. Kaufmann. Drawing high degree graphs with low bend numbers. In [Bra96], 254-266.

[GRS80] R. Graham, B.L. Rothschild and J. Spencer. *Ramsey Theory*. John Wiley, 1980.

[PT96] A. Papakostas and I. Tollis. High-degree orthogonal drawings with small gridsize and few bends. In *5th Workshop on Algorithms and Data Structures*, Springer Verlag LNCS, 1997. To appear.

[PT97] A. Papakostas and I. Tollis. Incremental orthogonal graph drawing in three dimensions. In these proceedings.

[Ros83] A. Rosenberg. Three-dimensional VLSI: A case study. *Journal of the Association of Computing Machinery*, **30** (3), 1983, 397-416.

[TT86] R. Tamassia and I. Tollis. A unified approach to visibility representations of planar graphs. *J. of Discrete and Computational Geometry* **1**, 1986, 321-341.

[Wis85] S. K. Wismath. Characterizing bar line-of-sight graphs. *Proc. of the 1st ACM Symp. on Computational Geometry*, Baltimore, Maryland, USA, 1985, 147-152.

[Yan89] M. Yannakakis, Embedding planar graphs in four pages. *18th Annual ACM Symposium on Theory of Computing*, J. Comput. System Sci. 38 (1989), no. 1, 36-67.

Finding the Best Viewpoints for Three-Dimensional Graph Drawings

Peter Eades, Michael E. Houle, and Richard Webber

Department of Computer Science and Software Engineering
University of Newcastle
Callaghan 2308, Australia
{eades,mike,richard}@cs.newcastle.edu.au

Abstract. In this paper we address the problem of finding the best viewpoints for three-dimensional straight-line graph drawings. We define goodness in terms of preserving the relational structure of the graph, and develop two continuous measures of goodness under orthographic parallel projection. We develop Voronoi variants to find the best viewpoints under these measures, and present results on the complexity of these diagrams.

1 Introduction

Three-dimensional drawings of graphs are being used in many areas of computing. For example: to visualise the structures of object-oriented [16,24] and parallel [22] software; for interactive information retrieval [11]; and to navigate the World-Wide Web [18,20].

There has been much work on creating graph drawings that are both easy to remember and easy to understand [8]. Most work has been to produce two-dimensional drawings, but recent work addresses three-dimensions [6]. Much of this work has concentrated on the problems of finding a good three-dimensional drawing for a graph. In this paper, we consider a problem in the process of rendering a three-dimensional graph drawing as a two-dimensional image: finding a good viewpoint.

There is some experimental evidence that three-dimensional graph drawings have advantages over their two-dimensional counterparts. One claimed advantage [23] is that three dimensions allow users to work with larger graphs – the natural three-dimensional actions of rotation and translation allow users to resolve ambiguities in large graphs while maintaining their overall mental map [9].

Ware et al. [23] consider the problem of finding connections in a three-dimensional graph drawing, with a variety of display and navigation combinations. Using a sequence of human experiments, they show that allowing users to navigate around the drawing results in a significant decrease in the error rate, but at the cost of increased decision time. Current navigation methods tend to require users to manually select their desired viewpoints. We conjecture that the increase in decision time is (partly) due to the time taken to manually select good viewpoints.

We propose the use of logical viewpoint selection – the users specify the part of the graph drawing on which they wish to focus, and the system automatically selects a (locally) good viewpoint. An additional use for logical viewpoint selection is when a three-dimensional graph drawing *must* be rendered as a static two-dimensional image. Using a good viewpoint under these circumstances can be critical, as was evident in last year's graph drawing contest [10].

Some existing systems already support simple versions of logical selection. The application ivview [21] navigates around three-dimensional scenes. It has a logical selection function that allows users to select a surface in the scene being viewed, then it moves the viewpoint to be perpendicular to the surface.

In Sec. 2 we briefly present some background material, and outline previous work specifically on the problem of finding good viewpoints [4,14]. In Sec. 3 we model good viewpoints in terms of the information conveyed by a drawing, then in Sec. 4 we use the techniques described by Bose, Gomez, Ramos and Toussaint [4] to find all bad viewpoints under our model.

In Sec. 5 we develop two new models that provide continuous measures of the goodness of a viewpoint. We introduce Voronoi variants to find all best viewpoints under these measures, and give proofs of their complexity. Finally, in Sec. 6 we outline future research possibilities based on this work.

2 Background

A three-dimensional straight-line drawing $D : V \to \mathbb{R}^3$ of an *abstract graph* $G = (V, E)$ associates a three-dimensional position $\langle x_i, y_i, z_i \rangle$ with each vertex $v_i \in V$. We consider a vertex and its position to be synonymous. Each edge e_{ij} is drawn as a line-segment $\overline{v_i v_j}$ between its end-points. A vertex with no incident edges is an *isolated vertex*. We denote the set of isolated vertices V_0.

The same abstract graph can be represented by many different drawings, and there are many techniques for automatically generating these drawings. For an extensive collection of graph drawing papers refer to the annotated bibliography by Di Battista et al. [8].

To convert a three-dimensional graph drawing into a two-dimensional image (for rendering onto a computer screen or paper) we use a *projection*. In computer graphics, the most common projections are *parallel*, *perspective*, and *fisheye-lens* projections. They approximate the way in which people (or fish) see the world around them.

Projections have parameters, the most important being the *viewpoint*. This has two components: the *position* of the viewer in three-dimensional space; and the *direction* that the viewer is facing. These projections map three-dimensional points onto a two-dimensional surface called the *projection surface*.

Under parallel projection (Fig. 1) the three-dimensional points are translated parallel to the viewpoint direction, forming two-dimensional points where they intersect the projection surface, which is a plane. In an *orthographic* parallel projection (also called an orthogonal projection, a perpendicular projection, or

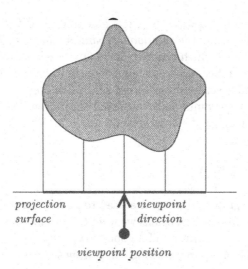

projection   viewpoint
surface   direction

viewpoint position

Fig. 1. Orthographic parallel projection.

simply "a projection") the viewpoint direction is perpendicular to the projection surface. A detailed explanation of projections can be found in [12].

If a projection maps two three-dimensional points to the same two-dimensional point, then there is an *occlusion*. We say the *front* point (the first one encountered when moving in the viewpoint direction) *occludes* the *rear* point.

Both a three-dimensional graph drawing and the two-dimensional result of its projection contain information. The preservation of this information under projection forms the basis for many models of *good viewpoints*.

Kamada and Kawai [14] model good viewpoints that preserve the shape information of a wire-frame drawing ($V_0 = \emptyset$), excluding viewpoints for which edges appear collinear. Bose et al. [4] preserve the depth-order of a wire-frame drawing, permitting only viewpoints that yield regular projections[1] [17].

In Sec. 3, we define good viewpoints to be those that preserve the abstract graph of a drawing, where a drawing may include isolated vertices. Our definition covers the significant *bad viewpoints* of Kamada and Kawai, and leads to faster algorithms than those in [4].

Bose et al. suggest an algorithm to describe all bad viewpoints in by building an arrangement [5,19] of curves consisting of bad viewpoints. Under their model this requires $O(|E|^3 \log|E| + k')$ time, where the parameter k' is $O(|E|^6)$ in the worst case. In Sec. 4 we employ the same technique with our simpler model to build an arrangement of bad viewpoints in $O(|\Psi| \log|\Psi| + k)$ time, where $|\Psi| = 2\binom{|V_0|}{2} + 2|V||E|$ and k is $O(|\Psi|^2)$ but can be much less.

Kamada and Kawai use the great-circle distance between a viewpoint and its nearest bad viewpoint as a continuous measure of *goodness*. We call this *rota-*

---

[1] A regular projection [17] excludes three three-dimensional points appearing as a single two-dimensional point. The end-points of edges count for two points each.

tional separation, and in Sec. 5.1 we show how to find all *best* viewpoints under rotational separation using the *rotational separation diagram*. This is a variant of the Voronoi diagram [2] which can be built in $O((|\Psi| + k)\log(|\Psi| + k))$ time, with $|\Psi|$ and k as above. In Sec. 5.2 we define goodness to be the *observed separation* after projection and describe the *observed separation diagram*, a Voronoi variant that finds all best viewpoints under observed separation. We also present some results on the complexity of this diagram.

3 Good Viewpoints

Three-dimensional graph drawings are used to convey information. When projecting a drawing into two dimensions we want to preserve this information. This forms the basis for our definition of a good viewpoint.

A three-dimensional graph drawing can convey several forms of information:

- Relational information: the vertices, edges and the incidences between them.
- Attribute information: attributes, such as shape and colour, of each element.
- Geometric information: the position of each element relative to some known point (origin), or more typically, the distance between a pair of elements.

Attribute and geometric information are features of a particular drawing of the graph. Attribute information is preserved by most useful projections. Geometric information is highly dependent on the projection and viewpoint used.

Relational information is dependent on the underlying abstract graph. Preservation of relational information motivates our definition for a good viewpoint:

Definition 1. *A good viewpoint is one for which the abstract graphs of the three-dimensional drawing and its two-dimensional projection appear the same.*

This in turn depends on how the projection maps each element. For now we make a few simplifying assumptions:

1. We use an orthographic parallel projection.
2. There is no clipping.
3. All vertices and edges are mathematical ideals, with zero width for the purpose of calculating occlusions.

Under an orthographic parallel projection we can parameterise a viewpoint by its normalised direction vector $\hat{d} \in \mathbf{S}^2$, where $\mathbf{S}^2$ is the unit sphere centred at the origin in $\mathbb{R}^3$.

The abstract graphs of a three-dimensional drawing and its two-dimensional projection appear the same under an orthographic parallel projection if and only if there are no occlusions between elements of the drawing. For two graph elements a and b we denote the set of viewpoints that generate occlusions of b by a (a is in front) by $\psi(a, b)$. We refer to the corresponding viewpoints as *bad viewpoints*.

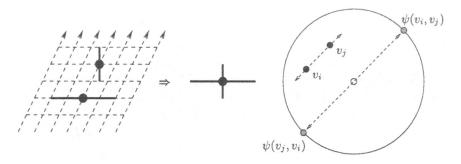

Fig. 2. Vertex-vertex occlusions.

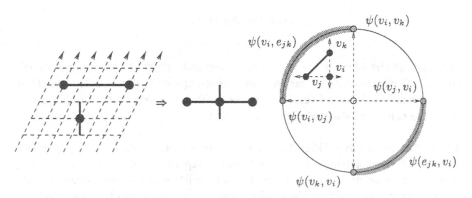

Fig. 3. Vertex-edge/edge-vertex occlusions.

There are four pairings of graph elements that generate occlusions:

Vertex-Vertex Occlusions (Fig. 2): A pair of three-dimensional vertices map to a single two-dimensional vertex. The abstract graph of the projection has a single vertex in place of the original two, and any edges incident to the original vertices now appear incident to the combined vertex. The set $\psi(v_i, v_j)$ contains one element, the normalised vector from v_i to v_j which we will denote $\hat{d}_{ij}$.

Vertex-Edge/Edge-Vertex Occlusions (Fig. 3): A three-dimensional vertex maps to an internal point of a two-dimensional edge. The abstract graph of the projection has two edges in place of the original edge, both incident to the vertex. The difference between vertex-edge and edge-vertex occlusions is simply which element is in front. In this case, $\psi(v_i, e_{jk})$ is the set of direction vectors on the great-circle arc between $\hat{d}_{ij}$ and $\hat{d}_{ik}$.

Notice that the set of bad viewpoints for a vertex-edge/edge-vertex occlusion is a superset of the bad viewpoints corresponding to the pair of vertex-vertex occlusions of the vertex and each end-point of the edge:

$$\psi(v_i, e_{jk}) \supseteq \psi(v_i, v_j) \cup \psi(v_i, v_k) \ .$$

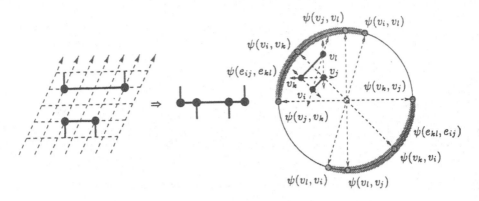

Fig. 4. Edge-edge occlusions.

We say that the vertex-edge/edge-vertex occlusion *covers* the vertex-vertex occlusions. If we are already checking for vertex-edge/edge-vertex occlusions then we only need to check for vertex-vertex occlusions between pairs of isolated vertices. We call these *isolated-vertex occlusions*.

Edge-Edge Occlusions (Fig. 4): There are two cases of edge-edge occlusions. Case 1 – a pair of three-dimensional edges map to a pair of two-dimensional edges that cross at a single internal point. With straight-line edges, we can (ideally) trace the paths of these edges, and the abstract graph is not effected.

Case 2, *significant edge-edge occlusions* – the two-dimensional edges share a continuous sequence of points. This only occurs when the three-dimensional edges are coplanar. The abstract graph of the projection has three edges in place of the original two. Some vertices incident to the original edges become incident to the new edge.

For significant edge-edge occlusions, $\psi(e_{ij}, e_{kl})$ is the union of (and hence covered by) four vertex-edge/edge-vertex occlusions:

$$\psi(e_{ij}, e_{kl}) = \psi(v_i, e_{kl}) \cup \psi(v_j, e_{kl}) \cup \psi(e_{ij}, v_k) \cup \psi(e_{ij}, v_l) \ .$$

If we are already checking for vertex-edge/edge-vertex occlusions then we do not need to check for edge-edge occlusions.

Lemma 1. *A good viewpoint is one that does not generate any isolated-vertex or vertex-edge/edge-vertex occlusions.*

4   Bad Viewpoints

For a given three-dimensional drawing of a graph $G = (V, E)$, we want to find all the bad viewpoints. Bose et al. [4] do this by constructing an arrangement of the sets of bad viewpoints under their model. They require $O(|E|^3 \log |E| + k')$ time and $O(|E|^3 + k')$ space, where k' is the number of intersections in the arrangement, which is $O(|E|^6)$ in the worst case. We apply the same technique to our model, obtaining a description in asymptotically less time and space.

Notice that, under an orthographic parallel projection with no clipping, the viewpoint directions $\hat{d}$ and $-\hat{d}$ yield the same two-dimensional projection. It follows that we only need to build the arrangement for one hemisphere.

An arrangement of bad viewpoints is an arrangement of a set Ψ of points and great-circle arcs on the unit sphere. We can transform this for one hemisphere to an arrangement of points and line-segments in the plane using a *central projection* [19]. We can now use a segment intersection algorithm in the plane [5,19] to form the arrangement, which leads us to the following result:

Result 1 (Bad Viewpoint Arrangement [4]). *For a given three-dimensional drawing of a graph $G = (V, E)$, we can build a bad viewpoint arrangement in $O(|\Psi| \log |\Psi| + k)$ time; $|\Psi| = 2\binom{|V_0|}{2} + 2|V||E|$ is the number of points and great-circle arcs in the arrangement; k is the number of intersections between them, which is $O(|\Psi|^2)$ in the worst case, but can be much less. The arrangement has $O(|\Psi| + k)$ size.*

5 Best Viewpoints

The bad viewpoint arrangement discussed above only distinguishes between good and bad viewpoints. We would prefer to have a continuous measure for the *goodness* of a viewpoint. In the following sections we discuss two such measures, and show how to find the *best* viewpoints for a given drawing.

5.1 Rotational Separation

Let the function $\delta(p, q)$ denote the great-circle distance between the points p and q on the unit sphere $\mathbb{S}^2$. The value of $\delta(p, q)$ is also the angle between the vector representations of p and q. We will use either interpretation as appropriate.

We will overload the notation by letting $\delta(\hat{d}, \psi(a, b))$ denote the minimum great-circle distance (angle) between $\hat{d}$ and the points (vectors) from the set $\psi(a, b)$. For a vertex-vertex occlusion $\psi(v_i, v_j)$ this is simply the great-circle distance between $\hat{d}$ and $\hat{d}_{ij}$. For a vertex-edge occlusion $\psi(v_i, e_{jk})$ and an edge-vertex occlusion $\psi(e_{jk}, v_i)$ it is defined as: the perpendicular great-circle distance between $\hat{d}$ and the great-circle through $\hat{d}_{ij}$ and $\hat{d}_{ik}$, when the perpendicular intersects the great-circle between $\hat{d}_{ij}$ and $\hat{d}_{ik}$; and $\min(\delta(\hat{d}, \hat{d}_{ij}), \delta(\hat{d}, \hat{d}_{ik}))$ otherwise.

We call $\delta(\hat{d}, \psi(a, b))$ the *rotational separation* of $\hat{d}$ from $\psi(a, b)$, and use its minimum over all occlusions to measure the goodness of $\hat{d}$. This is equivalent to the measure used by Kamada and Kawai [14].

Notice that $\delta(\hat{d}, \psi(a, b)) = 0$ implies $\hat{d} \in \psi(a, b)$; in other words, finding good viewpoints that preserve the relational information of a drawing also avoids the worst viewpoints under the rotational separation measure.

We can use the *rotational separation diagram (RSD)* to find the directions $\hat{d}$ for which a given occlusion $\psi(a,b)$ is the nearest bad viewpoint in terms of $\delta(\hat{d}, \psi(a,b))$. This is a variant of the *Voronoi diagram* [2], whose Voronoi sites S are the points and great-circle arcs of the bad viewpoint arrangement, resulting in *bisectors* that are great-circle arcs and spherical parabolas.

We can build the rotational separation diagram by modifying a planar algorithm for the Voronoi diagram of line-segments to work on the sphere $\mathbb{S}^2$. The similar problem of modifying Fortune's sweepline algorithm [13] for a cone has been described by Dehne and Klein [7].

The rotational separation diagram has $O(|S|)$ size and requires $O(|S|\log|S|)$ time to build, where $|S|$ is the size of the bad viewpoint arrangement that defines its Voronoi sites. The size $|S|$ is $O(|\Psi| + k)$; $|\Psi| = 2\binom{|V_0|}{2} + 2|V||E|$ and k is $O(|\Psi|^2)$ as before.

The goodness of a viewpoint increases as it moves away from its nearest bad viewpoint(s). This increase is maximised locally at the Voronoi vertices of the rotational separation diagram. A two-dimensional example of a best viewpoint for two occlusions under rotational separation is given in Fig. 5.

We can find best viewpoints with arbitrary locality criteria, by traversing the rotational separation diagram in time proportional to its size $O(|S|)$ (assuming the criteria can be tested in constant time). When "locally" means "nearest by great-circle distance", we repeat the process of building the rotational separation diagram, this time using the Voronoi vertices of the first diagram as the sites. This requires the same preprocessing time and space as before, after which we can apply a point-location algorithm [15,19], again with the same preprocessing time and space requirements. The resulting diagram can be used to find the nearest site (the locally best viewpoint) in $O(\log|S|)$ time.

Result 2. *The rotational separation diagram has $O(|S|)$ size and requires $O(|S|\log|S|)$ time to build, where the size $|S|$ of the corresponding bad viewpoints arrangement is $O(|\Psi| + k)$; $|\Psi| = 2\binom{|V_0|}{2} + 2|V||E|$ and k is $O(|\Psi|^2)$ in the worst, but can be much less.*

We can use the rotational separation diagram to find best viewpoints under rotational separation, with arbitrary locality criteria, in $O(|S|)$ time, and the nearest best viewpoint by great-circle distance in $O(\log|S|)$ time.

5.2 Observed Separation

Given two elements a and b from a three-dimensional graph drawing, and a viewpoint direction $\hat{d}$, we define $\Delta(\hat{d}, \psi(a,b))$ to be the shortest Euclidean distance between the projections of a and b, and call it the *observed separation* between a and b from $\hat{d}$.

Let the function $\rho(a,b)$ denote the shortest Euclidean distance between the two elements a and b in $\mathbb{R}^3$. For a vertex-vertex occlusion $\psi(v_i, v_j)$ this is simply the Euclidean distance $\|v_j - v_i\|$. For a vertex-edge occlusion $\psi(v_i, e_{jk})$ and an edge-vertex occlusion $\psi(e_{jk}, v_i)$ it is defined as: the perpendicular Euclidean distance between v_i and the line $\overleftrightarrow{v_j v_k}$ when the perpendicular intersects the line between v_j and v_k; and $\min(\|v_j - v_i\|, \|v_k - v_i\|)$ otherwise.

Now we can define the observed separation as:

$$\Delta(\hat{d}, \psi(a, b)) = \rho(a, b) \sin \delta(\hat{d}, \psi(a, b)) .$$

The proportion of the distance between a and b which is preserved after projection, when viewed from the direction $\hat{d}$, is given by the equation:

$$\frac{\Delta(\hat{d}, \psi(a, b))}{\rho(a, b)} = \sin \delta(\hat{d}, \psi(a, b)) .$$

As $0 \le \delta(\hat{d}, \psi(a, b)) \le \frac{\pi}{2}$ (by symmetry of the bad viewpoint arrangement), this ratio is maximised when $\delta(\hat{d}, \psi(a, b))$ is maximised. It follows that the best viewpoints under rotational separation maximise the minimum of these proportions preserved. Another interpretation is that best viewpoints under rotational separation maintain the relative distance between pairs of points under projection, a form of the geometric information listed in Sec. 3.

It is suggested that people are more sensitive to the boundary conditions of an aesthetic criterion than to the average, and prefer to maximise the minimum separation from these boundaries [14].

An alternate approach would be to use observed separation as a measure of goodness. The best viewpoints under this measure are those that maximise the minimum observed separation between pairs of elements in the two-dimensional projection. This has the effect of maximising the users' ability to resolve elements in the two-dimensional projection, which is proposed as an aesthetic criterion for two-dimensional graph drawings [8]. However, best viewpoints under observed separation sacrifice the relative distance information between pairs of points. A two-dimensional example of a best viewpoint for two occlusions under observed separation is given in Fig. 6.

The *observed separation diagram (OSD)* is a variant of the Voronoi diagram, used to find best viewpoints under the observed separation measure. Like the rotational separation diagram, its sites are the points and great-circle arcs of the bad viewpoint arrangement; however its bisectors are more complex, being defined by the equation $\Delta(\hat{d}, \psi(a, b)) = \Delta(\hat{d}, \psi(a', b'))$. This form is similar to the multiplicatively-weighted Voronoi variant [2,3], but with an extra sine factor.

Consider the restricted case where $E = \emptyset$. Each vertex-vertex occlusion $\psi(v_i, v_j)$ generates a point Voronoi site $s_a = \hat{d}_{ij}$ with a corresponding weight $w_a = \rho(v_i, v_j)$. By symmetry of the bad viewpoint arrangement, each site s_a has a corresponding site $-s_a$ of equal weight, and by the characteristics of the observed separation measure we can use either site interchangeably. The resulting diagram is symmetric about the origin.

When $w_a = w_b$, the bisector $B(a, b)$ of sites s_a and s_b is a pair of great circles, one equidistant from the points s_a and s_b, the other equidistant from the points s_a and $-s_b$, both intersecting at the poles of the great-circle through s_a and s_b. We divide $B(a, b)$ into two halves, consisting of the pairs of polar

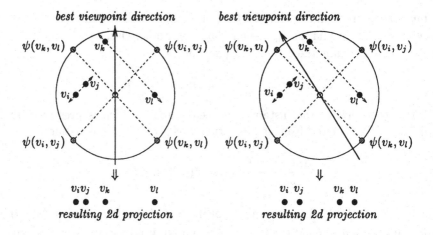

Fig. 5. Rotational separation. **Fig. 6.** Observed separation.

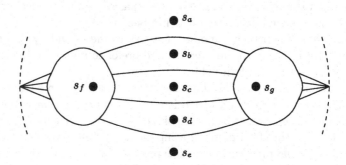

Fig. 7. $\Omega(|S|^2)$ sized example of an observed separation diagram.

great-semicircles that enclose s_a and $-s_a$ respectively. When $w_a > w_b$, the bisector $B(a, b)$ is a pair of ellipsoids, enclosing s_a and $-s_a$ respectively, formed by intersecting an elliptical double-cone with the unit sphere, both centred at the origin.

Lemma 2. *The worst case size complexity of the restricted observed separation diagram ($E = \emptyset$) on $|S|$ point sites lies within the bounds $\Omega(|S|^2)$ and $O(|S|^2 2^{\alpha(|S|)})$, where α is the inverse Ackermann function [1].*

Proof. An example of a diagram with $\Omega(|S|^2)$ size is given in Fig. 7. It is a simple adaption of Aurenhammer and Edelsbrunner's [3] worst case example for the multiplicatively-weighted Voronoi diagram of points in the plane.

To establish the upper worst case bound, consider an incremental construction of the diagram in non-descending order of weight. When the site s_t is inserted into OSD_{t-1}, the newly created region is convex about s_t. Label each segment of the new region's boundary by the region of OSD_{t-1} that it passes through. For example, in Fig. 7, the region of s_a is labelled *abcdedcb*.

For the sequence $a \ldots b$ to occur in the labelling of the region of s_t, the bisectors $B(t, a)$ and $B(t, b)$ must intersect at a single point. Solving the equations for intersecting a pair of bisectors yields at most 4 intersection points. It follows that the longest sequence of alternating as and bs within the labelling is $a \ldots b \ldots a \ldots b \ldots a$, and hence the labelling is an $(t, 4)$-Davenport-Schinzel sequences which has a maximum length of $O(t \, 2^{\alpha(t)})$ [1]. An observed separation diagram built from $|S|$ such regions has a worst case size of $O(|S|^2 2^{\alpha(|S|)})$.

For practical purposes, $\alpha(|S|)$ can be considered a constant, and these bounds converge. The unrestricted case is at least this complex.

6 Future Research

In this paper we have described techniques for finding best viewpoints for three-dimensional graph drawings. We have developed a model of good viewpoints and applied the technique of Bose et al. [4] to describe all bad viewpoints. We have extended our model and developed algorithms to find best viewpoints.

An important future step in this research is to perform experimental evaluation of our good viewpoint techniques against other methods for displaying and navigating three-dimensional graph drawings; in particular, to decide our initial conjecture that logical selection of good viewpoints decreases decision time in the experiments of Ware et al. [23].

Another avenue of research is to investigate good viewpoints under other projections, in particular under perspective and fisheye-lens projections. In our research (not detailed in this paper) we have established that the notions of abstract graphs, occlusions and a bad viewpoint arrangement, all hold under these projections. However, the current models of best viewpoints do not hold, and we are investigating other models for use under these projections.

Other possibilities for future research include: developing algorithms that trade-off query time against preprocessing complexity; developing fast (heuristic) solutions to find viewpoints that are "good enough"; removing the assumptions of mathematically ideal elements and no clipping; and investigating the implications of our techniques in higher dimensions.

References

1. P. Agarwal: *Intersection and Decomposition Algorithms for Planar Arrangements*, 1991; Cambridge University Press
2. F. Aurenhammer: "Voronoi Diagrams – A Survey of a Fundamental Geometric Data Structure" in *ACM Comp. Surveys*, Sep 1991; 23(3)
3. F. Aurenhammer, H. Edelsbrunner: "An Optimal Algorithm for Constructing the Weighted Voronoi Diagram in the Plane" in *Patt. Recog.*, 1984; 17(2):251-257
4. P. Bose, F. Gomez, P. Ramos, G. Toussaint: "Drawing Nice Projections of Objects in Space" in *Graph Drawing* (Sep 1995; Passau, Germany); pp. 52-63
5. B. Chazelle, H. Edelsbrunner: "An Optimal Algorithm for Intersecting Line Segments" in *IEEE Found. Comp. Sc.* (1988)

6. R. Cohen, P. Eades, T. Lin, F. Ruskey: "Three-Dimensional Graph Drawing" in *Algorithmica*, 1996; 17(2)

7. F. Dehne, R. Klein: "The Voronoi Diagram of Points on a Cone"; School of Computer Science, Carleton University, Ottawa

8. G. Di Battista, P. Eades, R. Tamassia, I. Tollis: "Algorithms for Drawing Graphs: An Annotated Bibliography", Jun 1994;
   http://www.cs.brown.edu/calendar/gd94/biblio.html

9. P. Eades, W. Lai, K. Misue, K. Sugiyama: "Preserving the Mental Map of a Diagram", 1991; Research Report IIAS-RR-91-16E, Fujitsu Laboratories Ltd., Japan

10. P. Eades, J. Marks, S. North: "Graph-Drawing Contest Report" in *Graph Drawing* (Sep 1996; Berkeley, U.S.A.); pp. 129–138

11. FADIVA, VIRI: "Actual Listing of Information Visualization Systems", 1995;
    http://www-cui.darmstadt.gmd.de:80/visit/Activities/Viri/visual.html

12. J. Foley, A. van Dam, S. Feiner, J. Hughes: *Computer Graphics: Principles and Practice, 2nd Ed.*, 1990; Addison-Wesley

13. S. Fortune: "A Sweepline Algorithm for Voronoi Diagrams" in *Algorithmica*, 1987; 2(2):153–174

14. T. Kamada, S. Kawai: "A Simple Method for Computing General Position in Displaying Three-Dimensional Objects" in *Comp. Vision, Graphics and Image Processing*, 1988; 41:43–56

15. D. Kirkpatrick: "Optimal Search in Planar Subdivisions" in *SIAM J. Comp.*, 1983; 12(1):28–35

16. H. Koike: "An Application of Three-Dimensional Visualization to Object-Oriented Programming" in *Proc. Adv. Visual Interfaces* (1992); pp. 180–192

17. C. Livingston: *Knot Theory*, 1993; Math. Assoc. America

18. T. Munzner, P. Burchard: "Visualizing the Structure of the World Wide Web in 3D Hyperbolic Space" in *VRML* (Dec 1995; San Diego, U.S.A.); pp. 33–38

19. F. Preparata, M. Shamos: *Computational Geometry: An Introduction*, 1985; Springer-Verlag

20. B. Regan: "Information Diagrams for the DOOMed Generation" in *Visual* (Feb 1996; Melbourne, Australia); pp. 557–566

21. Silicon Graphics Inc.: "ivview"; UNIX Manual Page

22. E. Trichina, B. Thomas: "3D Interactive Animation for Visualization of Parallel Design", 1995; Technical Report CIS-96-001, University of South Australia

23. C. Ware, G. Franck: "Evaluating Stereo and Motion Cues for Visualizing Information Nets in Three Dimensions" in *ACM Trans. Graphics*, Apr 1996; 15:121–140

24. C. Ware, D. Hui, G. Franck: "Visualizing Object Oriented Software in Three Dimensions" in *CASCON* (1993)

A Linear Algorithm for Optimal Orthogonal Drawings of Triconnected Cubic Plane Graphs

Md. Saidur Rahman, Shin-ichi Nakano and Takao Nishizeki *

Graduate School of Information Sciences
Tohoku University, Sendai 980-77, Japan.

Abstract. An orthogonal drawing of a plane graph G is a drawing of G in which each edge is drawn as a sequence of alternate horizontal and vertical line segments. In this paper we give a linear-time algorithm to find an orthogonal drawing of a given 3-connected cubic plane graph with the minimum number of bends. The best known algorithm takes time $O(n^{7/4}\sqrt{\log n})$ for any plane graph of n vertices.

1 Introduction

An *orthogonal drawing* of a plane graph G is a drawing of G with the given embedding in which each vertex is mapped to a point, each edge is drawn as a sequence of alternate horizontal and vertical line segments, and any two edges do not cross except at their common end. Orthogonal drawings have attracted much attention due to its numerous practical applications in circuit schematics, data flow diagrams, entity relationships diagrams, etc [B96, K96, T87]. In particular, we wish to find an orthogonal drawing with the minimum number of bends. For the plane graph in Fig. 1(a), the orthogonal drawing in Fig. 1(b) has the minimum number of bends, that is, seven bends.

For a given planar graph G, if it is allowed to choose its planar embedding, then finding an orthogonal drawing of G with the minimum number of bends is NP-complete [GT94]. However, Tamassia [T87] and Garg and Tamassia [GT96] presented algorithms which find an orthogonal drawing of a given plane graph G with the minimum number of bends in $O(n^2 \log n)$ and $O(n^{7/4}\sqrt{\log n})$ time respectively unless it is allowed to choose its planar embedding, where n is the number of vertices in G. They reduce the minimum-bend orthogonal drawing problem to a minimum cost flow problem. On the other hand, several linear-time algorithms are known for finding orthogonal drawings of plane graphs with a presumably small number of bends although they do not always find orthogonal drawings with the minimum number of bends [B96, K96].

In this paper we give a linear-time algorithm to find an orthogonal drawing of a 3-connected cubic plane graph with the minimum number of bends. To the best of our knowledge, our algorithm is the first linear-time algorithm to find an orthogonal drawing with the minimum number of bends for a fairly large class of graphs.

---

* E-mail: saidur@nishizeki.ecei.tohoku.ac.jp, {nakano, nishi}@ecei.tohoku.ac.jp

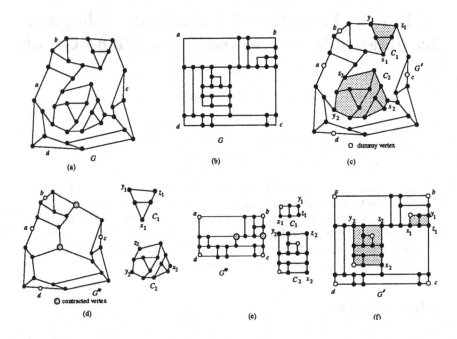

Fig. 1. A plane graph and its orthogonal drawing.

An orthogonal drawing in which there is no bend and each face is drawn as a rectangle is called a *rectangular drawing*. Linear-time algorithms have been known to find a rectangular drawing of a plane graph in which every vertex has degree three except four vertices of degree two on the outer boundary whenever such a graph has a rectangular drawing [KH94, RNN96]. The key idea of our algorithm is to reduce the orthogonal drawing problem to the rectangular drawing problem.

An outline of our algorithm is illustrated in Fig. 1. Given a plane graph as shown in Fig. 1(a), we first put four dummy vertices a, b, c and d of degree two on the outer boundary of G, and let G' be the resulting graph. The four dummy vertices are drawn by white circles in Fig. 1(c). We then contract each of some cycles $C_1, C_2, \cdots$ and their insides (shaded in Fig. 1(c)) into a single vertex as shown in Fig. 1(d) so that the resulting graph G'' has a rectangular drawing as shown in Fig. 1(e). We also find orthogonal drawings of those cycles $C_1, C_2, \cdots$ and their insides recursively (see Figs. 1(d) and (e)). Patching the obtained drawings, we get an orthogonal drawing of G' as shown in Fig. 1(f). Replacing the dummy vertices a, b, c and d in the drawing of G' with bends, we finally obtain an orthogonal drawing of G as shown in Fig. 1(b).

The rest of the paper is organized as follows. Section 2 gives some definitions and presents a known result. Section 3 presents an algorithm to find an orthogonal drawing in which the number of bends may not be the minimum but does not exceed the minimum number by more than four. Section 4 presents an algorithm to find an orthogonal drawing with the minimum number of bends, using the algorithm in Section 3. Finally Section 5 is a conclusion.

2 Preliminaries

In this section we give some definitions and present a known result.

Let G be a connected graph with n vertices and m edges. We denote the set of vertices of G by $V(G)$, and the set of edges of G by $E(G)$. The *degree* of a vertex v is the number of neighbors of v in G. If every vertex of G has degree three, then G is called a *cubic graph*. The *connectivity* $\kappa(G)$ of a graph G is the minimum number of vertices whose removal results in a disconnected graph or a single-vertex graph K_1. We say that G is *k-connected* if $\kappa(G) \geq k$.

A graph is *planar* if it can be embedded in the plane so that no two edges intersect geometrically except at a vertex to which they are both incident. A *plane* graph is a planar graph with a fixed embedding. A plane graph divides the plane into connected regions called *faces*. We regard the *contour* of a face as a clockwise cycle formed by the edges on the boundary of the face. We denote the contour of the outer face of graph G by $C_o(G)$.

For a simple cycle C in a plane graph G, we denote by $G(C)$ the plane subgraph of G inside C (including C). We say that cycles C_1 and C_2 in a plane graph G are *independent* if $G(C_1)$ and $G(C_2)$ have no common vertex. An edge which is incident to exactly one vertex of a simple cycle C and located outside of C is called a *leg* of the cycle C, and the vertex on C to which the leg is incident is called a *leg-vertex* of C. A simple cycle with exactly k legs is called a *k-legged cycle*.

An *orthogonal drawing* of a plane graph G is a drawing of G with the given embedding in which each vertex is mapped to a point, each edge is drawn as a sequence of alternate horizontal and vertical line segments, and any two edges do not cross except at their common end. A point where an edge changes its direction in a drawing is called a *bend*. We denote by $b(G)$ the minimum number of bends needed for an orthogonal drawing of G.

A *rectangular drawing* of a plane graph G is a drawing of G such that each edge is drawn as a horizontal or vertical line segment, and each face is drawn as a rectangle. Thus a rectangular drawing is an orthogonal drawing in which there is no bend and each face is drawn as a rectangle. The drawing of G'' in Fig. 1(e) is a rectangular drawing. The drawing of G' in Fig. 1(f) is not a rectangular drawing, but is an orthogonal drawing. The following result on rectangular drawings is known.

Lemma 1. *Let G be a connected plane graph such that all vertices have degree three except four vertices of degree two on $C_o(G)$. Then G has a rectangular drawing if and only if G has none of the following three types of simple cycles [T84]: (r1) 1-legged cycles, (r2) 2-legged cycles which contain at most one vertex of degree two, and (r3) 3-legged cycles which contain no vertex of degree two.*

Furthermore one can check in linear time whether G satisfies the condition above, and if G does then one can find a rectangular drawing of G in linear time [RNN96]. □

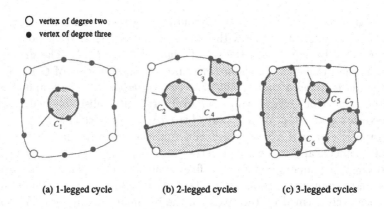

(a) 1-legged cycle   (b) 2-legged cycles   (c) 3-legged cycles

Fig. 2. Bad cycles C_1, C_2, C_3 and C_5, and non-bad cycles C_4, C_6 and C_7.

In a rectangular drawing of G, the four vertices of degree two are the four corners of the rectangle corresponding to $C_o(G)$. A cycle of type (r1), (r2) or (r3) is called a *bad cycle*. Figs. 2(a), (b) and (c) illustrate 1-legged, 2-legged and 3-legged cycles, respectively. Cycles C_1, C_2, C_3 and C_5 are bad cycles. On the other hand, cycles C_4, C_6 and C_7 are not bad cycles; C_4 is a 2-legged cycle but contains two vertices of degree two, and C_6 and C_7 are 3-legged cycles but contain one or two vertices of degree two.

Linear-time algorithms to find a rectangular drawing of a plane graph satisfying the condition in Lemma 1 have been obtained [KH94, RNN96]. Our orthogonal drawing algorithm uses the algorithm in [RNN96] which we call Rectangular-Draw in this paper.

3   Orthogonal Drawing

In this section we give a linear-time algorithm to find an orthogonal drawing of a 3-connected cubic plane graph G with at most $b(G) + 4$ bends. Thus there are at most four extra bends in a drawing produced by the algorithm.

Let G be a 3-connected cubic plane graph. Since G is 3-connected, G has no 1- or 2-legged cycle. A polygonal drawing of every cycle C in G has at least four convex corners, i.e., polygonal vertices of inner angle 90°, in any orthogonal drawing of G. Since G is cubic, such a corner must be a bend if it is not a leg-vertex of C. Thus we have the following facts for any orthogonal drawing of G.

Fact 2. *At least four bends must appear on $C_o(G)$.* □

Fact 3. *At least one bend must appear on each 3-legged cycle in G.* □

An outline of our algorithm is as follows.

Let G' be a graph obtained from G by adding four dummy vertices a, b, c and d of degree two on $C_o(G)$ as follows. If $|E(C_o(G))| \geq 4$, then add four dummy vertices on any four distinct edges on $C_o(G)$. If $|E(C_o(G))| = 3$, then add two dummy vertices on any two distinct edges on C_o and two dummy vertices on the remaining edge.

If the resulting graph G' has luckily no bad cycle, then by Lemma 1 G' has a rectangular drawing, in which the four dummy vertices become the corners of the rectangle corresponding to $C_o(G')$. From the rectangular drawing of G' one can immediately obtain an orthogonal drawing of G with exactly four bends by replacing the four dummy vertices with bends at the corners. By Fact 2 the orthogonal drawing of G has the minimum number of bends.

Thus we may assume that G' has a bad cycle. Since G has no 1- or 2-legged cycle, every bad cycle in G' is a 3-legged cycle containing no vertex of degree two like C_5 in Fig. 2(c). Among all bad cycles of G', let $C_1, C_2, \cdots, C_l$ be the "maximal" ones, that is, those that are not located inside of any other bad cycle. Let G'' be the graph obtained from G' by contracting the inside $G'(C_i)$ of each cycle C_i, $1 \leq i \leq l$, into a single vertex. We find a rectangular drawing of G'', and recursively find a "suitable" orthogonal drawing of $G(C_i), 1 \leq i \leq l$, called a *feasible drawing* and defined later, and finally patch them to get an orthogonal drawing of G. (See Fig. 1.)

Before describing the algorithm in details, we analyze a hierarchical structure of 3-legged cycles in G'. A 3-legged cycle C in G' is called a *descendant cycle* of $C_o(G')$ if C contains none of the four dummy vertices. We denote by $\mathcal{D}(G')$ the set of all descendant cycles of $C_o(G')$. A cycle C in $\mathcal{D}(G')$ is called a *child-cycle* of $C_o(G')$ if C is not located inside of any other cycle in $\mathcal{D}(G')$. Since G is a 3-connected cubic plane graph, all child-cycles of $C_o(G')$ are independent each other. For two distinct cycles C and C^* in $\mathcal{D}(G')$, if C^* is located in $G(C)$ but not located inside of any other 3-legged cycle in $G(C)$ except C, then C^* is called a *child-cycle* of C. For any cycle $C \in \mathcal{D}(G')$, all child-cycles of C are independent each other. Thus we get a hierarchical structure of cycles in $\mathcal{D}(G')$ represented by a "genealogical tree" T_g.

In Fig. 3(a) $\mathcal{D}(G') = \{C_1, C_2, C_3, C_4, C_5, C_6\}$. The 3-legged cycle C_7, indicated by a dashed line, is not in $\mathcal{D}(G')$ since it contains a dummy vertex a. Cycles C_1, C_2, C_3 and C_4 are the child-cycles of $C_o(G')$, and C_5 and C_6 are the child-cycles of C_4. The hierarchical structure of 3-legged cycles in G' in Fig. 3(a) is illustrated as a tree T_g in Fig. 3(b).

Using a method similar to one in [RNN96], one can find such a hierarchical structure T_g of 3-legged cycles in G' in linear time by traversing the contour of each face a constant number of times.

We assign the following information to each cycle in $\mathcal{D}(G')$ in a recursive manner from leafs to the root of T_g. Each cycle C in $\mathcal{D}(G')$ is divided into

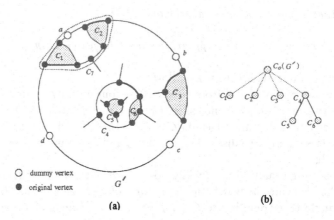

Fig. 3. (a) 3-legged cycles in G', and (b) a genealogical tree T_g.

three paths P_1, P_2 and P_3 by the three leg-vertices x, y and z of C. These three paths P_1, P_2 and P_3 are called *the contour paths* of C. Each contour path of C is classified as a *green path* or a *red path*. In addition, each cycle C in $\mathcal{D}(G')$ is assigned an integer $bc(C)$ called the *bend-count* of C. We will show later that $G(C)$ has an orthogonal drawing with $bc(C)$ bends and has no orthogonal drawing with fewer than $bc(C)$ bends, that is $b(G(C)) = bc(C)$. Furthermore we will show that, for any green path of C, $G(C)$ has an orthogonal drawing with $bc(C)$ bends including a bend on the green path. On the other hand, for any red path of C, $G(C)$ does not have any orthogonal drawing with $bc(C)$ bends including a bend on the red path. We do these classification and assignment by the following bottom-up computation on the tree T_g, as follows.

Assume that we have already done the classification and the assignment for all child-cycles $C_1, C_2, \cdots, C_l$ of a cycle $C \in \mathcal{D}(G')$ and are going to do them for C. There are three cases.

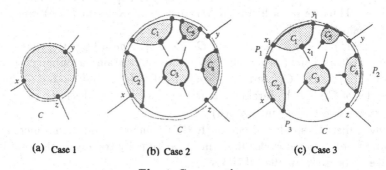

(a) Case 1   (b) Case 2   (c) Case 3

Fig. 4. Green paths.

Case 1: C has no child-cycle, that is, $l = 0$ and C is a leaf in T_g (see Fig. 4(a)).

In this case, we classify all the three contour paths of C as green paths, and set $bc(C) = 1$. (By Fact 3 we need at least one bend.) In Fig. 4(a) green paths of C are indicated by dotted lines.

Case 2: None of the child-cycles of C has a green path on C.

(In Fig. 4(b) the child-cycles of C are $C_1, C_2, \cdots, C_5$, and all green paths of them, drawn by thick lines, do not lie on C.) In this case, we classify all the three contour paths of C as green paths, and set $bc(C) = 1 + \sum_{i=1}^{l} bc(C_i)$. Since none of $C_1, C_2, \cdots, C_l$ and their descendant cycles in T_g has a green path on C, the orthogonal drawings of $G(C_1), G(C_2), \cdots, G(C_l)$ have no bend on C and hence we need to introduce a new bend on C in an orthogonal drawing of $G(C)$. In Fig. 4(b) the three green paths of C are indicated by dotted lines.

Case 3: Otherwise (see Fig. 4(c)).

In this case at least one of the child-cycles $C_1, C_2, \cdots, C_l$, for example C_1, C_4 and C_5 in Fig. 4(c), has a green path on C. Classify a contour path $P_i, 1 \le i \le 3$, of C as a green path if a child-cycle of C has its green path on P_i. Otherwise, classify P_i as a red path. Thus at least one of P_1, P_2 and P_3 is a green path. (In Fig. 4(c) P_1 and P_2 are green paths but P_3 is a red path.) We set $bc(C) = \sum_{i=1}^{l} bc(C_i)$. (For a cycle C_j having a green path on C, $G(C_j)$ has an orthogonal drawing with $bc(C_j)$ bends including a bend on the green path, and hence we need not to introduce any new bend on C.)

We have the following lemmas.

Lemma 4. *At least one of the three contour paths of every 3-legged cycle in* $\mathcal{D}(G')$ *is a green path under the classification above.* □

Lemma 5. *For every cycle C in $\mathcal{D}(G')$, $G(C)$ has at least $bc(C)$ vertex-disjoint 3-legged cycles in G' which contain no edge on red paths of C.*

Proof. It is easy to prove the proposition by an induction based on T_g. □

Lemma 6. *For every cycle C in $\mathcal{D}(G')$, $b(G(C)) \ge bc(C)$.*

Proof. By Fact 3 at least one bend must appear on each of the 3-legged cycles in $\mathcal{D}(G')$. By Lemma 5 $G(C)$ has at least $bc(C)$ vertex-disjoint 3-legged cycles in $\mathcal{D}(G')$. Therefore any orthogonal drawing of $G(C)$ has at least $bc(C)$ bends, that is, $b(G(C)) \ge bc(C)$. □

Conversely proving $b(G(C)) \le bc(C)$, we have $b(G(C)) = bc(C)$ for any cycle $C \in \mathcal{D}(G')$. Indeed we will prove a stronger claim later in Lemma 7 after introducing the following definition.

Let x, y and z be the three leg-vertices of C. One may assume that x, y and z appear on C in clockwise order. For a green path P with ends x and y on C, an orthogonal drawing of $G(C)$ is defined to be *feasible* for P if the drawing satisfies the following properties (p1)-(p3):

(p1) Exactly $bc(C)$ bends appear in the drawing of $G(C)$.

(p2) At least one bend appears on the green path P.

(p3) None of the the following six open halflines intersects the drawing of $G(C)$.

 – the vertical open halfline with the upper end at x.
 – the horizontal open halfline with the right end at x.
 – the vertical open halfline with the lower end at y.
 – the horizontal open halfline with the left end at y.
 – the vertical open halfline with the upper end at z.
 – the horizontal open halfline with the left end at z.

It should be noted that the starting line segment of each leg of C must lie on one of the six open halflines above in any orthogonal drawing of G extended from an orthogonal drawing of $G(C)$ feasible for P.

Lemma 7. *For any cycle C in $\mathcal{D}(G')$ and any green path P of C, $G(C)$ has an orthogonal drawing feasible for P.*

Proof. We give an algorithm to find a feasible orthogonal drawing of $G(C)$ recursively, as follows. There are three cases.

Case 1: C has no child-cycle.

In this case $bc(C) = 1$. We insert, as a bend, a dummy vertex t of degree two on an arbitrary edge on the green path P. Let F be the resulting graph. Every vertex of F has degree three except the three leg vertices x, y and z, and the dummy vertex t. Furthermore F has no bad cycle. Therefore by the algorithm Rectangular-Draw in [RNN96] one can find a rectangular drawing of F with four corners on x, y, z and t. The drawing of F immediately yields an orthogonal drawing of $G(C)$ with exactly one bend at t. Thus the drawing satisfies (p1)-(p3), and hence is feasible for P.

Case 2: None of the child-cycles of C has a green path on C.

Let $C_1, C_2, \cdots, C_l$ be the child-cycles of C. First, we find an orthogonal drawing $D(G(C_i))$ of $G(C_i)$ feasible for an arbitrary green path of C_i, for each $i = 1, 2, \cdots, l$, in a recursive manner.

Next, we construct a graph F from $G(C)$ by contracting each $G(C_i), i = 1, 2, \cdots, l$, to a single vertex v_i. We then construct a graph H from F by adding a dummy vertex t on any of the edges of P that remain in F. Thus there are exactly four vertices x, y, z and t of degree two on $C_o(H)$, and H has no bad cycle. Therefore, by the algorithm Rectangular-Draw, we can find a rectangular drawing $D(H)$ of H with four corners on x, y, z and t. Fig. 5(a) illustrates H for C in Fig. 4(b).

Finally, patching the drawings $D(G(C_1)), D(G(C_2)), \cdots, D(G(C_l))$ into $D(H)$, we can construct an orthogonal drawing of $G(C)$ with $bc(C) = 1 + \sum_{i=1}^{l} bc(C_i)$ bends (see Figs. 1 and 5). As illustrated in Fig. 6(b), there are twelve distinct embeddings of a contracted vertex v_i and the three legs incident to v_i, depending on the directions of the three legs. For each of the twelve cases, we can replace a contracted vertex v_i with a feasible orthogonal drawing of $G(C_i)$ or a rotated one shown in Fig. 6(c). Clearly t is a bend on P in the drawing of $G(C)$. Thus the drawing is feasible for P.

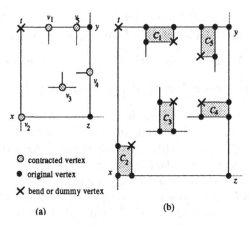

Fig. 5. (a) A rectangular drawing of H and (b) a feasible orthogonal drawing of $G(C)$ for Case 2.

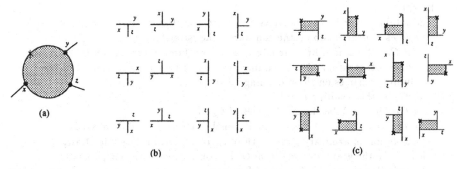

Fig. 6. (a) A 3-legged cycle, (b) twelve embeddings of a vertex v_i and three legs incident to v_i, and (c) twelve feasible orthogonal drawings of $G(C_i)$ and rotated ones.

Case 3: Otherwise.

Let $C_1, C_2, \cdots, C_l$ be the child-cycles of C, where $l \geq 1$. In this case, a child-cycle of C has a green path on C. One may assume without loss of generality that C_1 has a green path Q on the green path P of C, that the three leg-vertices x_1, y_1 and z_1 of C_1 appear on C_1 clockwise in this order, and that x_1 and y_1 are the ends of Q. We first construct a graph F from $G(C)$ by contracting each $G(C_i), 2 \leq i \leq l$, to a single vertex and by replacing $G(C_1)$ with a quadrangle $x_1 t y_1 z_1$, where t is a dummy vertex of degree two. Thus F has four vertices of degree two on $C_o(F)$, that is, t and the three leg-vertices x, y and z of C. By the algorithm Rectangular-Draw we then find a rectangular drawing $D(F)$ of F with four corners on x, y, z and t, in which the contour $x_1 t y_1 z_1$ of a face is drawn as a rectangle. We next find an orthogonal drawing $D(G(C_1))$ of $G(C_1)$ feasible for Q and an orthogonal drawing $D(G(C_i))$ of $G(C_i)$ feasible for an arbitrary green path of C_i for each $i = 2, 3, \cdots, l$ in a recursive manner. Finally, patching the drawings $D(G(C_1)), D(G(C_2)), \cdots, D(G(C_l))$ into $D(F)$, we can construct an orthogonal drawing of $G(C)$ feasible for P. Clearly $bc(C) = \sum_{i=1}^{l} bc(C_i)$ bends appear in the drawing of $G(C)$, and the bend t appears on Q and hence on P. □

Lemmas 6 and 7 immediately imply the following Lemma 8.

Lemma 8. *For every cycle C in $\mathcal{D}(G')$, $b(G(C)) = bc(C)$.* □

The algorithm for finding a feasible orthogonal drawing of $G(C)$ described in the proof of Lemma 7 above will be called Feasible-Draw from now on. Rectangular-Draw takes linear time. As mentioned before, one can find T_g in linear time by traversing every face boundary a constant number of times. Using T_g, one can easily compute $bc(C)$ and classify the three contour paths as green or red paths for all cycles $C \in \mathcal{D}(G')$ in linear time. Thus one can show that Feasible-Draw takes linear time. (The detail is omitted in this extended abstract.)

We are now ready to present our algorithm for orthogonal drawings of G.

Algorithm Orthogonal-Draw(G)
begin
1   add dummy vertices of degree two on $C_o(G)$ so that the resulting graph has four vertices of degree two on the contour of the outer face;
      {if $C_o(G)$ has four or more edges, add four dummy vertices on four distinct edges, otherwise, add two dummy vertices on two distinct edges and two dummy vertices on the remaining edge.}
2   let G' be the resulting graph;
3   let $C_1, C_2 \cdots, C_l$ be the child-cycles of $C_o(G')$;
4   for each $i = 1, 2, \cdots, l$, find an orthogonal drawing $D(G(C_i))$ of $G(C_i)$ feasible for an arbitrary green path of C_i by algorithm Feasible-Draw;
5   let G'' be the graph derived from G' by contracting each $G(C_i)$, $i = 1, 2, \cdots, l$, to a single vertex; {G'' has no bad cycle.}
6   find a rectangular drawing of $D(G'')$ of G'' by algorithm Rectangular-Draw;
7   patch the drawings $D(G(C_1)), D(G(C_2)), \cdots, D(G(C_l))$ into $D(G'')$ to get an orthogonal drawing of G
end.

We now have the following theorem.

Theorem 9. *For any 3-connected cubic plane graph G, an orthogonal drawing of G with at most $b(G) + 4$ bends can be found in linear time.*

Proof. (a) *Number of bends.*
There are two cases.
Case 1: $C_o(G)$ has no child-cycle.
In this case we have a drawing with exactly four bends. By Fact 2 it is a drawing with the minimum number of bends.
Case 2: Otherwise.
Let $C_1, C_2, \cdots, C_l$ be the child-cycles of G'. Then we have an orthogonal drawing of G with $4 + \sum_{i=1}^{l} bc(C_i)$ bends. By Lemma 5 G has $\sum_{i=1}^{l} bc(C_i)$ vertex-disjoint 3-legged cycles. Therefore Fact 3 implies that $\sum_{i=1}^{l} bc(C_i) \leq b(G)$. Thus the number of bends in the derived drawing is at most $b(G) + 4$.

(b) *Time complexity.*

Orthogonal-Draw calls Rectangular-Draw for G'' and Feasible-Draw for $G(C_i)$, $1 \leq i \leq l$. Both Rectangular-Draw and Feasible-Draw take linear time. Furthermore cycles $C_i, 1 \leq i \leq l$, are independent each other. Therefore Orthogonal-Draw takes linear time. □

4 Bend Minimization

The algorithm Orthogonal-Draw in the preceding section finds an orthogonal drawing of a 3-connected cubic plane graph G with at most $b(G) + 4$ bends. In this section we give a linear-time algorithm to find an orthogonal drawing with the minimum number $b(G)$ of bends, using Orthogonal-Draw.

In Section 3 we have defined a child-cycle for the cycle $C_o(G')$ and for cycles in $\mathcal{D}(G')$. In this section we define a child-cycle for any 3-legged cycle in G as follows. For two distinct 3-legged cycles C and C^* in G, if C^* is located in $G(C)$ but not located inside of any other 3-legged cycle in $G(C)$ except C, then C^* is called a *child-cycle* of C. We also extend the definitions of a contour path, a green path, a red path and the bend-count $bc(C)$ for any 3-legged cycle C in G.

Our idea is as follows. If G has a 3-legged cycle which has a green path on $C_o(G)$, then we can save one of the four bends mentioned in Fact 2, because a bend on the green path can play a role of a bend in Fact 2 and also a role of a bend in Fact 3. We therefore want to find such 3-legged cycles as many as possible and up to four. We define a *corner cycle* to be a 3-legged cycle that has a green path on $C_o(G)$ but none of whose child-cycles has a green path on $C_o(G)$. If G has independent corner cycles $C'_1, C'_2, \cdots, C'_k, k \leq 4$, then we can save k bends. We are now ready to give our algorithm to find an orthogonal drawing with the minimum number of bends.

Algorithm Minimum-Bend(G)
begin
1   find independent corner cycles $C'_1, C'_2, \cdots, C'_k$ of G as many as possible and up to four; {$k \leq 4$. The detail of this step is omitted in this extended abstract.}
2   let $P'_i, 1 \leq i \leq k$, be a green path of C'_i on $C_o(G)$;
3   let x_i, y_i and z_i be the leg-vertices of C'_i, and let the ends of P'_i be x_i and y_i;
4   replace each subgraph $G(C'_i), 1 \leq i \leq k$, in G with a quadrangle $x_i t_i y_i z_i$ where t_i is a dummy vertex of degree two;
5   let G' be the resulting graph;
6   Orthogonal-Draw(G');
7   {$t_1, t_2, \cdots, t_k$ are corners of the drawing of $C_o(G')$, and the quadrangle $x_i t_i y_i z_i$ is drawn as a rectangle for each i, $1 \leq i \leq k$.}
8   find an orthogonal drawing $D(G(C'_i))$ of $G(C'_i)$ feasible for P'_i for each $i = 1, 2, \cdots, k$;
9   patch the drawings $D(G(C'_1)), D(G(C'_2)), \cdots, D(G(C'_k))$ into the drawing of G' to get an orthogonal drawing of G
end.

We now have the following theorem.

Theorem 10. *Algorithm Minimum-Bend produces an orthogonal drawing of a 3-connected cubic plane graph G with the minimum number of bends in linear time.* □

5 Conclusions

In this paper we presented a linear-time algorithm to find an orthogonal drawing of a 3-connected cubic plane graph with the minimum number of bends. It is remained as a future work to find a linear-time algorithm for a larger class of graphs.

An orthogonal drawing is called an *orthogonal grid drawing* if all vertices and bends are located on integer grid points. Given an orthogonal drawing, one can find a *corresponding* (that means preserving the directions of edges) orthogonal grid drawing in linear time [T87]. Let W be the *width* of a grid, that is the number of vertical lines in the grid minus one, and H be the *height* of a grid. Let n be the number of vertices, and let m be the number of edges in a given graph. It is known that any orthogonal drawing using b bends has a corresponding orthogonal grid drawing on a grid such that $W + H \leq b + 2n - m - 2$ [B96]. It is also known that any 3-connected cubic plane graph has an orthogonal grid drawing using at most $\frac{n}{3} + 3$ bends on a grid such that $W \leq \frac{n}{2}$ and $H \leq \frac{n}{2}$ [B96, K96]. By using our algorithm and the algorithm in [T87], one can find in linear time an orthogonal grid drawing of a 3-connected cubic plane graph with exactly $b(G)$ bends on a grid such that $W \leq \frac{n}{2}$, $H \leq \frac{n}{2}$ and $W + H \leq b(G) + 2n - m - 2 = b(G) + \frac{1}{2}n - 2$.

References

[B96] T. C. Biedl, *Optimal orthogonal drawings of triconnected plane graphs*, Proc. of SWAT'96, LNCS 1097 (1996), pp. 333-344.

[GT94] A. Garg and R. Tamassia, *On the computational complexity of upward and rectilinear planarity testing*, Proc. of Graph Drawing'94, LNCS 894 (1995), pp. 286-297.

[GT96] A. Garg and R. Tamassia, *A new minimum cost flow algorithm with applications to graph drawing*, Proc. of Graph Drawing'96, LNCS 1190 (1997), pp. 201-226.

[K96] G. Kant, *Drawing planar graphs using the canonical ordering*, Algorithmica 16 (1996), pp. 4-32.

[KH94] G. Kant and X. He, *Two algorithms for finding rectangular duals of planar graphs*, Proc. of WG'93, LNCS 790 (1994), pp. 396-410.

[RNN96] M. S. Rahman, S. Nakano and T. Nishizeki, *Rectangular grid drawings of plane graphs*, Proc. of COCOON'96, LNCS 1090 (1996), pp. 92-105. Also, Comp. Geom. Theo. Appl. (submitted).

[T87] R. Tamassia, *On embedding a graph in the grid with the minimum number of bends*, SIAM J. Comput. 16 (1987), pp. 421-444.

[T84] C. Thomassen, *Plane representations of graphs*, (Eds.) J.A. Bondy and U.S.R. Murty, Progress in Graph Theory, Academic Press Canada, (1984), pp. 43-69.

Interactive Orthogonal Graph Drawing:
Algorithms and Bounds

Ulrich Fößmeier

Universität Tübingen, Wilhelm-Schickard-Institut, Sand 13, 72076 Tübingen, Germany, email: foessmei@informatik.uni-tuebingen.de

Abstract. Incremental graph drawing is a model gaining more and more importance in many applications. We present algorithms that allow insertions of new vertices into an existing drawing without changing the position of the objects drawn so far. We prove bounds for the quality of our drawings and considerably improve on previous bounds. Here the number of bends and the used area are our quality measures. Besides we discuss lower bounds for this problem.

1 Introduction

Representing objects and the relations between them is a task arising permanently in many applications of Computer Science. Therefore the concept of *graphs* is widely used to represent the data. A graph is a set V of *vertices* together with a set $E \subseteq V \times V$ of *edges*. Examples for the use of graphs are Petri nets, entity-relationship diagrams, production planning or data flow diagrams in Software Engineering (see e.g. [4,3]). For human processing it is necessary to *visualize* graphs; sometimes this visualization (called *graph drawing*) is even the desired output of some computation, e.g. in VLSI design [15]. In a graph drawing the vertices of the graph are represented by geometric objects (points, circles, squares, ...) and the edges by simple curves connecting them. The drawing must fulfill certain properties which depend on the specific application. Typical properties are: The edges must be straight lines, they must not cross each other (planar drawing) or they must consist only of horizontal and vertical segments (orthogonal drawing). Moreover the drawing should optimize some cost function; this could be to minimize the used area, to minimize the number of crossings or bends, to minimize the total edge length etc. A survey of graph drawing algorithms and applications can be found in [6].

In many cases not the total amount of data is given in advance, but the vertices of the graph are added one by one and after every insertion a drawing is required. This situation is called *incremental drawing*. Thus incremental drawing has its own importance, moreover it is an important step to the more general concept of *interactive drawing*, where also deletions of vertices are allowed. We investigate incremental drawings using the model of orthogonal drawings: Here graphs with a maximum vertex degree of at most four are drawn in such a way that the vertices are placed on grid points of an integer grid and the edges run along grid lines. If the drawings are destinated for human observers, inserting a new vertex should not change too much the 'old' drawing. In [11] the idea of the user's 'mental map' is introduced (the way how a human user 'understands' a drawing); adding a new vertex should not destroy the user's mental map. Thus an incremental drawing system must guarantee a) to maintain the user's mental map and b) to handle a single insertion in a reasonable time (we require constant time per insertion).

If adding a new vertex must not change at all the position of the objects drawn so far, we speak of the No-Change-Scenario (the No-Change-Scenario is the best way to support the maintenance of a mental map). Papakostas and Tollis [13] introduced this model and also investigated another one: The Relative-Coordinates-Scenario, where it is allowed to shift every drawn vertex by a constant number of units along the x- and y-axes when inserting a new vertex. They gave first results on the number of bends and the used area in both scenaria and presented some examples.

Obviously the Relative-Coordinates-Scenario is more powerful than the No-Change-Scenario. All drawings in the No-Change-Scenario maintain the conditions for the Relative-Coordinates-Scenario. Hence bounds for the No-Change-Scenario-drawings also hold for the other model.

In this paper we focus on the No-Change-Scenario; we refine the methods of [13] and considerably improve on their bounds for the number of bends and the used area; our algorithm even leads to better results (in general) than the Relative-Coordinates-drawings of Papakostas and Tollis.

The rest of the paper is organized as follows: In Section 2 we give some definitions and describe the basic algorithm. In Section 3 we introduce our technique how we bound the value of some cost function (number of bends), and in the following two sections we present new methods how to improve on these bounds. Section 6 is devoted to another important cost function (used area) and in Section 7 we discuss lower bounds for the problem. Finally Section 8 illustrates our algorithms with some examples.

2 First Results

Papakostas and Tollis [13] gave a first algorithm that computes drawings in the No-Change-Scenario and proved $\frac{8}{3} \cdot n \approx 2.66 \cdot n$ (n being the number of vertices drawn so far) as an upper bound for the number of bends produced by their algorithm. In [12] statistics about the practical behaviour of this algorithm are given. We present a more general approach that can be used to minimize any cost function and analyse it for the example of the number of bends being the cost criterion.

We start with some definitions:

A graph with a maximum vertex degree of at most four is called a *4-graph*.

The *local degree ld(v)* of a vertex v is the number of incident edges at the moment when v is inserted.

A vertex v which is inserted at step t is called *new* at this time; in all further steps $t' > t$ it will be called *old*.

An old vertex v has a *free direction* to the top, if there is no other vertex v' having the same x-coordinate as v and a larger y-coordinate than v **and** if there is no vertical edge segment above of v in the same column (this definition implies that there is no edge being incident to v at its top side). Free directions to the bottom, left and right are defined analogously.

A half straight line starting at an old vertex v and pointing into a free direction of v is called a *free line* (of v).

A vertex v is called *critical* if it has exactly one free direction and *semi-critical* if it has exactly two free directions.

The *bounding box* of a drawing is the smallest rectangle containing the whole drawing.

The four gridlines with the minimum distance to the bounding box without intersecting or touching it are called *basic gridlines*.

To prove the correctness of our algorithms we use the following

Incremental Invariant: Every vertex v of degree $deg(v)$ $(deg(v) \leq 4)$ in the current drawing has $4 - deg(v)$ free directions.

Lemma 1. *An algorithm that maintains the incremental invariant can compute an orthogonal drawing of any 4-graph.*

Proof. Let v be a vertex that has to be added to the drawing and $v_1, \ldots, v_{ld(v)}$ its neighbours in the current drawing. The incremental invariant implies that we can draw straight lines starting at $v_1, \ldots, v_{ld(v)}$ that do not cross forbidden objects (other vertices, parallel edges); after leaving the bounding box these lines may have some bends in order to join at a geometric point where we place v. $\Diamond$

The strategy of our first algorithm is the following: Place a new vertex v at one of the following places: Either at the insertion of two basic gridlines ('at the corner' of the bounding box), or at the intersection of a basic gridline with a free line of v' (v' being a neighbour of v). In this case v and v' will be connected by a straight line. Since v has at most four neighbours and every neighbour has at most four free directions, there are at most 20 possible places for v. Also the bends along the new edges have to be placed at these positions. Fig. 1a) shows the possible places for a new vertex v having the neighbours v_1, v_2, v_3.

Our strategy is to choose the place that optimizes the main criterion (e.g. the number of bends) *for this insertion*. Fig. 1b) shows two different possibilities, the first one minimizing the number of bends and the second one minimizing edge crossings.

In [7] we give a simple proof for the following

Lemma 2. *An algorithm following the strategy described above maintains the incremental invariant.*

Next we analyse the behaviour of the algorithm with respect to the number of produced bends.

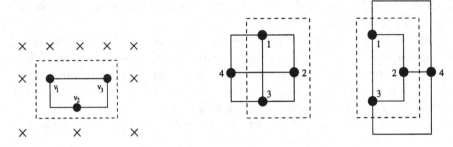

Fig. 1. a) The possible places for vertices and bends b) Minimizing bends or crossings

3   Bounds on the Number of Bends

We subdivide the vertices into different classes according to their local degree and to the number of bends they produce when they are inserted.
B_{ij} denotes the class of vertices with local degree i producing j new bends and $b_{ij} = |B_{ij}|$.

In [13] the graph is demanded to be *connected at any time*, since otherwise the bounds could not be proved. We do not want to restrict our algorithm more than necessary, so we allow insertions with local degree 0; such an insertion (called *ld0-insertion*) creates a new connected component of the graph (which can join with other connected components in the course of the algorithm). Surprisingly, although an insertion of local degree 0 does not cause any bends, the upper bound on the number of bends increases compared to the model where the graph is always connected. This observation encouraged us to distinguish between two classes of problems: With and without ld0-insertions. The analysis of this section forbids vertices with local degree 0 in order to have an easy and understandable proof for the bound.

We distinguish the disjoint classes B_{48}, B_{47}, B_{46}, B_{34}, B_{33}, B_{23}, B_{22}, B_{10}, and characterize the most important classes (B_{46}, B_{33} and B_{22} mean that *at most* 6, 3 respectively 2 bends are needed): Obviously a vertex v with local degree one can always be inserted without any new bends.

The first line of Fig. 2 shows the three different cases how a new vertex v of local degree two can be connected to its neighbours: The two new edges can leave the bounding box at the same side, at different but not opposite sides or at opposite sides (all other situations can be obtained by rotating and reflecting one of the pictures in Fig. 2). According to these cases such an insertion requires one, two or three bends. Thus a vertex $v \in B_{23}$ has two neighbours requiring the incident edges to run in opposite directions; this implies that both neighbours must be critical, otherwise one of the other (cheaper) cases would work.

In the lower part of Fig. 2 the different cases for insertions with local degree three and four can be seen.

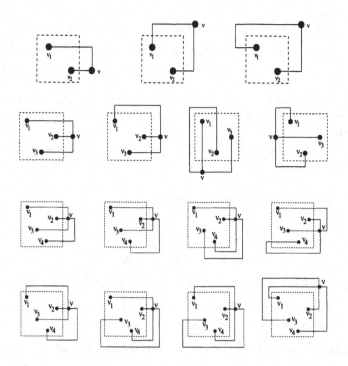

Fig. 2. Insertions of a vertex with local degree two and more

Table 1 shows necessary conditions what kind of neighbours a vertex v must have in order to (possibly) belong to a class B_{ij}.

Table 1

Class	B_{48}	B_{47}	B_{34}	B_{23}
critical neighbours	4	2	1	2

We shortly prove this fact for the example of the class B_{34} (the proof for the other classes is similar): Let v_1, v_2 and v_3 be the neighbours of v; if none of them is critical then there is a pair of them (say v_1 and v_2) having a common free direction; in this case the edge connecting v and v_3 cannot be forced to leave the bounding box in the opposite direction, since v_3 is not critical, and thus the conditions for case B_{34} cannot be fulfilled.

In the following we try to construct a worst case (i.e. a case with a maximum number of bends) obeying the rules of Table 1. This case gives an upper bound for the number of bends needed by our algorithm (in [12] a similar technique was used to prove upper bounds). We use linear programming and maximize the number of bends under certain restrictions. A first Linear Program (LP) could be

$$max \quad 8 \cdot b_{48} + 7 \cdot b_{47} + 6 \cdot b_{46} + 4 \cdot b_{34} + 3 \cdot b_{33} + 3 \cdot b_{23} + 2 \cdot b_{22}$$

$$due\ to\ \ b_{48} + b_{47} + b_{46} + b_{34} + b_{33} + b_{23} + b_{22} + b_{10} = n$$

n is the number of vertices inserted so far. Since the solution of this LP does not give a non-trivial result we must add some more restrictive constraints:

- We evaluate the data from Table 1 and take into account that only vertices that are inserted with a local degree of at most three can ever be or become critical.
- We bound the total number of edges (which is equal to the sum of all local degrees) by $2 \cdot n$.

This leads to the following constraints:

$$crit \leq b_{34} + b_{33} + b_{23} + b_{22} + b_{10}$$

$$4 \cdot b_{48} + 2 \cdot b_{47} + b_{34} + 2 \cdot b_{23} \leq crit$$

$$4 \cdot b_{48} + 4 \cdot b_{47} + 4 \cdot b_{46} + 3 \cdot b_{34} + 3 \cdot b_{33} + 2 \cdot b_{23} + 2 \cdot b_{22} + b_{10} \leq 2 \cdot n$$

A solution of this LP gives $2.5 \cdot n$ as the value of the objective function with the variables $b_{23} = b_{22} = 0.5 \cdot n$, $crit = n$.

As a consequence of all facts discussed in this section we state

Theorem 1. *Our algorithm incrementally computes orthogonal drawings of any 4-graph without ld0-insertions producing at most $2.5 \cdot n$ bends (where n is the number of vertices inserted so far); every insertion can be done in constant time.*

Fig. 3 a) shows an easy example for a drawing that exactly realizes this bound: Vertices 1, 2, 3, 4 are already drawn. Vertex 5 has to be connected with 1 and 3, vertex 6 with 2 and 4, vertex 7 with 1 and 4 and vertex 8 with 2 and 3. In the resulting drawing the vertices 5 and 7 play the role of the vertices 1 and 2 and the vertices 6 and 8 play the role of the vertices 3 and 4 and thus we can iterate.

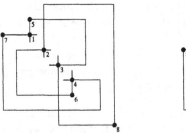 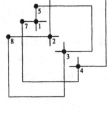

Fig. 3. a) An example with $2.5 \cdot n$ bends   b) A drawing with less bends

4 A First Improvement

In the drawing of Fig. 3 a) there are many insertions of type B_{23} using critical vertices having their free direction at opposite sides. If all critical vertices would have the same shape this bad example could not arise. Thus our strategy is to produce always the same type of critical and semi-critical vertices, if possible. In the example of Fig. 3 a) this means that the insertions of type B_{22} (e.g. 5 and 6) create *the same* type of critical vertices, say vertices being free to the left or to the bottom. Following this strategy the vertices 5, 6, 7 and 8 could be inserted with 8 bends (instead of 10 bends) and the resulting drawing has four semi-critical vertices of the same shape, thus making following insertions even cheaper. Following the iteration the graph can be drawn with n bends instead of $2.5 \cdot n$ bends (see Fig. 3 b)).

This effect can also be proved theoretically: We take into account that e.g. an insertion of type B_{23} not only needs two critical vertices but two critical vertices having the free direction at opposite sides. Thus in the LP two types of B_{23} insertions are distinguished. One of them consumes a critical vertex with free direction to the left and a critical vertex with free direction to the right, and the other one consumes each a critical vertex being free to the top and to the bottom. Both types of insertion create a semi-critical vertex. Here it is a free choice of the algorithm whether this semi-critical vertex is free to a) the left and to the top, b) the top and to the right, c) the right and to the bottom or d) the bottom and to the left. If e.g. the two critical neighbours of the new vertex u are free to the left (v_1) resp. to the right (v_2), the algorithm can realize case a) by placing u in the left upper corner of the bounding box, case b) by placing u at the right upper corner of the bounding box etc.

In the LP we establish a variable for every type of insertion (e.g. new vertex with local degree two, both neighbours are critical, one of them being free to the left, the other to the right), and an equality for every type of vertex (e.g. critical being free to the right). These equalities ensure that the same number of vertices of some type is consumed and created. The resulting LP consists of 16 equalities and 3073 variables and can be constructed automatically (in order to avoid mistakes that are inevitable when dealing with so many data by hand).

In order to create vertices of the same shape we give a cost to every vertex type: For critical vertices this cost function prefers vertices being free to the left and increases monotonically for vertices being free to the bottom, to the top and to the right (in this order). It is important that the cheapest and the most expensive type are free in opposite directions because having vertices of both types makes insertions with many bends possible. Similarly the cost function for semi-critical vertices increases monotonically from vertices being free to the left and to the bottom via the other shapes until the most expensive vertices being free to the right and to the top. Note that these costs are favourable compared with a cost function where vertices being free e.g. to the bottom and to the right are cheap because such vertices could become expensive critical vertices with the free direction to the right after future insertions, whereas the cheapest semi-critical vertex of our cost function always becomes a relatively cheap critical vertex in the future. This consideration indicates that the chosen cost function is optimal with respect to a worst-case analysis. For practical examples also a randomized variant could be interesting.

Using the optimizing system CPLEX [5] the LP can be solved in about half a second; the value of the objective function (i.e. the number of bends) is $\frac{92}{41} \cdot n \approx$ $2.24 \cdot n$ if the graph is always connected and $\frac{3333}{1204} \cdot n \approx 2.77 \cdot n$ if insertions with local degree 0 are allowed.

Using this method we get not only an upper bound but also a family of examples realizing this bound (in the case of an always connected graph this example consists of a series of 41 insertions that can be iterated and need 92 bends per iteration). So we are sure that the upper bound is tight with respect to the algorithm.

5 Placing Vertices inside of the Bounding Box

Our final method to improve on the upper bounds for cost functions is very powerful in practice and even leads to an improvement of the theoretic bounds: We allow a new vertex to be placed *inside* of the bounding box if this placement does not violate the incremental invariant.

Lemma 3. *Placing a new vertex v at the intersection of free lines of two neighbours v_1 and v_2 of v does not violate the incremental invariant.*

Proof. Since v will be placed at the intersection point of two free lines there are no other vertices on these gridlines in the used directions. Thus v cannot destroy a free direction of other vertices than v_1 and v_2. After the insertion v is free in the two directions in which v_1 and v_2 were free before the insertion, so the invariant holds for v. $\diamond$

An insertion of a vertex having local degree two now in some cases can be done without producing a new bend, while this is impossible following the insertion rules of Section 2 and 4. Vertices 10 and 13 in Fig. 6 are examples for such insertions. The enormous practical profit of this idea is clear: Especially large graphs with many vertices having a degree of less than four define a large number of such intersection points, such that the probability that the new idea can save bends is very high. Applying the new insertion method to improve on the theoretic bound is tricky: The number of intersection points does not depend only of the number and shape of the vertices drawn so far, but also of their geometric position: If there are e.g. a critical vertex v_1 having its free direction to the left and a critical vertex v_2 having its free direction to the top, then they only define an intersection point if v_1 is placed to the right of and above v_2. These conditions cannot be expressed by constraints of a linear program.

But there are situations where two vertices *always* define an intersection point: If vertex v_1 is free to the left and to the right and vertex v_2 is free to the top and to the bottom, the corresponding free lines always intersect independently of the placement of the vertices. Thus many insertions using two such vertices as neighbours get cheaper when placing the new vertex inside of the bounding box.

Our algorithm tries to create semi-critical vertices having their free directions at opposite sides whenever possible, in order to increase the number of intersection points. For example the insertion of a vertex v of local degree one can always create a semi-critical vertex of the desired shape if the neighbour of v has degree one. Thus the cost function introduced in Section 4 must be slightly modified for semi-critical vertices. Of course vertices of degree one often define intersection points, too.

The number of variables of the LP increases considerably because the reciprocal geometric position of the vertices now plays a role: The placement of the intersection points determines the shape of the new vertices (see Fig. 4 for an example). After the insertion the vertices in question are free to the left, to the bottom and (semi-critical) to the top and right (left drawing) or to the top, to the right and (semi-critical) to the left and bottom (right drawing).

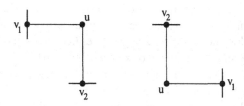

Fig. 4. Placing a new vertex u inside of the bounding box creates different situations.

The large number of cases (especially when the new vertex has local degree four) makes a complete analysis of the problem extraordenarily difficult. Inserting only the necessary rules by hand (i.e. the rules concerning types of insertions that arise in worst examples of the always-connected case) leads to an LP with 3119 variables (without ld0-insertions).

Theorem 2. *Every 4-graph without ld0-insertions allows an incremental orthogonal drawing where the number of bends is bounded by $\frac{100}{47} \cdot n \approx 2.22 \cdot n$ at any time, n being the actual number of vertices. Every insertion can be done in constant time.*

Proof. The value of the bound is the solution of the new Linear Program. The time bound follows from the fact that the number of intersection points is bounded by 12: There are at most four vertical gridlines that can define intersection points (the x-coordinates of the at most four neighbours of the new vertex), and on each of them there can arise at most three intersection points (having the y-coordinates of the other three neighbours). Thus the algorithm must check at most $20 + 12 = 32$ positions where the new vertex can be placed (see [9] for an exact case analysis). $\diamond$

When allowing ld0-insertions the number of cases is so high that we did not succeed in establishing a corresponding LP by hand. We conjecture that the insertions of isolated vertices are always cheap such that the value of the objective function should be very close to the case of an always connected graph.

6 Bounds for the Used Area

We can bound the area used by our drawings using a method similar to the bend minimizing strategy: The criterion where to place a new vertex now is not the number of new bends but the number of new gridlines. We bound the total number of gridlines used by the drawing and argue that a drawing using k gridlines cannot require more than $\left(\frac{k}{2}\right)^2$ area. This is because the bounding box of the drawing has a circumference of $2 \cdot k$ and the rectangle with the largest area having a circumference of $2 \cdot k$ is the square with a side length of $\frac{k}{2}$.

In our LP we only have to change the objective function (which now must count the number of new gridlines instead of the number of new bends) and get a result of $0.937 \cdot n^2$ resp. $1.057 \cdot n^2$ (allowing ld0-insertions or not), whereas Papakostas and Tollis only can guarantee an area of at most $1.77 \cdot n^2$. The profit of the idea described in Section 5 for the used area is obvious: a new vertex placed inside of the bounding box does not require new gridlines because it uses old ones. For the area it is not surprising that insertions of vertices with local degree 0 increase the upper bound because such an insertion always needs two new gridlines in order to guarantee four free directions of the new vertex, whereas an insertion of a vertex of local degree 1 requires only one new gridline. Again the bounds are tight for the algorithm.

In Section 7 we present the results on the area when using the different approaches desribed in this paper.

7 Lower Bounds and Offline Bounds

A classification of our results requires the comparison with existing results about upper bounds for the offline-version of the corresponding problems as well as the comparison with lower bounds. Table 2 shows some results on the number of bends and on the used area:

Table 2

	Bends	Area
Upper bound offline	$1.9 \cdot n$	$0.8 \cdot n^2$
Lower bound offline	$1.83 \cdot n$	$\Omega(n^2)$
Papakostas and Tollis	$2.66 \cdot n$	$1.77 \cdot n^2$
Basic algorithm	$2.5 \cdot n$	$1.44 \cdot n^2$
Section 4 always connected	$2.24 \cdot n$	$0.937 \cdot n^2$
Section 5 always connected	$2.22 \cdot n$	$0.937 \cdot n^2$
Section 4 general	$2.77 \cdot n$	$1.057 \cdot n^2$
Section 5 general	??	??

The lower bound for the number of bends given in Table 2 is a lower bound for offline drawings and is established in [1]. For the area no good lower bound is known. In [15] a lower bound of $\Omega(n^2)$ is proved and in [2] the author precises the constant of this bound to be about 10^{-6}. Thus this bound is not very useful for comparisons. The offline upper bounds in Table 2 are established in [14]. It is worth to mention that the minimization problems for the number of bends and for the used area are both $\mathcal{NP}$-complete [8,10].

The upper bounds for the area and the number of bends in incremental drawings are not much worse than the corresponding offline-bound. In [7] we prove a lower bound of $2 \cdot n$ for the number of bends in incremental drawings. This result makes the upper bound of $2.22 \cdot n$ even more valuable.

For the last line of the table the exact values are not known, but we conjecture that they are very close to the always-connected case.

8   Examples and Practical Results

To estimate the practical significance of the algorithm we compared some drawings with
a) drawings got by the algorithm in [13];
b) drawings got by an offline-algorithm.

Having a closer look at the results in [13] and [12] it can be seen that the No-Change-Scenario was mainly developped in order to apply an easy and effective technique of analysis, whereas the Relative-Coordinate-Scenario was developped for practical applications. This is the reason for the bad upper bounds of the Relative-Coordinates-algorithm ($3 \cdot n$ for the number of bends, $2.25 \cdot n^2$ for the area) compared to the weaker model. Thus a comparison of our drawings with No-Change-drawings of Papakostas and Tollis would not be fair. But our drawings are even better than the Relative-Coordinates-drawings of Papakostas and Tollis in most cases, although we guarantee the rules of the No-Change-Scenario. Fig. 5 and 6 show three drawings of the same graph produced by the three algorithms. The No-Change-drawing needs 16 bends and 13 × 13 area, the Relative-Coordinates-drawing 12 bends and 11 × 11 area and our algorithm produces only 9 bends on an area of 9 × 10. Further tests have shown that these numbers show a typical relation between the three algorithms.

9   Extensions

The algorithms described so far are all implemented in our system [9]. The methods are very flexible such that it is easily possible to think about extensions in order to have algorithms that can be used in a large field of applications.

In [7] a detailed discussion about possible extensions can be found. The goals are

- to provide interactive drawings, i.e. also deletions of vertices and edges;
- to optimize other cost functions;

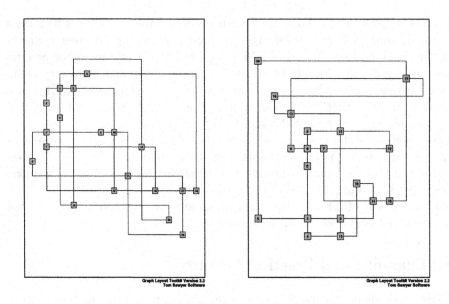

Fig. 5. No-Change Papakostas/Tollis and Relative-Coordinates Papakostas/Tollis

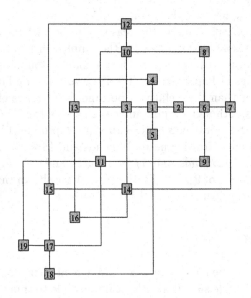

Fig. 6. No-Change new approach

- to realize the ideas of Relative-Coordinates;
- to draw graphs with an arbitrary vertex degree.

Especially the last point is important in order to have an algorithm that is universally applicable to arbitrary graphs.

Acknowledgement: I wish to thank Michael Kaufmann and Frank Jahrmarkt for useful discussions, Frank Jahrmarkt for the implementation and Achilleas Papakostas, Janet Six and Ioannis Tollis for getting some drawings of their algorithms for comparisons.

References

1. Biedl, T.C., *New Lower Bounds for Orthogonal Graph Drawings*, Proceedings on GD'95, Passau, 28-39, 1995.
2. Biedl, T.C., *Orthogonal Graph Drawing, Algorithms and Lower Bounds*, Diploma Thesis TU Berlin, 1995.
3. Batini, C., E. Nardelli and R. Tamassia, *A Layout Algorithm for Data-Flow Diagrams*, IEEE Trans. on Software Engineering, Vol. SE-12 (4), 538-546, 1986.
4. Batini, C., M. Talamo and R. Tamassia, *Computer Aided Layout of Entity-Relationship Diagrams*, The Journal of Systems and Software, Vol. 4, 163-173, 1984.
5. CPLEX optimization, Inc. *Using the CPLEX Base System*. CPLEX Optimization, Inc., 1995.
6. DiBattista G., P. Eades, R. Tamassia and I.G. Tollis, *Algorithms for Drawing Graphs: An Annotated Bibliography*, Computational Geometry: Theory and Applications, vol. 4, no 5, 235-282, 1994.
7. Fößmeier U., *Interactive Orthogonal Graph Drawing: Algorithms and Bounds*, Technical Report WSI-97-12, University of Tübingen.
8. Garg, A., R. Tamassia, *On the computational complexity of upward and rectilinear planarity testing*, Proceedings of GD'94, Princeton, 286-297, 1994.
9. Jahrmarkt, F., *Knickminimierende Verfahren für interaktives orthogonales Graphenzeichnen*, Diplomarbeit Universität Tübingen, 1997 (in German language).
10. Kramer M.R., J. van Leeuwen, *The complexity of wire routing and finding minimum area layouts for arbitrary VLSI circuits*, Advances in Computer Research, Vol. 2: VLSI Theory, Jai Press, Reading, MA, 129-146, 1992.
11. Misue, K., P. Eades, W. Lai and K. Sugiyama, *Layout Adjustment and the Mental Map*, Journal of Visual Languages and Computing, vol. 6, 183 - 210, 1995.
12. Papakostas, A., J. M. Six, I. G. Tollis, *Experimental and Theoretical Results in Interactive Orthogonal Graph Drawing*, Proceedings on GD'96, Berkeley, 371-386, 1996.
13. Papakostas, A. and I.G. Tollis, *Issues in Interactive Orthogonal Graph Drawing*, Proceedings on GD'95, Passau, 419-430, 1995.
14. Papakostas, A. and I.G. Tollis, *Improved Algorithms and Bounds for Orthogonal Drawings*, Proceedings on GD'94, Princeton, 40-51, 1994.
15. Valiant, L. G., *Universality considerations in VLSI circuits*, IEEE Trans. Comput., C-30, 135-140, 1981.

Embedding a Graph in the Grid of a Surface with the Minimum Number of Bends Is NP-hard

M. A. Garrido and A. Márquez

Dept. Matemática Aplicada I, Universidad de Sevilla, 41012-Sevilla (Spain)
email: vizuete@cica.es and almar@cica.es

Abstract. This paper is devoted to the study of graph embeddings in the grid of non-planar surfaces. We provide an adequate model for those embeddings and we study the complexity of minimizing the number of bends. In particular, we prove that testing whether a graph admits a rectilinear (without bends) embedding essentially equivalent to a given embedding, and that given a graph, testing if there exists a surface such that the graph admits a rectilinear embedding in that surface are NP-complete problems and hence the corresponding optimization problems are NP-hard.

1 Introduction

Drawings of graphs on the rectilinear grid have been studied because of their applications in VLSI planning. In this context, three important optimization subtasks related to theoretical problems in Graph Theory arise in VLSI design

- To minimize the number of holes to be made in the board in order to avoid inappropiate intersection.
- To minimize the number of bends.
- To minimize the area.

This paper is related to diverse aspects of the second problem. The first task is one of the most important problems in Topological Graph Theory, where it is known as the genus of a graph, and, in 1990 C. Thomassen proved that this problem is NP-hard [8]. Nonetheless, several heuristics have been presented obtaining good aproximations to the optimal solution (see [6]). The other two problems are NP-hard problems even if there exists no fixed embedding of the graph (see [4] and [1]). In practice, the heuristics mentioned above provide actual embeddings of the graphs in certain surfaces. Thus, it remains as an open question to find a polynomial algorithm providing the minimum number of bends when a fixed embedding in a surface is given. In this sense, Tamassia proves that if the embedding is in the plane that algorithm exists [7], but nothing is known in other surfaces. And, actually in most practical situations, the graph that modelizes the design is non-planar. In this paper, we establish the basis to the study of grid drawings of graphs in orientable surfaces defining a good model (a grid in each surface of genus n) that allows to considerate the same topics as in the plane, and we prove that the two following related problems are NP-complete:

- Testing whether a graph admits a rectilinear (without bends) embedding essentially equivalent to a given embedding.
- Given a graph, testing if there exist a surface such that the graph admits a rectilinear embedding in that surface.

Hence, the corresponding bend-minimization problems are NP-hard.

In this paper, we assume standard concepts on NP-completeness [3].

2 A model to study grid embeddings in non-planar surfaces

In order to consider grid embeddings of graphs in non-planar surfaces, we need to give an appropiate representation of a surface of genus n. Of course, the first attemp to achieve that goal must be to use the well-known Classification of Surfaces (see, for instance [2]).

Theorem 1. *[2] Every closed, connected, orientable surface is homeomorphic to one of the standard polygonal surfaces.*

The standard polygonal surfaces are obtained by identifying the sides of a $4n$-gone following the scheme described in Figure 1.

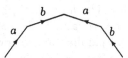

Fig. 1. One of the handles in a surfaces of genus n.

On the other hand, we can observe that in our problem the existence of an orthogonal system is fundamental. Nevertheless, the standard polygonal representation of surfaces lacks that system. However, it is possible to give another representation of surfaces. As a consequence of Theorem 1 we get the following classification theorem.

Theorem 2. *Every closed, connected, orientable surface is homeomorphic to one of the standard orthogonal surfaces.*

Where the standard orthogonal surface of genus n is obtained by identifying the sides of a rectangle with $n-1$ small rectangles deleted in its interior following the scheme described in Figure 2.

Observe that in the standard orthogonal surface of genus n there exist two distinguish directions (horizontal and vertical, those directions parallel to the

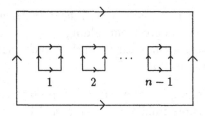

Fig. 2. The standard orthogonal surface of genus n.

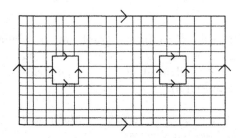

Fig. 3. A grid of the surface of genus 3.

sides of the external rectangle). The horizontal direction will be called the direction of the *parallels* and the vertical direction will be called the direction of the *meridians*. Thus, any set of parallels and meridians will be called a *grid* of the standard orthogonal surface. Figure 3 shows an example.

Now, it is possible to say that a *grid drawing* (or embedding) of a graph in a surface of genus n maps each edge to a chain of segments parallel to the sides of the external rectangle in the standard orthogonal surface and no two edges intersect except at a common vertex. A change of direction (from a parallel to a meridian) in an edge is called a *bend*. A *rectilinear* drawing is a grid drawing where each edge is either a horizontal or a vertical segment (i. e. a grid embedding without bends).

Figure 4 shows a grid drawing of a graph on the standard orthogonal surface of genus 3.

As usual (see [5]), it is possible to identify an embedding with a rotation system (a cyclic permutation of the edges incident to each vertex) and two embeddings are said to be *equivalent* if their rotations systems are the same up to cyclic permutation. Two embeddings are said to be *essentially equivalent* if the two subgraphs constituted, one with the blocks with essential cycles (those "wrapping around" the surface or, more precisely, those homotopically non-trivial) and the other with the blocks without essential cycles are pairwise equivalent. Figures 5 and 6 show some examples.

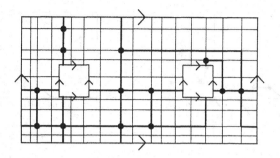

Fig. 4. A grid drawing of a graph on the standard orthogonal surface of genus 3.

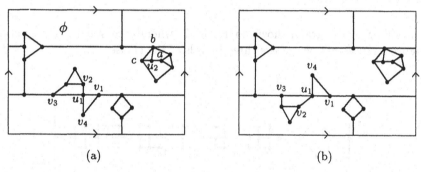

(a)   (b)

Fig. 5. (a) ϕ is an embedding on the torus of a graph with two blocks, one with essential cycles and the other without it. (b) This embedding is not essentially equivalent to ϕ because the embedding of the block with essential cycles is not equivalent.

3   Essentially equivalent rectilinear drawing

In this section, we prove that the following problem is NP-complete:
ESSENTIALLY EQUIVALENT RECTILINEAR DRAWING (EERD):
INSTANCE: Surface S, graph G, embedding $\phi : G \longrightarrow S$.
QUESTION: Does there exist a rectilinear drawing of G in S essentially equivalent to ϕ?

Theorem 3. EERD *is NP-complete.*

Proof. (SKETCH) EERD is easily seen to be in NP. For the second part of the proof, we transform 3-SATISFIABILITY (3SAT) to EERD. Let $U = \{u_1, u_2, \ldots, u_n\}$ be a set of variables and $C = \{c_1, c_2, \ldots, c_m\}$ be a set of clauses making up an arbitrary instance S of 3SAT. Starting from S, we construct an embedding ϕ_S of a graph G_S in a suitable surface, as follows. We associate to each clause the

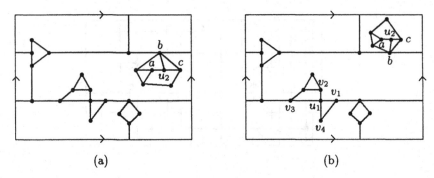

(a)   (b)

Fig. 6. (a) This embedding is not essentially equivalent to ϕ of Figure 5.(a) because the embedding of the block without essential cycles is not equivalent. (b) This embedding is essentially equivalent to ϕ of Figure 5.(a).

embedding of the graph shown in Figure 7, where each subgraph despicted in Figure 8 represents one of the three literals of the clause.

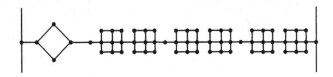

Fig. 7. Embedding associated to each clause.

Fig. 8. Embedding associated to each literal.

Finally, we join the literals of the different clauses as it is shown in Figure 9, and we add enough handles to avoid crossings. (Figure 10 shows a complete example for the set of variables $U = \{u_1, u_2, u_3, u_4, u_5\}$ and the clauses $C = \{c_1, c_2, c_3\}$ with $c_1 = \{u_2, u_3, u_5\}$, $c_2 = \{u_1, \overline{u_4}, u_5\}$ and $c_3 = \{u_2, u_3, \overline{u_5}\}$).

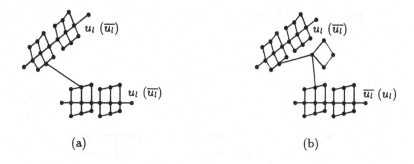

Fig. 9. (a) Union between embeddings associated to equal literals.
(b) Union between embeddings associated to opposite literals.

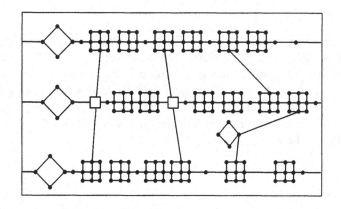

Fig. 10. An example for three clauses.

We claim that S is satisfiable if and only if there exists a rectilinear drawing of G_S essentially equivalent to ϕ_S.

In a rectilinear drawing of G_S essentially equivalent to ϕ_S, we say that one of the subgraph associated to a literal of a clause is vertical if the two vertices that have degree one in the subgraph are in the same meridian. It is horizontal if they are in the same parallel (those are the only possible options). Then, it is easy to see the following two facts:

1. Two literals corresponding to the same variable in different clauses are one of them vertical and the other horizontal if and only if one is the negation of the other.

2. In all graphs corresponding to clauses of S there exists at least one vertical subgraph associated to one of its literals (this happens because in all rectilinear embeddings of the graph associated to a clause, the subgraph that

is not part of a literal has its vertices of degree one at different heights, see Figure 11).

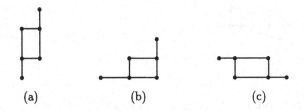

(a)   (b)   (c)

Fig. 11. Rectilinear embeddings of the graph associated to a clause.

Therefore, if we assign the value *True* to vertical literals and *False* to horizontal ones we get a satisfying truth assignment for S.

Reciprocally, the assignation given above leads to a rectilinear drawing of G_S from a satisfying truth assignment for S. In such a way that true literals correspond to vertical subgraphs, those vertical subgraphs are always drawn upward if we follow the graph representing the clause from left to right as it is shown in Figure 12.

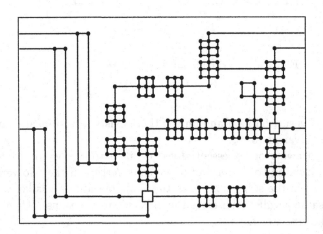

Fig. 12. Rectilinear embedding of the graph in the Figure 6.

As a consequence of Theorem 3, the corresponding optimization problem related with EERD is NP-hard, then

Corollary 4. *Given an embedding* $\phi : G \longrightarrow S$ *of a graph* G *in a surface* S, *computing a grid drawing of* G *in* S *essentially equivalent to* ϕ *with the minimum number of bends is NP-hard.*

4 Rectilinear drawing of graphs

As it has been said in the introduction, Garg and Tamassia [4] proved that testing whether a graph is rectilinear planar is an NP-complete problem. In this section we study a problem related in some way with that, which we can establish in the following terms.

RECTILINEAR DRAWING IN SURFACE (RDS):

INSTANCE: Graph G.

QUESTION: Does there exist a surface S and a rectilinear drawing of G in S?

Theorem 5. RDS *is NP-complete.*

Proof. (SKETCH) Again, RDS is easily seen to be in NP. For the second part of the proof, we will transform the following well-known NP-complete problem:

NOT-ALL-EQUAL-3SAT (NE3SAT):

INSTANCE: Set U of variables, collection C of clauses over U such that each clause $c \in C$ has $|c| = 3$.

QUESTION: Is there a truth assignment for U such that each clause in C has at least one true literal and at least one false literal?

In the proof of Theorem 3, we used the fact that in any rectilinear embedding of a given graph, there are two points that we know that are not over the same parallel, here we will exploit that in any vertex of degree 3 in a rectilinear embedding at least one of its incident edges must be vertical and at least one of its incident edges must be horizontal.

Let $U = \{u_1, u_2, \ldots, u_n\}$ be a set of variables and $C = \{c_1, c_2, \ldots, c_m\}$ be a set of clauses making up an arbitrary instance $\mathcal{S}$ of NE3SAT. Starting from $\mathcal{S}$, we construct a graph $G_{\mathcal{S}}$ in the following way:

Each variable u_i, $i = 1, \ldots, n$ is represented by a sequence of cycles C_3 of length 3 adjacent by different vertices to the its anterior and to its posterior. The cycles in an odd position in the sequence will represent the variable in its affirmative form and those in even position will represent the variable in its negative form. The length of this sequence is determined by the number of literals where the variable appears in such a way that we will have enough vertices of degree 2 to join with new vertices that will represent the clauses (increasing, in this way the degree of some of those vertices of degree 2 up to degree 3). Then vertices representing clauses will have degree 3 (Figure 13 shows the graph corresponding to the set of variables $U = \{u_1, u_2, u_3, u_4\}$ and the clauses $C = \{c_1, c_2, c_3\}$ with $c_1 = \{u_1, \overline{u_2}, u_4\}$, $c_2 = \{\overline{u_1}, u_3, u_4\}$ and $c_3 = \{u_2, u_3, u_4\}$).

Observe that in a rectilinear drawing of $G_{\mathcal{S}}$ it is easy to check the following three facts:

1. A subgraph associated to a literal is always either vertical or horizontal.

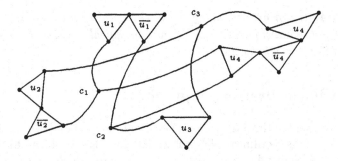

Fig. 13. An example for three clauses.

2. In the case that two literals correspond to the same variable, one of them is vertical and the other is horizontal, if and only if one is the negation of the other.
3. Each vertex corresponding to a clause is adjacent to at least one vertical literal and at least to one horizontal literal.

Therefore, if we assign the value *True* to vertical literals and *False* to horizontal ones we get a satisfying truth assignment for $\mathcal{S}$.

Reciprocally, the assignation given above leads to a rectilinear drawing of $G_\mathcal{S}$ from a satisfying truth assignment for $\mathcal{S}$. In such a way that true literals correspond to vertical subgraphs (see Figure 14).

As above, the corresponding optimization problem is NP-hard

Corollary 6. *Given a graph G, finding a surface S and an embedding ϕ of G in S such that ϕ minimizes the number of bends among all embeddings in any surface of G is NP-hard.*

5 Conclusions and Open Problems

In this paper we give an adequate model for studying embeddings of graphs in the grid of non-planar surfaces, proving, moreover, that two natural problems arising in such embeddings are NP-complete problems, namely, testing whether a graph admits a rectilinear (without bends) embedding essentially equivalent to a given embedding; and given a graph, testing if there exist a surface such that the graph admits a rectilinear embedding in that surface. Hence, the corresponding optimization problems are NP-hard.

Of course, the main open problem related to our work is deciding if testing whether a graph admits a rectilinear embedding equivalent (without the restriction of being essentially equivalent) to a given embedding is an NP-complete

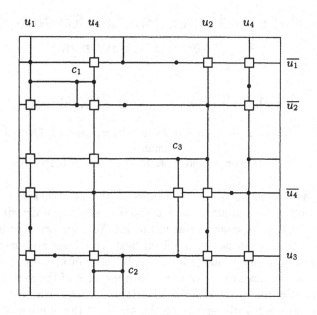

Fig. 14. Rectilinear embedding of the graph in the Figure 9.

problem. Moreover, it can be interesting to study the existence of approximations to the optimal solutions of the corresponding optimization problems.

References

1. D. Dolev and R. Trickey. On linear area embedding of planar graphs. Technical Report STAN-CS-81-876, 1981.
2. M. Fréchet and Ky Fan. *Initiation to Combinatorial Topology*. Weber and Schmidt, 1967.
3. M. R. Garey and D. S. Johnson. *Computers and Intractability: a guide to the theory of NP-completeness*. Freeman, 1979.
4. A. Garg and R. Tamassia. On the computational complexity of upward and rectilinear planarity testing. In R. Tamassia and I. G. Tollis, editors, *Graph Drawing 94, Lectures Notes in Computer Science*, pages 286–297. Springer-Verlag, 1994.
5. J. L. Gross and T. W. Tucker. *Topological graph theory*. John Wiley & Sons, 1987.
6. T. A. J. Nicholson. Permutation procedure for minimizing the number of crossings in a network. *Proc. Inst. Elec. Eng.*, 115:21–26, 1968.
7. R. Tamassia. On embedding a graph in the grid with the minimum number of bends. *Siam J. Comput.*, 16(3):421–444, 1987.
8. C. Thomassen. The graph genus problem is NP-complete. *Journal of Algorithms*, 10:568–576, 1989.

Algorithms and Area Bounds for Nonplanar Orthogonal Drawings

Ulrich Fößmeier   Michael Kaufmann

Universität Tübingen, Wilhelm-Schickard-Institut, Sand 13, 72076 Tübingen,
Germany,
email: { foessmei / mk } @informatik.uni-tuebingen.de

Abstract. We report on some extensions of the **Kandinsky** model: A new
and highly nontrivial technique to incorporate nonplanar drawings into the
Kandinsky model in the same way as in the **GIOTTO** approach is presented. This
means a major step towards the practical usability of our approach. The used
technique even gives new insights for the solvability of network flow problems.
Another variant of **Kandinsky** ensures a minimal size of the vertices removing
the requirement of uniform size of each vertex. We present a new technique to
evaluate our approach with respect to the area and the number of bends, and
to perform a reasonable comparison with the **GIOTTO** approach.

1 Introduction

1.1 Previous Approaches: GIOTTO and Kandinsky

In the last decade many different ways were examined how to draw graphs in
the plane in order to get a clear and understandable visualization of the data.
Further many algorithms were developed which compute drawings in these models (see [4] for an overview). One of the most important ways, maybe even the
most appealing one to draw a graph is to produce *orthogonal drawings*. Here
the vertices of the graph are placed on grid points of a rectilinear grid and the
edges run along gridlines. This approach leads to very good results, especially
if the graph is planar. In this case the famous algorithm of Tamassia [14] computes a planar drawing with the minimum number of bends while preserving a
given topological embedding of the graph. The basic idea is to compute a related network from the underlying topological embedding, to solve a min-cost
flow problem on this network, and to obtain a bend-minimum orthogonal representation of the embedded graph that just includes topological information like
sequences of bends, angles, etc. In a final compaction step, the missing geometric
data are filled in and exact coordinates for all objects are computed.

This algorithm is restricted to work on graphs with a maximum vertex degree
of at most four, so other methods had to be found in order to have drawings
for high-degree graphs. GIOTTO [15], a further extension of Tamassia's algorithm
solves this problem by allowing the vertices to grow in order to have enough
place for the incident edges. GIOTTO replaces a vertex v of degree d ($d \geq 5$) by

a rectangle R. The horizontal sides of R consist of $\lfloor \frac{d}{4} \rfloor$ vertices and the vertical sides of $\lceil \frac{d}{4} \rceil$ vertices. The d edges incident to v are connected to these new vertices such that the new vertices have degree four or three depending of whether the new vertex is a corner of R or not. Using this strategy the edges are distributed uniformly around v. In the final drawing R is represented by a large rectangle. Unfortunately the size of the vertices cannot be bounded even for graphs having only one vertex of degree five and all others have degree one or two.

An experimental comparison of three different approaches [5] clearly proves the great practical performance of the bend minimizing approach. The algorithm presented in [8] – now called **Kandinsky** – proposes a more flexible way to draw high-degree planar graphs guaranteeing that each vertex has the same size (and this size is not (much) larger than necessary). The model requires the uniform-sized vertices to lie on a (sparse) grid whereas the edges may run between them along the gridlines of a dense grid. One of the most important consequences of these rules is that every vertex v may have at most one straight edge (without bends) leaving v at each side, and among all the edge segments leaving v at the same side, the segment of the straight edge is the longest. Fig. 1 shows an example for a drawing in this model. Again the number of bends is minimized.

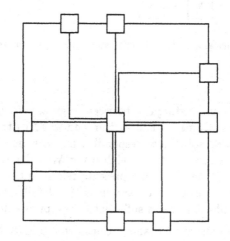

Fig. 1. An example for a Kandinsky-drawing.

All these algorithms are restricted to work on *planar* graphs. This restriction is very unfavourable for most of the practical applications. GIOTTO provides a simple but powerful way to handle nonplanar graphs: GIOTTO computes for a given graph a corresponding topological embedding and planarizes it by replacing each crossing by an artificial vertex (*crossing vertices*). After the computation of a drawing for this graph these artificial vertices are re-replaced by crossings.

Applying this strategy to **Kandinsky** would lead to drawings like shown in the left drawing of Fig. 2 because each but one of the edges $(2,5)$, $(2,7)$, $(2,9)$ must have a bend before reaching the crossing. The right drawing of Fig. 2 is a better alternative because it produces less bends and avoids the additional crossing vertices of large size. More precisely: We allow parallel edges to have no bend if we can assure that these edges lie close to each other thus not increasing the size of the vertex. This can only be assured if the face between them has zero area with respect to the sparse grid. The faces defined by the vertex 2 and the crossing vertices in Fig. 2 show an example for this situation.

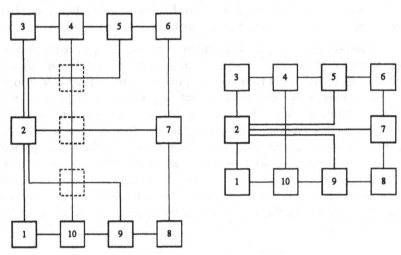

Fig. 2. Kandinsky-drawings without (left) and with (right) modification of the alg.

1.2 This Paper

The new algorithm must distinguish between original vertices and crossing vertices. **Kandinsky** (like Tamassia's algorithm and GIOTTO) transforms the graph into a directed network, solves a corresponding min-cost flow problem and transforms the solution back to an orthogonal drawing. We will neglect the compaction step in the following. The network represents relations between every vertex v of the graph and the faces occuring around v (for details see [14] and [8]). This 'local exchange' of informations is sufficient to compute a drawing.

Using **Kandinsky** this situation changes: Consider two edges incident at the same side of some vertex. At least one of them must bend, but not necessarily before the first crossing vertex. Thus the bends in the right drawing of Fig. 2 are related to the vertex 2 without being part of one of the faces occuring around of 2. In Section 2 we will present a new network approach and an alternative technique to solve these problems of 'long-distance'-exchange of informations.

The rest of the paper is devoted to the aspect of the used area. In Section 3 we present a variant of the **Kandinsky** model, where each of the vertices has minimal size. When enhanced by a reasonable compaction, the new algorithm produces drawings that are not only bend-minimal but also of good quality in terms of area. In Section 4 we will derive area bounds for the drawings of the new variant of **Kandinsky** and compare them with corresponding bounds for GIOTTO-drawings. Finally, we compare the two approaches using some examples.

2 Nonplanar Orthogonal Drawings

2.1 The New Network

We follow the idea to reduce nonplanar graphs to the planar case by replacing the crossings by crossing vertices. An intuitive view to the networks in all algorithms mentioned so far shows that the faces of the planar graph are nodes in the network and a bend corresponds to a flow unit between two faces having a common edge. The crucial point in the model suggested above is that some arcs in the network are not allowed to have a positive flow *simultaneously*. If e.g. one of the bends in the right drawing of Fig. 2 is related to vertex 2, the edge segments between this bend and vertex 2 are not allowed to have other bends, i.e. the flow related to these segments must be zero if there is a positive flow representing the bend. We call such a set of arcs *forbidden combinations*.

If generally there are k arcs in some network N having the restriction that only one of them may have a positive flow the following construction models this situation: Let $a_1 = (v_1, v_1'), \ldots, a_k = (v_k, v_k')$ be these arcs (each having capacity 1); construct a new network N^* by establishing k additional nodes $\bar{v}_1, \ldots, \bar{v}_k$ in the network N and connecting them by a simple cycle $(\bar{v}_1, \bar{v}_2), \ldots, (\bar{v}_{k-1}, \bar{v}_k), (\bar{v}_k, \bar{v}_1)$ of directed edges of capacity 1 and cost $-c$ for some constant c. Replace a_i by two arcs $(v_i, \bar{v}_i)$ and $(\bar{v}_i, v_i')$ both having capacity 1. The cost of $(v_i, \bar{v}_i)$ is set to $c \cdot k + cost(a_i)$ and the cost of $(\bar{v}_i, v_i')$ is 0. A flow on an arc $a_i = (v_i, v_i')$ in N corresponds to a flow on the path $(v_i, \bar{v}_i), (\bar{v}_i, \bar{v}_{i+1}), \ldots, (\bar{v}_{i-1}, \bar{v}_i), (\bar{v}_i, v_i')$ (mod k) in N^* having the same cost (see Fig. 3 for an example). But the realization of a flow on this path uses all edges along the cycle, thus another flow unit on an arc $a_j = (v_j, v_j')$ in N cannot be represented in N^* in this way because the arcs of the cycle are saturated. Thus the flow on a_j would cause $c \cdot k + cost(a_j)$ cost in N^* by just using the arcs $(v_j, \bar{v}_j)$ and $(\bar{v}_j, v_j')$. If c is chosen large enough, e.g. $c = \sum_{a \in N} cost(a)$, we can easily prove the following

Lemma 1. *There is a flow x of size r in N and without any forbidden combinations if and only if there is a flow x^* in N^* with $|x^*| = r$ and $cost(x^*) < c$.*

The size of the new network depends on the number of arcs defining forbidden combinations and can be quadratic in the size of N.

2.2 Solving the New Min-Cost Flow Problem

The common way to solve min-cost flow problems is to use an augmenting algorithm [12] that iteratively augments the flow along shortest paths from s to t. But the existance of negative cycles in N^* makes this strategy unprofitable because solving shortest-path problems in graphs with negative cycles is $\mathcal{NP}$-hard [10]. A better way is to transform the min-cost flow problem into a linear program (LP) and to solve this LP. This can be done using any polynomial time algorithm for solving such problems. Since all coefficients of the LP (= capacities in N^*) have integer values, the convex polyhedron representing the feasible

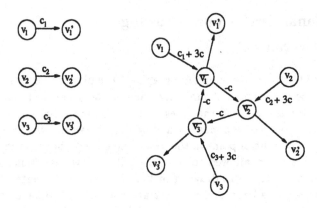

Fig. 3. The construction of N^* if at most one of the arcs (v_1, v_1'), (v_2, v_2') and (v_3, v_3') is allowed to have positive flow.

solutions of the LP only has extreme points with integer coordinates [11]. Thus an integer solution of the LP will be found and we get an integer solution for the flow in N^* which corresponds to a drawing of the graph. Since the objective function of the LP minimizes the cost of flows in N^*, the total number of bends of this drawing is minimized.

For practical implementations we can use the LP solving package CPLEX [6] which leads to an very short running time for this part of the algorithm.

If the way to solve the problem of computing flows with additional constraints of the form $\sum_{i \in I} x_i \leq 1$ leads to a complicated large network which is then transformed into a linear program, an obvious idea is to solve an LP representing the simpler network N to which we add these constraints explicitly. This idea leads to excellent results because N can be *constructed* easier.

2.3 The Drawing Algorithm

We describe the 'simple' network N together with the additional constraints. The transformation either to a network N^* really modeling the drawing problem or to a linear program can easily be done using the techniques presented in the last sections.

Let $G = (V, E)$ be the (nonplanar) embedded graph to be drawn and $G^* = (V \cup V^*, E^*)$ the corresponding planarized version of G, where V^* are the crossing vertices and E^* is the set of edges of G^* as defined above. Let F be the faces of G^* (due to the given drawing); $F_v' \subseteq F$ is the set of faces consisting of three edges e_i, e_j and e_l, two of them (say e_i and e_j) being incident to the same vertex $v \in V$ and the third one (e_l) being incident to two crossing vertices. These faces are called *critical triangles with respect to v*. Define $F' = \bigcup_{v \in V} F_v'$. $F_v^* \subseteq F$ is the set of faces consisting of exactly four edges, all of them being incident to two crossing vertices, *and* having the following property: One of the faces being neighboured to a face $f \in F_v^*$ lies in F_v' or F_v^*. The faces in F_v^* are called *critical quadrangles with respect to v*. Define $F^* = \bigcup_{v \in V} F_v^*$. The set $F_n = F \setminus (F' \cup F^*)$ contains the *normal faces*. Fig. 4 shows an example: f_1 and f_{10} are critical triangles with respect to v; f_6 is a critical triangle with respect to w; f_{11} is a critical triangle with respect to u; f_2 and f_9 are critical quadrangles with respect to v; f_{12} is a critical quadrangle with respect to u; f_8 is a critical quadrangle with respect both to v and to u. It is easy to see that only critical faces have the property that a pair of parallel edges uses two gridlines (of the dense grid) lying close to each other (like described in Section 1.1). If we would define e.g. face f_5 in Fig. 4 as critical, too, the edges incident to w could not run parallel without having bends but using neighboured gridlines: The crossing vertex lying opposite to w would need an additional gridline to be placed on between these edges. This gives an intuition for the definition of the critical triangles as the only faces where parallel edges without bends are allowed.

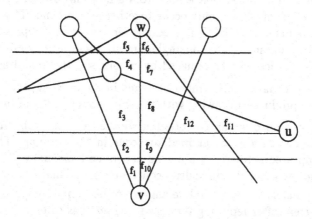

Fig. 4. Critical faces.

The network $N = (U, A, s, t, b, c)$ is defined as follows:

Nodes of N: $U = U_V \cup U_F \cup \{s, t\}$ where $U_V = \{u_v : v \in V\}$ and $U_F = \{u_f : f \in F\}$.

Arcs of N: $A = A_{SV} \cup A_{SF} \cup A_{VT} \cup A_{FT} \cup A_{VF} \cup A_{FFV} \cup A_{FF}$ where

A_{SV} contains arcs from s to $u_v \in U_V$ if degree$(v) \leq 3$; the arcs have cost 0 and capacity $4 - $ degree(v).

A_{SF} contains arcs from s to $u_f \in U_F$ if f is not the external face of G^* and if f has at most three edges; the arcs have cost 0 and capacity $4 - (\#$ edges of $f)$.

A_{VT} contains arcs from $u_v \in U_V$ to t if degree$(v) \geq 5$; the arcs have cost 0 and capacity degree$(v) - 4$.

A_{FT} contains arcs from $u_f \in U_F$ to t if f is not the external face of G^* and if f consists of at least five edges; the arcs have cost 0 and capacity $(\#$ edges of $f) - 4$. Moreover there is an arc from u_f to t in A_{FT} if f is the external face of G^*; this arc has cost 0 and capacity $(\#$ edges of $f) + 4$.

A_{VF} contains arcs from $u_v \in U_V$ to $u_f \in U_F$ if v lies at the border of f; the arcs have cost 0 and capacity 3.

A_{FFV} contains arcs $(u_f, u_v)_e$ from $u_f \in U_F$ to $u_v \in U_V$, namely one for every edge segment e if either e lies at the border of f and is incident to v, or g is a critical quadrangle with respect to v, and f and g have e as a common edge; these arcs have cost 1 and capacity 1.

A_{FF} contains arcs from $u_f \in U_F$ to $u_g \in U_F$ if f and g have at least one common edge; these arcs have cost 1 and capacity ∞.

Let $x : A \to \mathbb{R}^+$ be a feasible integer flow in N with maximum $|x|$; a drawing can be constructed as follows: the angle at a vertex v between two incident edges having the face f between them is set to $(x((u_v, u_f)) - x((u_f, u_v)) + 1) \cdot 90°$. Each edge e has $|x((u_f, u_g))|$ bends if f and g have e as common edge.

Before these angles and bends are computed the arcs in A_{FFV} have to be transformed: An arc $(u_f, u_v)_e \in A_{FFV}$ stands for a combination of an arc $(u_f, u_g) \in A_{FF}$ and an arc $(u_g, u_v) \in A_{FV}$ if f and g have e as common edge. In this way it is guaranteed that a flow from a face into a vertex (which corresponds to an angle of $0°$) can only occur together with a bend. This situation is described more detailed in [8]. The crossing vertices are not represented in the network, so it is guaranteed that all angles around these vertices are equal to $90°$ and thus these vertices can be replaced by crossings in the final drawing step.

The cost of x is equal to the number of bends in the drawing, thus a flow with minimum cost produces a drawing with the minimum number of bends.

The network described so far cannot simulate the drawing problem: Obviously several combinations are not allowed to occur in the drawing. The first two conditions appeared similarly in previous papers on **Kandinsky** [8,9], while the third condition reflexed the difficulties concerning nonplanarity.

- It is not possible to have negative angles; so the sum of the flows from a face into a vertex (after replacing the arcs in A_{FFV}) has to be at most one.
- Each bend belonging to a $0°$-angle at a vertex v must lie close to v (i.e. between the bend and v there are no other bends); thus an edge cannot bend in different directions corresponding to $0°$-angles at the same vertex.
- Further if b is a bend on an edge belonging to a critical quadrangle with respect to v and b corresponds to a $0°$-angle at v then all critical faces between b and v have to be drawn without bends (see the right drawing in Fig. 2 for an example: the faces defined by vertex 2 and the crossing vertices do not have any bends at their edge segments).

These conditions can be guaranteed by applying one of the techniques described in the previous section. There is one weak point in this concept that does not allow to have an unbounded number of normal bends on a single edge. Details can be found in the full paper. We summarize the results of this section in the following

Theorem 1. *For a given drawing of a graph $G = (V, E)$ an embedding preserving drawing in the Kandinsky-model with the minimum number of bends can be computed in polynomial time under the restriction that each edge of a critical triangle or of a critical quadrangle has at most k bends (for a constant k).*

3 Vertices of Minimal Size

The requirement of vertices of uniform size sometimes seems artifical. Often it is more important to represent the vertices just as big that the incident edges can be drawn separately. In the full version, we show how to modify the **Kandinsky** model such that the vertices may have different size and a detailed analysis of the area can be done.

In original **Kandinsky** drawings, the vertices of uniform squarish size sit on the intersections of a sparse uniform grid. In between the sparse grid lines there is a finer grid supporting the routing of the edges [8]. Now we resolve the two grids such that only one grid remains. The spacing between neighbored parallel grid lines is set to be the minimal distance of two parallel edges.

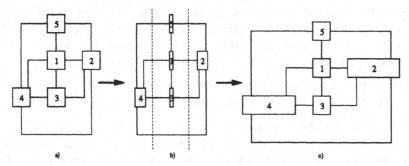

Fig. 5. Minimizing the size of the vertices: 3 snaphots of the algorithm.

4 Bounds for the Area of Orthogonal Drawings

In this section we give upper bounds for the area used by drawings produced by **GIOTTO** and **Kandinsky**. We consider the **Kandinsky** only with shrunken vertices, since otherwise a fair comparision is not possible because of different underlying grids. Recently two area bound have been given for general graphs [13] and [3] ($\frac{3}{2}m$ and $m + n$ gridlines respectively) but the approaches are completely different, and the algorithms produce highly nonplanar drawings with relatively many bends. Note that our bounds are not restricted to planar drawings.

4.1 Graphs of Small Degree

For the **GIOTTO** model, the following easy observation [2] helps us to develop an upper bound for the area. Later, we will extend it to the **Kandinsky** model.

Assume that each vertex has degree at most 4. If each edge has exactly one bend, each gridline in horizontal and vertical direction carries at most one vertex in the worst case. The number of gridlines is $2n$ in this case. If one edge has no bends, the corresponding vertices share a gridline, if an edge has b bends, there are $b - 1$ segments not incident to any vertex. Each of the segments might need a separate gridline. The number of gridlines is linear in the number of bends.

Hence the total number of gridlines is at most $2n - m + k_G$, where k_G is the number of bends produced by GIOTTO. Note that the argument does not use planarity but only that the degree is at most 4 and the edges extend to all sides.

In **Kandinsky** drawings (even for graphs with a vertex degree of at most four) there may arise larger vertices. The vertices use at most $2 + \#(0°\text{-angles})$ gridlines because every $0°$-angle may force the vertex to grow by one unit. This gives a bound of $2n - m + k_K + \#(0°\text{-angles})$ where k_K denotes the number of bends.

In order to compare this formula with the bound for GIOTTO we analyse the number of $0°$-angles. A $0°$-angle always arises together with a bend; thus $\#(0°\text{-angles}) \leq k_K$. Clearly, $k_K \leq k_G$, and if there are $0°$-angles then $k_K < k_G$. We can show

$$k_G \leq k_K + \#(0°\text{-angles}) \leq 2 \cdot k_G$$

, since GIOTTO can simulate a $0°$-angle by two 'normal' bends and the second inequality follows from the observations above. Usually $k_G \approx k_K + \#(0°\text{-angles})$ and both algorithms produce drawings of the same size, but there are lower bound examples where the two terms are a factor of two off.

As a consequence the upper bound for the area of GIOTTO drawings is never larger than the upper bound for **Kandinsky** drawings.

4.2 Graphs of Higher Degree

To apply the technique of the previous section to graphs with an arbitrary vertex degree we have to formulate it in a more general way. A vertex v is said to have a *free pin* at every position where an additional edge could be connected to v. 'freepins(v)' denotes the number of free pins of vertex v and 'freepins' $= \sum_v$ freepins(v). Thus a vertex v uses $\frac{\deg(v) + \text{freepins}(v)}{2}$ gridlines. Therefore the total area can be bounded by $m + \frac{\text{freepins}}{2}$ if every edge has exactly one bend. With the argument of the previous section an upper bound for the area in the general case is $m + \frac{\text{freepins}}{2} + k - m = k + \frac{\text{freepins}}{2}$ where k is the number of bends.

In **Kandinsky** drawings we have to take into account that the edges being incident to some vertex v might extend all towards the same direction. In this case freepins$(v) = \deg(v) + 2$ and the total area can only be bounded by $k_K + \frac{1}{2}\sum_v(\deg(v) + 2) = k_K + m + n$. But we can do better: Using some properties for bend minimum drawings excludes most of the vertices from the sum:

Theorem 2. *A drawing in the* **Kandinsky** *model with different vertex sizes uses at most* $m + k_K + |\{v : \deg(v) = 1\}|$ *gridlines,* k_K *being the number of bends.*

Note that we did not take into account that there may be many pairs of edges incident to opposite sides of the same vertex and using the same gridline. In all examples in the rest of this paper there is not a single vertex having the bad property that all its incident edges run into the same direction.

Analyzing GIOTTO is easier because the edges are forced to be uniformly distributed to the vertex sides. Similarily as above, we can show:

Theorem 3. *A drawing in the* GIOTTO *model uses at most* $k_G + \frac{1}{2}(|\{v : deg(v) \text{ is } odd\}|)$ *+* $|\{v : deg(v) \leq 2\}|$ *gridlines,* k_G *being the number of bends.*

Orthogonal drawings usually need smaller area than due to the formulas given in this section. The formulas induce that all small vertices of the auxiliar representation for a high degree vertex that lie on the same gridline are connected by straight edges. If a gridline is used by r such connected sequences of vertices the gridline is counted r times. To count this effect we replace the large vertices by the auxiliary construction for GIOTTO (a rectangle of small vertices) and look for gridlines that are used by more than one connected sequence (a connected sequence also can consist of zero vertices if there is an edge segment connecting two bends). We can show

Theorem 4. *An orthogonal drawing uses exactly* $(k + 1/2(freepins) - saved-lines)$ *gridlines where* k *is the number of bends and savedlines is the number of multiple use of gridlines as observed above.*

An example for this analysis is given in the next section.

5 Examples

Our first example shows the graph G_3 of the graph drawing contest of GD'94 and we took the embedding found by Petra Mutzel for her winning drawing. The results are shown in Fig. 6. GIOTTO changed the embedding. Observe the marked edge. It has 10 crossings and 4 bends. A simple rerouting of this edge would reduce the size by 2 gridlines, save 6 crossings and 2 bends. In total we have size 23×20, 32 bends and 27 crossings. Some vertices are much larger than necessary, e.g. 13, 11, 2.

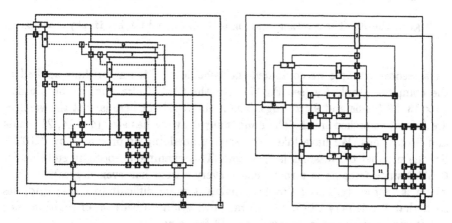

Fig. 6. The resulting drawings by GIOTTO (left) and by Kandinsky (right) for G_3.

For the **Kandinsky** drawing, the situation is better: We obtain size 21 × 20, 29 bends and 20 crossings. Note, that we have maintained the original embedding given in [7].

We derive the used area of the drawings using the formula from Theorem 4: In the **GIOTTO** drawing of Fig. 6 two gridlines are multiple used: The 8th leftmost column and the 7th bollowmmost row are doubly used. Thus the drawing should save two gridlines compared to the formula. Indeed, the drawing uses 23+20 = 43 gridlines, while the formula bounds the area by 32 (bends) + 12 (24 vertices have odd degree) + 1 (one vertex with degree two) = 45.

In the **Kandinsky** drawing of Fig. 6 we can save 10 gridlines compared to the formula by multiple use of gridlines. The number of free pins is 44, so the formula states an area of 29 (bends) + 22 (1/2 ·(freepins)) - 10 (savedlines) = 41. This is conform to the needed area of (21 × 20).

We leave the question open for future research if the better re-use of gridlines is a provable quality of the **Kandinsky** model.

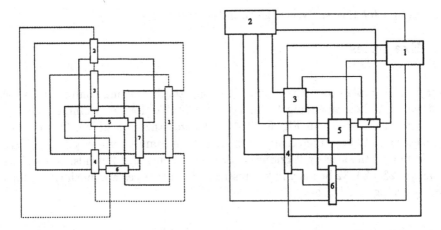

Fig. 7. The K_7: The resulting drawing by **GIOTTO**, and the **Kandinsky** drawing.

As a second example we compared the behaviour of the algorithms applied to the complete graph K_7 (see Fig. 7). For the **GIOTTO** drawing, we obtain size 12 × 13, 26 bends and 9 crossings. The main drawback here is that the vertices are (often) stretched much. The **Kandinsky** drawing has size 14 × 14, 21 bends and 9 crossings. We have taken the same embedding as produced by **GIOTTO**. So, the number of crossings is the same. This example is specially suitable for **GIOTTO** since freepins is zero (because every vertex has even degree) and the graph is extremely dense; this fact makes it more difficult to re-use gridlines such that **Kandinsky** can only subtract two gridlines because of multiple use, but it suffers from the large number of 18 free pins.

More examples can be found in the full version of the paper.

6 Conclusion

Summarizing the results of the comparison done in this paper we state that

- **GIOTTO**
  - \+ gives good theoretical bounds for the used area
  - \+ performs well in practice w.r.t. the number of bends and to the used area
  - − does not define a model of orthogonal drawings for high-degree graphs (it is not clear under which conditions a vertex may be large)
  - − area/bends/size of vertices depend on the chosen embedding [1] but also on the distribution of the edges around the vertices

- **Kandinsky**
  - \+ defines a model for orthogonal drawings
  - \+ minimizes a cost function (number of bends) in this model
  - \+ behaves well in practice w.r.t. the area (multiple use of gridlines)
  - − needs an embedding as part of the input
  - − does not allow an easy analysis of the used area.

References

1. Bertolazzi, P., G. Di Battista, W. Didimo, *Computing Orthogonal Drawings with the Minimum Number of Bends*, Proc. of WADS'97, to appear.
2. Biedl, T.C., *Orthogonal Graph Drawing, Algorithms and Lower Bounds*, Diplomarbeit TU Berlin, 1995.
3. Biedl, T.C., M. Kaufmann, *Area-Efficient Static and Incremental Drawings of High-Degree Graphs*. to appear in ESA 1997.
4. Di Battista, G., P. Eades, R. Tamassia, I.G.Tollis, *Algorithms for Drawing Graphs: An Annotated Bibliography*, Computational Geometry: Theory & Applications, 235-282, 1994.
5. Di Battista, G., A. Garg, G. Liotta, R. Tamassia, E. Tassinari, F. Vargiu, *An Experimental Comparison of Three Graph Drawing Algorithms*, Computational Geometry: Theory & Applications, 1996.
6. CPLEX optimization, Inc., *Using the CPLEX Base System*.
7. Eades, P., J. Marks, *Graph-Drawing Contest Report*, Proceedings on GD'94, Princeton, LNCS 894, 143-146, 1995.
8. Fößmeier U., M. Kaufmann, *Drawing High Degree Graphs with Low Bend Numbers*, Proceedings on GD'95, Passau, LNCS 1027, 254-266, 1995.
9. Fößmeier U., G. Kant, M. Kaufmann, *2-Visibility Drawings of Planar Graphs*, Proceedings on GD'96, Berkeley, LNCS 1190, 155-168, 1996.
10. Garey, M.R., D.S. Johnson, *Computers and Intractability: A Guide to the Theory of $\mathcal{NP}$-Completeness*, W.H. Freeman and Company, New York, 1979.
11. Kleinschmidt, P., *personal communication*.
12. Lawler, E.L., *Combinatorial Optimization: Networks and Matroids*, Holt, Rinehart and Winston, New York, 1976, Chapter 4.
13. Papakostas A., I.G. Tollis, *High-degree orthogonal drawings with small grid-size and few bends*. Technical report, University of Texas at Dallas, Richardson, TX 75083, December 1996.
14. Tamassia, R. *On Embedding a Graph in the Grid with the Minimum Number of Bends*, SIAM Journal of Computing, vol. 16, No. 3, 421-444, 1987.
15. Tamassia, R., G. Di Battista, C. Batini, *Automatic Graph Drawing and Readability of Diagrams*, IEEE Trans. Systems, Man and Cybernetics, 61 - 79, 1988.

Drawing Clustered Graphs on an Orthogonal Grid

Peter Eades[1] and Qing-Wen Feng[2*]

[1] Department of Computer Science and Software Engineering, University of Newcastle, NSW 2308, Australia.
[2] Tom Sawyer Software, 804 Hearst Avenue, Berkeley, CA 94710, USA.

(extended abstract)

Abstract. Clustered graphs are graphs with recursive clustering structures over the vertices. For graphical representation, the clustering structure is represented by a simple region that contains the drawing of all the vertices which belong to that cluster. In this paper, we present an algorithm which produces planar drawings of clustered graphs in a convention known as *orthogonal grid rectangular cluster drawings*. We present an algorithm which produces such drawings with $O(n^2)$ area and with at most 3 bends in each edge. This result is as good as existing results for classical planar graphs. Further, we show that our algorithm is optimal in terms of the number of bends in each edge.

1 Introduction

Clustered graphs are graphs with recursive clustering structures over the vertices (see Figure 1). This type of clustering structure appears in many systems. Examples include CASE tools [39], management information systems [19], and VLSI design tools [15]. For graphical representation, the clustering structure is represented by a simple region that contains the drawing of all the vertices which belong to that cluster. Algorithms for automatically drawing of clustered graphs are difficult. Heuristic methods for drawing similar structures have been developed by Sugiyama and Misue [22, 30], North [23], and by Madden et al. [18]. Algorithms for constructing straight-line drawings of clustered graphs are given in [8, 11]; note, however, that straight-line drawings of clustered graphs can require exponential area [11]. In this paper, we present an algorithm which produce planar drawings of clustered graphs in a convention called "orthogonal grid rectangular cluster drawings". We apply a technique to order the clusters of the graph recursively, and we use the visibility representation for directed graphs to produce our drawings.

The orthogonal grid drawing convention appears in a number of applications, such as VLSI circuit design [20, 21, 37, 38] and diagrammatic interfaces for relational information systems [1, 2, 24, 27, 31]. Under the orthogonal grid drawing convention, minimizing the number of bends and minimizing the area are the

---

* The work described in this paper was performed when the author was studying at the Department of Computer Science and Software Engineering of the University of Newcastle.

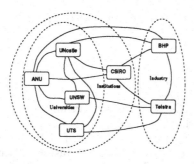

Fig. 1. An example of a clustered graph.

main criteria both for diagram readability and for VLSI design applications. For classical graphs, several basic results regarding planar orthogonal grid drawings have appeared in the literature. It has been shown by Valiant [38] that any planar graph of degree at most 4 admits a planar orthogonal grid drawing with area $O(n^2)$; further, there are graphs which need quadratic area. Tamassia [32] presented an $O(n^2 \log n)$ time algorithm that computes a planar orthogonal grid drawing with a given planar embedding so that the number of bends is minimized. Garg and Tamassia [14] have shown that if the planar embedding is not given, then the problem is NP-hard.

Several linear time algorithms for planar orthogonal grid drawings of classical graphs have been developed. Tamassia and Tollis [34, 35] have presented an algorithm that outputs drawings with $O(n^2)$ area, where n is the number of vertices of the graph. If the graph is biconnected, then there are at most $2n + 4$ bends in the drawing; otherwise, there are at most $2.4n + 2$ bends. Further, there are at most 4 bends in each edge; if the graph is biconnected, then all but 2 edges have at most 2 bends. Kant [16, 17] has presented an algorithm which improves the result of Tamassia and Tollis in some cases. For triconnected graphs, Kant's algorithm draws on an $n \times n$ grid with at most 2 bends per edge (if $n > 6$), and the total number of bends is no more than $\lceil 3n/2 \rceil + 4$. If the graph is connected with degree at most 3, then the algorithm draws on an $\lfloor n/2 \rfloor \times \lfloor n/2 \rfloor$ grid with at most 2 bends in each edge and no more than $\lfloor n/2 \rfloor + 1$ bends in total. Even and Granot [9] have presented an algorithm such that for any planar graph with degree at most 4, the drawing has $O(n^2)$ area, and there are at most 3 bends in each edge. Lower bounds on the area and the number of bends for planar orthogonal drawings of graphs have been presented by Tamassia, Tollis and Vitter [36], and by Biedl [5]. Another useful representation for planar graphs is the *visibility representation* [28, 33]. Visibility representation is related to orthogonal drawing in that it is often used as a basis for constructing an orthogonal drawing. Several orthogonal drawing algorithms [9, 34, 35] first construct a visibility representation of the graph, then transform it to an orthogonal drawing.

In this paper, we present an algorithm for planar drawing of clustered graphs using the same approach. Our algorithm produces orthogonal grid rectangular cluster drawings with $O(n^2)$ area and with at most 3 bends in each edge. This result is as good as the results for classical planar graphs [9, 17, 35]. Further, we give an example which shows that the bend per edge performance of our algorithm is optimal, that is, there is a class of graphs which require $\Omega(n)$ edges each

2 Terminology

A *clustered graph* $C = (G, T)$ consists of an undirected graph G and a rooted tree T such that the leaves of T are exactly the vertices of G. Each node ν of T represents a *cluster* $V(\nu)$ of the vertices of G that are leaves of the subtree rooted at ν. Note that tree T describes an inclusion relation between clusters. In a *drawing* of a clustered graph $C = (G, T)$, graph G is drawn as points and curves as usual. For each node ν of T, the cluster is drawn as simple closed region R that contains the drawing of $G(\nu)$, such that: (1) the regions for all sub-clusters of ν are completely contained in the interior of R; (2) the regions for all other clusters are completely contained in the exterior of R; (3) if there is an edge e between two vertices of $V(\nu)$ then the drawing of e is completely contained in R. We say that the drawing of edge e and region R have an *edge-region crossing* if the drawing of e crosses the boundary of R more than once. A drawing of a clustered graph is *c-planar* if there are no edge crossings or edge-region crossings. If a clustered graph C has a c-planar drawing then we say that it is *c-planar*. An edge is said to be *incident* to a cluster $V(\nu)$ if one end of the edge is a vertex of that cluster but the other endpoint is not in $V(\nu)$. An *embedding* of a clustered graph consists of the circular ordering of edges around each cluster which are incident to that cluster. A clustered graph $C = (G, T)$ is a *connected clustered graph* if each cluster induces a connected subgraph of G. The following results from [12] characterize c-planarity in a way which can be exploited by our drawing algorithms.

Theorem 1. *A connected clustered graph $C = (G, T)$ is c-planar if and only if graph G is planar and there exists a planar drawing $\mathcal{D}$ of G, such that for each node ν of T, all the vertices and edges of $G - G(\nu)$ are in the external face of the drawing of $G(\nu)$.*

Theorem 2. *A clustered graph $C = (G, T)$ is c-planar if and only if it is a sub-clustered graph of a connected and c-planar clustered graph.*

From Theorem 2, we can assume that we are given a connected clustered graph when drawing a c-planar clustered graph. In the rest of the paper we further assume that in a clustered graph $C = (G, T)$, every nonleaf node of T has at least two children.

Our techniques use the concept of planar st-graphs [3]. A planar st-graph is a planar directed graph with one source s and one sink t; and both source and sink above can be embedded on the boundary of the same face, say the external face.

3 Orthogonal Drawings for C-planar Clustered Graphs

An *orthogonal grid rectangular cluster drawing (OGRC drawing)* of a clustered graph maps the graph onto a grid, where edges are drawn as sequences of horizontal and vertical segments, vertices are drawn on grid points, and region boundaries for clusters are drawn as rectangles. In this section, we present an algorithm that produces c-planar OGRC drawings for clustered graphs. We assume that we are given a connected clustered graph C with n vertices, and with a c-planar embedding; further, C has a sub-clustered-graph C' with n vertices, and every vertex of C' has degree no more than 4. Our algorithm outputs c-planar OGRC drawings of C' with $O(n^2)$ area, and there are at most 3 bends in each edge. We show that there is a class of c-planar embedded clustered graphs of n vertices, for which every c-planar OGRC drawing requires $\Omega(n)$ edges bent more than twice. Our algorithm consists of three phases: visibility representation, orthogonalization, and bend reduction.

The result is summarized below.

Theorem 3. *Let $C = (G, T)$ be an n vertex connected clustered graph with a c-planar embedding, and let C' be an n vertex sub-clustered-graph of C with degree at most 4. One can construct a c-planar OGRC drawing of C' with $O(n^2)$ area, and with at most 3 bends in each edge, in $O(n^2)$ time.*

Visibility Representation To produce a visibility representation, we first transform the clustered graph to a planar st-graph, taking into account the clustering structure. Then, we make use of the algorithm by di Battista, Tamassia and Tollis [4] which produces *constrained* visibility representations, that is, edges of a selected path are vertically aligned. We use this property to form rectangles for clusters. When transforming the clustered graph to a planar st-graph, we need to consider the clustering structure so that the visibility representation that we produce respects the clustering constraints. This is achieved by computing an st-numbering of the vertices of G such that the vertices that belong to the same cluster are numbered consecutively. We call this numbering *c-st numbering*. The details of the construction of the c-st numbering are omitted in this extended abstract.

The *Constrained-Visibility* algorithm described in [4] takes as inputs a planar st-graph G; set Π of paths in G and produces a visibility drawing of G so that the x-coordinates of $\Gamma(e)$ and $\Gamma(e')$ are the same whenever e and e' are the edges in the same path in Π. The set Π is restricted to "nonintersecting paths" in the following sense. Two paths Π_1 and Π_2 of G are said to be "nonintersecting" if they are edge disjoint and do not *cross* at common vertices, i.e., there is no vertex v of G with edges e_1, e_2, e_3 and e_4 incident in this clockwise order around v, such that e_1 and e_3 are in Π_1 and e_2 and e_4 are in Π_2.

As discussed above, with *c-st* numbering, each cluster is assigned a vertical range. To obtain the 4 bounding sides of the rectangle for a cluster ν, we construct 4 dummy vertices denoted by $t(\nu)$ (top), $b(\nu)$ (bottom), $l(\nu)$ (left) and $r(\nu)$ (right); each represents one side of the rectangle. We modify G recursively from bottom to top of the tree T, to include the dummy vertices into G. At each nonleaf node ν of T, we "bunch together" all the edges going out of $G(\nu)$ at the dummy vertex $t(\nu)$ in the external face; similarly, we bunch together all the edges coming into $G(\nu)$ at the dummy vertex $b(\nu)$ in the external face (see Figure 2). Note that there are clusters that do not have incoming or outgoing edges for the following reason. Suppose that s is the single source of G; consider the path from s to the root γ in T. The construction of the c-st numbering ensures that for each node ν on this path, $G(\nu)$ has no incoming edges. In this case, we add an edge $(b(\nu), b(\sigma))$ to $G(\nu)$, where σ is the child of ν on the path

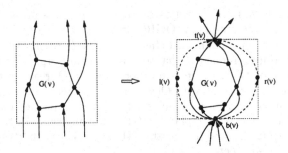

Fig. 2. Modify $G(\nu)$, adding dummy vertices to represent the rectangle.

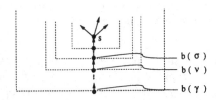

Fig. 3. Connect the dummy vertices to the source.

of T between the root γ and s (see Figure 3). Note that the dummy vertex $b(\sigma)$ has already been included into $G(\sigma)$ by recursion. If $\sigma = s$, then we add edge $(b(\nu), s)$ instead. Similarly, if $G(\nu)$ does not have an outgoing edge, then we add edge $(t(\tau), t(\nu))$ to $G(\nu)$, where τ is the child of ν on the path of T between the root γ and t. If $\tau = t$, then we add edge $(t, t(\nu))$ instead. After this, $G(\nu)$ becomes a planar subgraph with a single source $b(\nu)$ and single sink $t(\nu)$. Further, we add two dummy paths $(b(\nu), l(\nu), t(\nu))$ and $(b(\nu), r(\nu), t(\nu))$ in the left face and the right face of $G(\nu)$ respectively (see Figure 2). Clearly, the modified $G(\nu)$ is a planar st-subgraph with source $b(\nu)$ and sink $t(\nu)$, and with external face bounded by two directed paths $(b(\nu), l(\nu), t(\nu))$ and $(b(\nu), r(\nu), t(\nu))$. We modify every subgraph $G(\nu)$ recursively as above, obtaining an extended graph F that includes the dummy vertices for the rectangles. Clearly, by our construction, the resulting graph F is a planar st-graph. Assuming that every nonleaf node of T has at least two children, then the tree T has less than $2n$ nodes. Since we add 4 dummy vertices for every nonleaf node of T, the resulting graph F has $O(n)$ vertices. Note that each intercluster edge in G is replaced by a path in F, with vertices on the path representing the region boundaries it crosses (see Figure 4). Therefore, if G has n vertices and hence $O(n)$ edges, then graph F has $O(n^2)$ edges (F may have multiple edges).

We specify the following alignment requirements in F for our visibility representation. For each nonleaf node ν of T, we require that the edges on each of the dummy paths $(b(\nu), l(\nu), t(\nu))$ and $(b(\nu), r(\nu), t(\nu))$ are aligned. For each path of F that represents an intercluster edge, we require that all the edges on the path are aligned. To avoid introducing unnecessary bends in the orthogo-

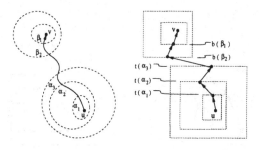

Fig. 4. An edge in G is replaced by a path.

(a)   (b)

Fig. 5. (a) Alignment requirement for a vertex v. (b) Edge alignment rules.

nalization phase, we require that some edges around a vertex are aligned. For example, suppose that there is a vertex v that has two incoming edges and two outgoing edges, then we require that the right incoming edge is aligned with the left outgoing edge (see Figure 5(a)). Note that we are drawing a sub-clustered-graph C' which has degree at most 4, some of the edges in G will be removed in the final drawing. Here we only need to consider those edges that are in the sub-clustered-graph, and note that each vertex has at most 4 edges incident to it in the sub-clustered-graph. Figure 5(b) illustrates all the cases for our alignment requirements; the edges that are marked by thick lines are required to be aligned (edges that do not belong to the sub-clustered-graph are not shown here). All the above alignment requirements together form a complete specification of the paths that are to be aligned for our visibility representation. Although some of these paths share common vertices, they do not intersect with each other. This is because there is at most one path going through each original (nondummy) vertex of G, and at every dummy vertex, the paths originate from distinct edges of G and therefore do not intersect. We apply the algorithm *Constrained_Visibility_Draw* to F with the above requirements for alignment. Thus we obtain a visibility representation Γ of F with the vertices and edges of each cluster ν drawn within a rectangle formed by $\Gamma(t(\nu))$, $\Gamma(b(\nu), l(\nu), t(\nu))$, $\Gamma(b(\nu))$ and $\Gamma(b(\nu), r(\nu), t(\nu))$ as in Figure 6. The extended graph F has $O(n)$ vertices and $O(n^2)$ edges and thus this step takes time $O(n^2)$; it produces a visibility drawing on a grid of size $O(n) \times O(n)$.

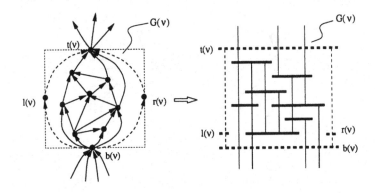

Fig. 6. Forming a rectangle for a cluster.

Orthogonalization In this phase, we first remove those edges which do not belong to the sub-clustered-graph, but we retain the dummy vertices and the dummy paths for the rectangles. To transform the visibility representation to an OGRC drawing, we only need to perform some local operations at each vertex, transforming a horizontal segment to a point. These local operations are illustrated in Figure 7; symmetric cases for (a), (c), (d), (e) and (f) are omitted for simplicity. Note that Figure 7 covers all the cases that can appear, since we have required that certain edges around a vertex are aligned. Further, note that a new row is added at every source or sink of degree 4 (see Figure 7(h) and (i)).

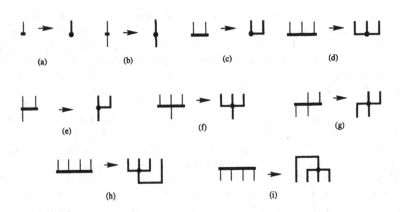

Fig. 7. Orthogonalization rules.

Bend Reduction In the OGRC drawing obtained from the previous phase, an edge is bent only near its endpoints. Hence every edge can have at most 3

Fig. 8. (a) An edge with 4 bends. (b) Reduce a bend on edge e.

bends except in the following case: the edge is between a source of degree 4 and a sink of degree 4 and it has 2 bends near each endpoint (see Figure 8(a)). We show that this 4 bend case can be eliminated by making some adjustments in the drawing. Note that for a source or sink u of degree 4, either the leftmost edge or the rightmost edge gets 2 bends near it, but not both. We call the leftmost or the rightmost edge of u an *extreme edge* of u. Suppose that there is an edge $e = (u, v)$ that has 2 bends near u and another 2 bends v. Then we fix one end u and rotate the edges around the other end v, letting the edge e bend only once near v (see Figure 8(b)). After this, we say that the edge e is "processed", and the vertices u and v are "visited". However, this operation creates a new bend in the other extreme edge $e' = (v, w)$ of v. This may cause some problems if vertex w is a source or sink of degree 4 and e' is one of its extreme edges. In this case, we continue to process edge e' with the visited endpoint v fixed. We repeat this until we meet an edge $e' = (v, w)$ such that either e' is not an extreme edge of w, or w has already been visited. Clearly, if e' is not an extreme edge of w, then it has at most 3 bends. If w has already been visited, then there is a processed edge incident to w that gets two bends near w, therefore e' has at most 1 bend near w and hence can have at most 3 bends. We keep performing this until there are no 4 bend edges exist.

The bend reduction step takes $O(n^2)$ time. Details are omitted from this extended abstract.

4 A Lower Bound for Bends

In this section, we present a class of c-planar embedded clustered graphs of n vertices, for which every c-planar OGRC drawing requires $\Omega(n)$ edges each bent more than twice. This shows that our algorithm *Orthogonal_Grid_Rectangular_Draw* is optimal in the worst case (in terms of the number of bends in each edge). To prove our result, let us first consider a small sub-clustered-graph I (see Figure 9), which serves as the building block of our cluster graph. There are two clusters A and B (see Figure 9) in the sub-clustered-graph I. Cluster A contains vertices $a_1, a_2, \ldots, a_7$; cluster B contains vertices $b_1, b_2, \ldots, b_4$. We assume that I has a fixed c-planar embedding; the orderings of the edges around cluster A and cluster B are shown in Figure 10. The drawings that we discuss in the rest of this section are all consistent with this embedding. We prove the following lemma.

Lemma 4. *In every c-planar OGRC drawing of I, there is at least one edge bent more than twice.*

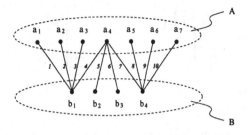

Fig. 9. The sub-clustered-graph I.

Fig. 10. The embedding for clusters A and B.

Proof. In a c-planar OGRC drawing of I, cluster A and B are drawn as disjoint rectangular regions. Without loss of generality, we assume that the rectangle for A is drawn above the rectangle for B, and there is a horizontal line ℓ separating them. Note that all the edges between A and B have to cross this horizontal line ℓ, and they cross the line ℓ in the order shown in Figure 10. Consider the edge (a_4, b_1) and the edge (a_4, b_4) both incident to a_4. We show that at least one of these two edges has more than one bend above the line ℓ. If the edge (a_4, b_1) has no bend above the line ℓ, then the other edge (a_4, b_4) must have at least 3 bends above the line ℓ; if the edge (a_4, b_1) has one bend above the line ℓ, then the other edge (a_4, b_4) must have at least 2 bends above the line ℓ. With out loss of generality, let us assume that the edge (a_4, b_4) has more than one bend above the line ℓ. Now consider the edge (a_4, b_4) together with the edge (a_7, b_4). We show that at least one of them has a total of more than two bends. If the edge (a_4, b_4) has at least one bend below the line ℓ, then it has more than two bends in total; if the edge (a_4, b_4) has no bend below the line ℓ, then the other edge (a_7, b_4) must have more than two bends below the line ℓ. Therefore, we have that there is at least one edge in I that has a total of more than two bends. □

Now we define a class of clustered graphs Φ_n $(n = 1, 2, \ldots)$ with sub-clustered-graph I as the building block. Clustered graph Φ_n consists of a sequence of n copies of the sub-clustered-graph I (see Figure 11). The vertex a_7 of a previous copy of I also serves as the vertex a_1 of the next copy of I. Clustered graph Φ_n has two clusters A_n and B_n. Cluster A_n contains the vertices in the cluster A of each sub-clustered-graph I; cluster B_n contains the vertices in the cluster B of each sub-clustered-graph I. Clearly, Φ_n has $10n + 1$ vertices. By Lemma 4, we have the following theorem.

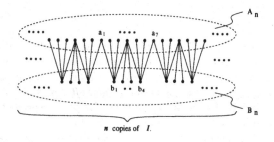

Fig. 11. The construction of the clustered graph Φ_n.

Theorem 5. *In every c-planar OGRC drawing of Φ_n ($n = 1, 2, \ldots$), there are at least n edges bent more than twice.*

5 Remarks

In this paper, we present an algorithm *Orthogonal_Grid_Rectangular_Draw* that produces c-planar OGRC drawings with $O(n^2)$ area and with at most 3 bends in each edge. This result is as good as the results for classical planar graphs [9, 17, 35]. Lower bounds for the area of orthogonal drawings of classical graphs [38] imply that the area of the drawing produced by our algorithm is asymptotically optimal. Further, we show that the bend per edge performance of our algorithm is optimal. Nevertheless, some open problems remain:

- Although the height and the width of our output drawings are both $O(n)$, our algorithm does not guarantee a good aspect ratio. In practice, our algorithm may produce drawings which clearly prefer one dimension against the other; this is because we use a visibility representation which is biased to one dimension. Recently, some study has been done on the problem of "2-dimensional visibility representations" [7, 13] of planar graphs. In a 2-dimensional visibility representation, each vertex is represented by a box and each edge is represented by a horizontal or vertical segment between the sides of the boxes. It would be interesting to study how to use 2-dimensional visibility representations to improve our algorithm, especially in terms of the aspect ratio of the drawing.
- Even and Granot [10] have presented some algorithms for grid layout of block diagrams. Although the drawing requirements there are different from the requirements of drawing clustered graphs, it would be worthwhile to investigate whether we can borrow some of the techniques there and use them in drawing clustered graphs.
- In recent years, many results have been achieved for orthogonal drawings of non planar graphs [6, 5, 25, 26]. It seems very profitable to use these results to extend our algorithm to non planar clustered graphs.

References

1. C. Batini, L. Furlani, and E. Nardelli. What is a good diagram? a pragmatic approach. In *Proc. 4th Int. Conf. on the Entity Relationship Approach*, 1985.
2. C. Batini, M. Talamo, and R. Tamassia. Computer aided layout of entity-relationship diagrams. *Journal of Systems and Software*, 4:163–173, 1984.
3. G. di Battista and R. Tamassia. Algorithms for plane representations of acyclic digraphs. *Theoretical Computer Science*, 61:175–198, 1988.
4. G. di Battista, R. Tamassia, and I.G. Tollis. Constrained visibility representations of graphs. *Information Processing Letters*, 41:1–7, 1992.
5. T. Biedl. New lower bounds for orthogonal graph drawings. In Franz J. Brandenburg, editor, *GD'95*, volume 1027 of *Lecture Notes in Computer Science*, pages 28–39. Springer-Verlag, 1995.
6. T. Biedl and G. Kant. A better heuristic for orthogonal graph drawings. In *ESA'94*, volume 855 of *Lecture Notes in Computer Science*, pages 24–35. Springer-Verlag, 1994.
7. P. Bose, A. Dean, J. Hutchinson., and T. Shermer. On rectangle visibility graphs. In Stephen C. North, editor, *GD'96*, volume 1190 of *Lecture Notes in Computer Science*, pages 25–44. Springer-Verlag, 1997.
8. P. Eades, Q. Feng, and X. Lin. Straight-line drawing algorithms for hierarchical graphs and clustered graphs. In Stephen C. North, editor, *GD'96*, volume 1190 of *Lecture Notes in Computer Science*, pages 113–128. Springer-Verlag, 1997.
9. S. Even and G. Granot. Rectilinear planar drawings with few bends in each edge. Technical Report 797, Computer Science Department, Technion, Israel Institute of Technology, 1994.
10. S. Even and G. Granot. Grid layout of block diagrams - bounding the number of bends in each connection. In R. Tamassia and I. G. Tollis, editors, *GD'94*, volume 894 of *Lecture Notes in Computer Science*, pages 64–75. Springer-Verlag, 1995.
11. Q. Feng, R. Cohen, and P. Eades. How to draw a planar clustered graph. In *COCOON'95*, volume 959 of *Lecture Notes in Computer Science*, pages 21–31. Springer-Verlag, 1995.
12. Q. Feng, R. Cohen, and P. Eades. Planarity for clustered graphs. In *ESA'95*, volume 979 of *Lecture Notes in Computer Science*, pages 213–226. Springer-Verlag, 1995.
13. U. Fößmeier, G. Kant, and M. Kaufmann. 2-visibility drawings of planar graphs. In Stephen C. North, editor, *GD'96*, volume 1190 of *Lecture Notes in Computer Science*, pages 155–168. Springer-Verlag, 1997.
14. A. Garg and R. amassia. On the computational complexity of upward and rectilinear planarity testing. In R. Tamassia and I. G. Tollis, editors, *GD'94*, volume 894 of *Lecture Notes in Computer Science*, pages 286–297. Springer-Verlag, 1995.
15. D. Harel. On visual formalisms. *Communications of the ACM*, 31(5):514–530, 1988.
16. G. Kant. Drawing planar graphs using the lmc-ordering. In *Proc. 33th IEEE Symp. on Foundations of Computer Science*, pages 101–110, 1992.
17. G. Kant. Drawing planar graphs using the canonical ordering. *Algorithmica*, 16:4–32, 1996.
18. G. Kar, B.P. Madden, and R.S. Gilbert. Heuristic layout algorithms for network management presentation services. *IEEE Network*, pages 29–36, November 1988.
19. J. Kawakita. The KJ method – a scientific approach to problem solving. Technical report, Kawakita Research Institute, Tokyo, 1975.

20. M.R. Kramer and J. van Leeuwen. The complexity of wire-routing and finding minimum area layouts for arbitrary VLSI circuits. In F.P. Preparata, editor, *Advances in Computing Research*, volume 2, pages 129–146. JAI Press, Greenwich, Conn., 1985.

21. C. E. Leiserson. Area-efficient graph layouts (for VLSI). In *Proceedings of the IEEE Symposium on the Foundations of Computer Science*, pages 270 – 281, 1980.

22. K. Misue and K. Sugiyama. An overview of diagram based idea organizer: D-abductor. Technical Report IIAS-RR-93-3E, ISIS, Fujitsu Laboratories, 1993.

23. S. North. Drawing ranked digraphs with recursive clusters. In *Proc. ALCOM Workshop on Graph Drawing '93*, September 1993.

24. J. Nummenmaa and J. Tuomi. Constructing layouts for er-diagrams from visibility representations. In *Proc. 9th Int. Conf. on Entity-Relationship Approach*, pages 303–317, 1990.

25. A. Papakostas and I. G. Tollis. Improved algorithms and bounds for orthogonal drawings. In R. Tamassia and I. G. Tollis, editors, *GD'94*, volume 894 of *Lecture Notes in Computer Science*, pages 40–51. Springer-Verlag, 1994.

26. A. Papakostas and I. G. Tollis. A pairing technique for area-efficient orthogonal drawings. In Stephen C. North, editor, *GD'96*, volume 1190 of *Lecture Notes in Computer Science*, pages 355–370. Springer-Verlag, 1997.

27. D. Reiner, G. Brown, M. Friedell, J. Lehman, R. McKee, P. Rheingans, and A. Rosenthal. A database designer's workbench. In S. Spaccapietra, editor, *Entity-Relationship Approach: Proc. 5th Int. Conf. on Entity-Relationship Approach (Dijon France 1987)*, pages 347–360, New York, N.Y., 1987. North-Holland.

28. P. Rosenstiehl and R.E. Tarjan. Rectilinear planar layouts and bipolar orientations of planar graphs. *Discrete and Computational Geometry*, 1(4):343–353, 1986.

29. J.A. Storer. On minimal node-cost planar embeddings. *Networks*, 14:181–212, 1984.

30. K. Sugiyama and K. Misue. Visualization of structural information: Automatic drawing of compound digraphs. *IEEE Transactions on Software Engineering*, 21(4):876–892, 1991.

31. R. Tamassia. New layout techniques for entity-relationship diagrams. In *Proc. 4th Int. Conf. on Entity-Relationship Approach*, pages 304–311, 1985.

32. R. Tamassia. On embedding a graph in the grid with the minimum number of bends. *SIAM J. Computing*, 16(3):421–444, 1987.

33. R. Tamassia and I.G. Tollis. A unified approach to visibility representations of planar graphs. *Discrete and Computational Geometry*, 1(4):321–341, 1986.

34. R. Tamassia and I.G. Tollis. Efficient embedding of planar graphs in linear time. In *Proc. IEEE Int. Symp. on Circuits and Systems*, pages 495–498, 1987.

35. R. Tamassia and I.G. Tollis. Planar grid embedding in linear time. *IEEE Trans. on Circuits and Systems*, CAS-36(9):1230–1234, 1989.

36. R. Tamassia, I.G. Tollis, and J.S. Vitter. Lower bounds for planar orthogonal drawings of graphs. *Information Processing Letters*, 39:35–40, 1991.

37. J.D. Ullman. *Computational Aspects of VLSI*. Principles of Computer Science. Computer Science Press, Rockville, Md., 1984.

38. L. Valiant. Universality considerations in VLSI circuits. *IEEE Transactions on Computers*, C-30(2):135–140, 1981.

39. C. Williams, J. Rasure, and C. Hansen. The state of the art of visual languages for visualization. In *Visualization 92*, pages 202 – 209, 1992.

Graph Clustering I: Cycles of Cliques

(Extended Abstract)*

F. J. Brandenburg

Lehrstuhl für Informatik
Universität Passau
94030 Passau, Germany
email: brandenb@informatik.uni-passau.de

Abstract. A graph is a cycle of cliques, if its set of vertices can be partitioned into clusters, such that each cluster is a clique and the cliques form a cycle. Then there is a partition of the set of edges into inner edges of the cliques and interconnection edges between the clusters. Cycles of cliques are a special instance of two-level clustered graphs. Such graphs are drawn by a two phase method: draw the top level graph and then browse into the clusters. In general, it is NP-hard whether or not a graph is a two-level clustered graph of a particular type, e.g. a clique or a planar graph or a triangle of cliques. However, it is efficiently solvable whether or not a graph is a path of cliques or is a large cycle of cliques.

Introduction

Graph clustering is a new direction in graph drawing. Several winners of last year's graph drawing competition have used this technique [9]. The Circular Library in the Graph Layout Toolkit of Tom Sawyer Software [5] clusters graphs and then displays them e.g. as a circle of circles.

There is a general need for clustering techniques. As the amount of information to be visualized becomes larger, more structure is needed on top of the classical graph model. There is the need for abstraction and reduction. Flat graphs no longer suffice. When graphs become huge, classical graph drawing algorithms behave poorly or even fail. The further difficulty comes from the inherent complexity of large graphs. The computational complexity of the graph drawing algorithms is directly effected by the size of the graphs. There is a need for efficient algorithms in graph drawing. Here, graph clustering brings us a step forward. Now, time consuming graph drawing algorithms can be applied to small portions only, and the overall running time still remains satisfactory. However, this can work only, if the partition into clusters can be computed efficiently.

* Partially supported by the Deutsche Forschungsgemeinschaft, Forschungsschwerpunkt "Effiziente Algorithmen für diskrete Probleme und ihre Anwendungen".

Several recursive clusterings have been proposed recently, such as higraphs [12], compound graphs [23], hierarchical and clustered graphs [7] and [8] or Hgraphs [14]. For directed acyclic graphs special hierarchical decompositions have been introduced in [11,20], or in [22]. These techniques impose a tree structure on top of the graphs. Ultimately, such approaches lead to graph grammars, where the derivation trees define a hierarchical structure and give raise to clusterings based on a finite set of rules, see [3,18].

Graph partitioning is a classical topic in graph theory (see e.g. [2]) and an important issue in the design of networks, VLSI, and parallel algorithms [18]. There are vertex and edge partitions of graphs, e.g. partitions into triangles, cliques or isomorphic graphs, and there are partitioning algorithms aiming at small sets of separators see, e.g., [1,18]. Partition is a classical NP-hard domain. Many of the graph partition problems stated above are well-known in the theory of NP-completeness, see GT10-GT16 in [10]. Generalized versions of NP-complete graph partition problems habe been studied in [2,6,13,15].

Our approach towards two-level clustered graphs considers graph partitions in more detail and aims at structural properties of the clusters and of the cluster graph. This graph is generally ignored. A graph is a two-level clustered graphs, or more precisely, an X-graph of Y-graphs, if the clusters from the partition of the vertices induce a Y-graph, and if the graph of the clusters is an X-graph. Examples are cycles of cliques or path of cliques. Every bipartite graph with a nonempty set of edges is an edge of two discrete graphs, and every rectangular grid is a (horizontal) path of (vertical) paths. Two-level clustered graphs are defined by graph projections. A graph projection is a mapping between graphs, which is different from other common mappings on graphs, such as graph embeddings and graph homomorphisms. In terms of generalized embeddings, graph projections have a high load, dilation one, and a shrinking expansion. As graph homomorphisms, projections are onto for the vertices and they require a membership property for the clusters.

Two-level clustered graphs have been defined in a syntactic way by the $\odot$-operation of Kratochvíl et. al. [17]. For two disjoint graphs G and G' let $G \odot G'$ be a graph obtained by the union of the sets of vertices and edges and adding an arbitrary set of matching edges between the two sets of vertices. Thus, the $\odot$-operation is nondeterministic and does not define $G \odot G'$ uniquely. In their "Stringing Lemma", Kratochvíl et. al. [17] have applied their $\odot$-operation to complete graphs, such that the graph of clusters is planar. In our terminology these graphs are the planar graphs of cliques. Planar graphs of cliques seem to be inherently nondeterministic. Their synthesis uses the nondeterministic $\odot$-operation and their recognition is NP-complete [16].

Obviously, the recognition and representation problems are the major problems for two-level clustered graphs. Given a graph G and two classes of X-graphs and Y-graphs. Then the recognition problem is, whether G is an X-graph of Y-graphs. Thus, one must find an X-graph G' and a projection of G onto G' such that every cluster is a Y-graph. However, these problems are NP-hard, in general. This is somewhat expected, since many classical graph partition problems can be defined in these terms, such as partition into triangles, partition into cliques, etc., see [10]. Here the top-level X-graphs are arbitrary graphs, and the Y-graphs are triangles or cliques etc. Kratochvíl's result [16] restrict the top-level to the planar graphs, but still there is NP-completeness. We add some further NP-completeness results, such as the recognition of a triangle of cliques or a clique of cliques. However, there are $O(n^3)$ resp. $O(n^2)$ algorithms to decide whether or not a graph is a path of cliques or is a cycle of cliques of length at least six. The latter looks surprising, since it is NP-hard whether a graph is a triangle of cliques, i.e. a cycle of length three.

This is our first paper on graph clustering. In forthcoming papers we'll investige all relevant classes of graphs, including paths, cycles, trees, grids, planar graphs, and cliques, and vary the depth of the clustering to more than two levels.

Basic Notions

First, we recall some basic notions on graphs and establish our notation.

A graph $G = (V, E)$ consists of a finite set of vertices V and a finite set of edges E. Edges are denoted as pairs of vertices (u, v). We consider simple, undirected graphs without self-loops and multiple edges. Such edges are erased by a clean-up procedure.

The *neighbors* of a vertex u are the vertices $neigh(u) = \{v \in V | (u, v) \in E\}$. $neigh(u)$ does not include u itself. Let $neigh[u] = \{u\} \cup neigh(u)$. Accordingly, for a set of vertices U, let $neigh(U) = \{v \in V - U | (u, v) \in E$ for some $u \in U\}$ and $neigh[U] = U \cup neigh(U)$. The subgraph *induced* by a set of vertices U is denoted by $G_{|U}$. The *complement* of G is $\bar{G} = (V, \bar{E})$ with $\bar{E} = \{(u, v) | u, v \in V, u \neq v, (u, v) \notin E\}$.

A graph $G = (V, E)$ with $V = \{v_0, \ldots, v_{n-1}\}$ is a *path*, if $E = \{(v_i, v_{i+1}) | i = 0, \ldots, n - 2\}$. G is a *cycle*, if $E = \{(v_i, v_{i+1}) | i = 0, \ldots, n - 1 \bmod n\}$ and is the *clique* K_n, if $E = \{(v_i, v_j) | i \neq j\}$.

The size of a graph is $size(G) = n$, where $n = |V|$. For convenience, we shall assume the neighbors of a vertex and the set of neighbors neigh(U) can be computed in time proportional to the sizes of U and neigh(U). Usually, the computation time also takes the number of edges into account. Then a scan of G takes $O(n^2)$ instead of $O(n)$.

We now come to the central definition of our approach. A graph projection is a mapping between two graphs. However, it is different from other well-known mappings between graphs and imposes a two-level structure on the graphs, defined by the host graph and by the induced subgraphs.

Definition

A *projection* of a graph $G = (V, E)$ onto a graph $G' = (V', E')$ is a mapping $f : G \to G'$ such that $f : V \to V'$ is many-to-one and onto and there is an edge $(u', v') \in E'$ if and only if there are vertices $u, v \in V$ with $f(u) = u'$, $f(v) = v'$ and $(u, v) \in E$.

Since f is onto, every vertex $v' \in V'$ induces a subgraph of G defined by the set of vertices $f^{-1}(v')$. Every vertex v' resp. the set $f^{-1}(v')$ is called a *supervertex* of G and the induced subgraph $G_{|v'} = (f^{-1}(v'), E \cap (f^{-1}(v') \times f^{-1}(v')))$ is called a *cluster*. The supervertices define a partition the set of vertices V into pairwise disjoint sets. The set of edges E is partitioned into inner and interconnection edges. An edge $e = (u, v) \in E$ is an *inner* edge, if $f(u) = f(v)$. Then $f(e)$ becomes a self-loop, and is omitted by our convention. An edge e is an *interconnection* edge, if $f(u) \neq f(v)$. Then there is an edge $(f(u), f(v)) \in E'$ connecting the supervertices. Conversely, for every edge $(u', v') \in E'$ there is an interconnection edge $e = (u, v)$ such that $f(u) = u'$ and $f(v) = v'$. Conversely, every partition of the set of vertices V defines a mapping between graphs, which is a projection, if it is onto and has a representative for every interconnection edge.

Definition

Let $f : G \to G'$ be a projection of G onto G'. If G' is a graph from a class X and for every $v' \in V'$, the clustered graph $G_{|v'}$ is from some class Y', then G is *a an X-graph of Y-graphs*. More generally, G is called a *two-level clustered graph*.

Two-level clustered graphs can be defined in a syntactic way by the nondeterministic $\odot$-operation introduced by Kratochvíl, et al. [17]. For graphs $G_1 = (V_1, E_1)$ and $G_2 = (V_2, E_2)$ with $V_1 \cap V_2 = \emptyset$ let $G_1 \odot G_2 = \{V_1 \cup V_2, E_1 \cup E_2 \cup M\}$, where $M \subseteq V_1 \cup V_2$ is an arbitrary matching. $G_1 \odot G_2$ is not uniquely determined. This operation is extended to tuples of graphs by defining $G_1 \odot (G_2 \odot \ldots \odot G_n)$.

Kratochvíl et al. have considered graphs of the form $G = G_1 \odot G_2 \odot \ldots \odot G_n$, where each G_i is a complete graph, and the graph of representatives $\tilde{G}$ is planar. $\tilde{G} = (\{G_1, \ldots, G_n\}| \tilde{E})$ and there is an edge (G_i, G_j) in $\tilde{E}$ if and only if $i \neq j$ and there are vertices $v_i \in V_i$ and $v_j \in V_j$ such that $\{v_i, v_j\}$ is an edge in G. Kratochvíl (personal communication) has observed, that graphs of the above form are exactly the planar graphs of cliques. In [17] it has been shown that the complete bipartite graph $K_{5,5}$ is not a planar graph of cliques and that every such graph is a string graph, i.e. the intersection graph of curves. In [16] Kratochvíl has proved the NP-completeness of the recognition problem for planar graphs of cliques. This results shall we improved to triangles of cliques.

How shall we draw two-level clustered graphs? How shall we draw a an X-graph of Y-graphs? This comes naturally with the definition. On the top level, draw the graph G' by your favorite algorithm for X-graphs. The algorithm may draw G' with variable size nodes reflecting the sizes of the induced subgraphs. We do not display the graph G as a whole; instead we browse into the vertices of G' and display the clusters by drawing the induced subgraphs as Y-graphs. This can be extended to pairs of supervertices $u', v' \in V'$ and then displaying the clusters as Y-graphs together with the interconnection edges between the vertices of the clusters. For a nice display, the clusters should be drawn such that the endpoints of interconnection edges appear on the outer boundary of the drawn Y-graphs. Graph drawing algorithms of that kind are not yet around.

For a graph $G = (V, E)$ and graph classes X and Y the recognition problem is, whether or not G is an X-graph of Y-graphs, and to find a corresponding representation. What is the complexity of this decision problem? A solution of this question requires a partition of the vertices $V = V_1 \cup \ldots \cup V_k$ with $V_i \cap V_j = \emptyset$ such that every induced subgraph $G_{|V_i}$ is a Y-graph. The sets $\{V_1, \ldots, V_k\}$ become the vertices of the graph G'. The edges of G' must have representatives in the graph G. Hence, the set of edges E is partitioned into the sets of edges of the induced subgraphs and the set of edges representing the edges of G'. This partition is many-to-one and onto and is directed by the properties of the graph classes X and Y.

The question whether or not a graph G is an X graph of Y graphs is NP-hard for very many instances of graphs G and classes X and Y. This is particularly true, when X is the class of all graphs. In that case, the host graph G' has no particular structure and the problem of the existence of a certain projection reduces to a classical decomposition problem for graphs or to the existence of a special subgraph. The interconnection edges can be ignored, and known NP-hardness results apply, see [6, 10]. In particular, the well-known graph partition problems into triangles, isomorphic subgraphs, Hamiltonian subgraphs, forests, cliques, or perfect matchings and graph colorability (see GT11-GT16 in [10]) can be described as a graph projection onto arbitrary graphs.

Similary, if Y is the set of graphs with a single vertex, then nothing changes and all NP-hard problems remain as they were.

However, these classes of graphs seem too general and are not interesting to us. Our inspiration came from the circular layouts of [5] and we first consider cycles and paths of cliques. Surprisingly, the recognition problem can be solved in efficiently.

A further consequence of our approach will be a richer classification of graphs, particularly for graph drawing. So far, graph drawing distinguishes four classes: general graphs, directed acyclic graphs, planar graphs and trees. Other classes of graphs are not of general interest, because either they seem too special or there are no appropriate graph drawing algorithms, yet. Our approach changes this situation. And it opens a wide field for further investigations.

Cycles of Cliques

For clarity we recall the definition of cycles of cliques.

Definition

A graph $G = (V, E)$ is a *cycle of cliques* (resp. *path of cliques*), if there is a cycle (a path) $G' = (V', E')$ and a projection $f : G \to G'$ of G onto G' such that for every $v' \in V'$, the subgraph induced by $f^{-1}(v')$ is a clique. G is a k-cycle of cliques, if G' is a cycle of size (or length) k.

Hence, there is a partition of the set of vertices $V = V_0 \cup \ldots \cup V_{k-1}$ with $V_i \cap V_j = \emptyset$ for $i \neq j$ such that for every $i, 0 = 1, \ldots, k - 1$, the subgraph induced by V_i is a clique. Moreover, with the proper ordering there are vertices u, v with $u \in V_i$, $v \in V_{i+1}$ (mod k) and $(u, v) \in E$, and conversely, for every edge $(u, v) \in E, u \in V_i$ implies $v \in V_i$ or $v \in V_{i\pm1}$ (mod k).

For the decision problem it must be checked whether or not the set of edges can be partitioned into clique edges and cycle edges, where the clique edges are one-to-one and the cycle edges are many-to-one.

Cycles of cliques have a particular structure. The deletion of a supervertex cuts the cycle and leaves a path of cliques. Now, every supervertex except at the ends is a clique separator. However, not every clique separator can be used as a supervertex. There may be some nondeterminism, which cannot be decided locally. The circular saw with two defects in Fig. 1 is an example of a cycle of cliques. However, the triangles cannot be chosen as clusters. Because of the quadrilateral on top the left resp. right sides must be taken.

Figure 1

A partition of a graph $G = (V, E)$ into cliques is directly related to a coloring of the complement graph $\bar{G}$. A graph coloring partitions the vertices such that there are edges only between vertices from different clusters. A partition of a graph into a k-cycle of cliques implies a k-coloring of the complement graph, such that every pair of non-adjacent clusters is a complete bipartite graph and two adjacent clusters are not complete bipartite.

It should be noted that the Circular Layouts from Tom Sawyer's Graph Layout Tool [5] are not fully captured by our definition of cycles of cliques. Their examples allow almost cycles of almost cliques. However, there is no reasonable way to define "almost" and to handle it efficiently.

A vertex and an edge of cliques can be recognized easily. The case $k = 3$ is different. Then a k-cycle is a triangle. Triangles of cliques are dense graphs, and there may be an edge between any pair of vertices. This explains why they are hard to recognize. Our NP result improves the result of Kratochvíl [16] on the NP-completeness of planar graphs of cliques.

Theorem 1
It is NP-complete, whether or not a graph G is a triangle of cliques.

Proof.
G is a triangle of cliques if and only if the complement graph $\bar{G} = (V, \bar{E})$ is 3-colorable and the induced partition $V = V_1 \cup V_2 \cup V_3$ is connected in G.
The latter holds e.g. by the reduction from 3-SAT given in [19]. Hence, a 3-SAT expression is satisfiable if and only if G is a triangle of cliques.

Using k-colorability, this result can be generalized to cliques of cliques. Notice that a clique is a clique of cliques, but not conversely.

Theorem 2
For $k \geq 3$ it is NP-complete, whether or not a graph G is a k-clique of cliques.

These results suggest, that the recognition problems for two-level clustered graphs are NP-hard, except for some trivial cases. *This is false!* There are polynomial time algorithms for paths of cliques and also for cycles of cliques, when triangles are excluded.

Theorem 3
There is an $O(n^3)$ algorithm to decide whether or not a graph G is a path of cliques and to compute the corresponding representation.

Proof.
(sketch). Suppose that $G = (V, E)$ is a path of cliques with the supervertices $V_0, \ldots, V_{k-1}$ and $V_i \subseteq V$ for $i = 0, \ldots, k-1$. Each supervertex is a clique, and there are further edges between vertices of G if and only if they are in adjacent supervertices. Thus, we order the vertices of G from V_0 to V_{k-1} and orient the edges accordingly.

By induction, suppose that a path of supervertices $V_0, \ldots, V_{i-1}$ has been computed, and a subset C_i of the next supervertex V_i is known. The vertices from $V_0, \ldots, V_{i-1}$ have been marked as "old". Then the next supervertex V_i contains the vertices from C_i and is contained in neigh$[C_i]$. The remaining vertices in neigh$[C_i]$ - V_i are a subset of the next supervertex. So we define an orientation of the edges from a marked to a yet unmarked vertex, i.e. from V_{i-1} to V_i. The vertices from V_i are marked as "old" and we proceed to the next stage and compute the next supervertex V_{i+1}.

For a set of vertices $U \subseteq V$ define $next(U) = \{w \in V | w \in neigh(U)$ and w is not marked$\}$. $next(U)$ is the set of new neighbors of U. A set of vertices $U \subseteq V$ forms a clique, if the induced subgraph $G_{|U}$ is a clique. Two disjoint cliques U_1 and U_2 form a clique, if $U_1 \cup U_2$ does. Otherwise, for every $u_1 \in U_1$ there is some $u_2 \in U_2$ such that the edge (u_1, u_2) is missing in G.

Consider a nonempty set of vertices C_i which forms a clique. C_i is taken as a subset of the current supervertex V_i. The first goal is to expand C_i into V_i. However, as we shall see, this is not unique and a decision cannot be made by local computations. See Figure 1. There the triangles do not form the proper supervertices. Hence, we compute several candidates $V_{i,j}$ for the supervertex V_i, and keep track of them and check which of them will eventually succeed to represent G as a cycle of cliques. A candidate fails, if it induces a set of vertices as a supervertex, which is not a clique. Candidates can be identified, if they behave similarly and induce cliques which differ not essentially. For example, if a vertex lies between V_i and V_{i+1} and forms a clique with both, then it can be added to either of them.

Let $C_i \subseteq V_i$ and let the vertices from $V_0, \ldots, V_{i-1}$ be marked. Then $C_i \subseteq V_i \cup next(C_i)$ and $\emptyset \neq next(C_i) \subseteq V_i \cup V_{i+1}$. Hence, the vertices from $next(C_i)$ must be partitioned into V_i and V_{i+1}. Consider the set $next(C_i)$. Since $next(C_i)$ is covered by at most two cliques, it partitions into four mutually disjoint sets C_{i+1}, R, B and Y. Some but not all of them may be empty. Each of them forms a clique. C_{i+1} and C_i do not form a clique. I.e. for every $u \in C_{i+1}$ there is some $v \in C_i$ such that (u, v) is not an edge in G. Hence, the vertices from C_{i+1} must belong to the next supervertex V_{i+1}. The sets R, B, and Y each form a clique with each of C_i and C_{i+1}. Y forms a clique with each of R and B, whereas R and B do not form a clique. A pair (u, v) is in $R \times B$ iff the edge (u, v) is missing in G. The computation of the sets C_{i+1}, R, B and Y is easily done by coloring the complement graph $\bar{G}$. Consider the subgraph of G induced by $C_i \cup next(C_i)$. Its complement must be 2-colorable, such that all vertices from C_i are colored by the same color. Otherwise, $C_i \cup next(C_i)$ must be partitioned into at least three supervertices. These are either disconneced or they branch and then cannot form a cycle. The complement graph restricted to $C_i \cup next(C_i)$ is almost discrete. However, every vertex from C_{i+1} is connected to some, but not all vertices from C_i, the "yellow" vertices from Y are isolated and the vertices from $R \cup B$ are bipartite subgraphs in $\bar{G}$, i.e. they are colored "red" and "blue". If there are several components, R or B are chosen arbitrarily.

Next, we decide the vertices from $R \cup B$, and move them into C_i and C_{i+1}. Mark the vertices from $C_i \cup next(C_i)$. For every pair u, v from $R \cup B$ with $u \in R$ and

$v \in B$, compute $next(u)$ and $next(v)$. If $next(u) \subset next(v)$, then add u to C_i and $v \cup next(u)$ to C_{i+1}. Since u and v must be in distinct supervertices, there is no other choice. If $next(u) = next(v)$, then arbitrarily add u to C_i and v and $next(u)$ to C_{i+1}. u and v can be interchanged. Finally, if $next(u)$ and $next(v)$ are pairwise incomparable, there is an error. The chosen C_i is false, and shall be ignored. If there is no alternative, then the graph G is not a path of cliques.

Now, consider the vertices from Y. Y forms a clique with each of C_i and C_{i+1}. Thus, for the vertices of Y there is a choice between the supervertices V_i and V_{i+1}. This nondeterminism cannot be solved by local computations. However, the vertices from Y can be ordered.

Mark the vertices from $C_i \cup next(C_i)$. For vertices y and y' from Y, if $next(y) = next(y')$, then identify y and y'. If $next(y)$ and $next(y')$ are pairwise incomparable, then y and y' must belong to the same supervertex, V_i or V_{i+1}. Then merge them into y'' with $next(y'') = next(y) \cup next(y')$. Finally, if $next(y) \subset next(y')$, then $y' \in C_i$ implies $y \in C_i$, otherwise, the next supervertex is not a clique. Hence, we can order the (subsets of) elements of Y according to the strict inclusion of their sets of next neighbors. Let $Y = (y_1, \ldots y_r)$ with $next(y_j) \subset next(y_{j+1})$. Since Y forms a clique with C_i, r is smaller than the cliquesize of G.

For the supervertex V_i we have $r + 1$ candidates, namely $V_{i,0} = C_i$ and $V_{i,j} = C_i \cup \{y_1, \ldots, y_s | s \leq j, j = 1, \ldots, r\}$. Each of them is taken into consideration and is tested in parallel in the next phases. However, if $r \leq 2$, then every $V_{i,j}$ uniquely determines its next supervertex $V_{i+1,j}$, or the number of candidates reduces, since a candidate fails or is merged with another candidate. Every candidate can be associated with a vertex of G or a supervertex. Hence there are at most $O(n)$ candidates, and each of them takes at most $O(n)$ computation steps, when we assume that neighbors can main be computed in linear time in the numbers of vertices. Thus, the decision procedure takes $O(n^2)$ time, which is $O(|V| \cdot |E|)$, if edges are taken into account.

It remains to compute a candidate for the first supervertex. Suppose that G is a k-path of cliques. The cases $k = 1$ and $k = 2$ are obvious. For $k \geq 3$ consider the set of endpoints of paths defining the diameter of G. Let C_0 be one of these sets, i.e. $C_0 = \{u|$ there is some vertex v such that the shortest path between u and v has length $diam(G)\}$, where $diam(G)$ is the diameter of G. The diameter of a graph and the set C_0 can be computed by using an "all-pairs shortest path" algorithm, and this takes $O(n^3)$.

Finally, we consider cycles of cliques for sufficiently long cycles.

Lemma
If G is a k-cycle of cliques, then the diameter of G is bounded by $diam(G) \leq 2 \cdot \lfloor k/2 \rfloor + 1$.

Notice that the graphs used for the NP-completeness result of Theorem 2 have diameter two.

Theorem 4

If $diam(G) \geq 4$, then there is an $O(n^2)$ algorithm to decide whether or not a graph G is a cycle of cliques and to compute the representation of G as a cycle of cliques.

Proof.

Suppose that G is a k-cycle of cliques with the supervertices $V_0, V_1 \ldots, V_{k-1}$ in that cyclic order. Then $k \geq 4$. The idea of the proof is to cut the cycle, say at a supervertex V_0. Then we obtain a path of cycles, and can use Theorem 5. Thus, we must compute a subset C_0 of the first supervertex V_0, and we must mark the vertices from V_{k-1} in order to move through G in circular order. These marks are crucial, because then the set of next neighbors of a supervertex is 2-colorable in the complement graph. Otherwise, and in particular for triangles, three or more colors are needed. Since $diam(G) \geq 4$, there is a shortest path of length four $v_0 \rightarrow v_1 \rightarrow v_2 \rightarrow v_3 \rightarrow v_4$. Mark (or temporarily delete) v_0 and its neighbors neigh(v_0). If G is cycle of cliques, then the rest graph $G - (neigh[v'_0])$ is a path of at least three cliques. Now use the construction from Theorem 3 with the vertex (v_2) as subset of the first supervertex. When the supervertices including (v_4) are compputed, reinsert neigh$[v_0]$.

When the cycle is closed it is checked, that the candidate chosen as the first supervertex is feasible.

The algorithm takes time $O(n^2)$, since the paths of length at least 4, the computations of C_0 and the main decision procedure can each be done in quadratic time with the assumptions made above. If a scan through a graph $G = (V, E)$ takes $O(|V| + |E|)$ time, then we obtain $O(n^3)$.

Conclusion

Graph projections and the induced graph clusterings are a field for new investigations. The partitions into cycles of cliques are a very first step. Other classes of interest are trees and planar graphs at the top level. For trees, there may be a similarity to the tree-decomposition of graphs; for planar graphs we take a look at planar graphs of planar graphs as a new attempt towards "almost planar graphs". E.g. the K_{16} can be decomposed into four K_4's and thus is a planar graph of planar graphs; however, the K_{17} does not admit such a projection.

References

1. C.J. Alpert and A.B. Kahng: Recent directions in netlist partitioning: a survey. Integration, the VLSI J. **19** (2995), 1-18.
2. J. Bosik: Decompositions of Graphs. Kluvwer Academic Publishers, Dordrecht (1990).

168

3. F.J. Brandenburg: Designing graph drawings by layout graph grammars. Graph Drawing 94, LNCS 894 (1995), 416-427.
4. R.C. Brewster, P. Hell and G. MacGillivray: The complexity of restricted graph homomorphisms. Discrete Mathematics **167/168** (1997), 145-154.
5. U. Dogrusöz, B. Madden and P. Madden: Circular layout in the graph layout toolkit. Graph Drawing 96, LNCS 1190 (1997), 92-100.
6. D. Dor and M Tarsi: Graph decomposition is NP-complete: proof of Holyer's conjecture. SIAM J. Comput. **26** (1997), 1166-1187.
7. P. Eades and Q.W. Feng: Multilevel visualization of clustered graphs. Graph Drawing 96, LNCS 1190 (1997), 101-112.
8. P. Eades, Q.W. Feng and X. Lin: Straight-line drawing algorithms for hierarchical graphs and clustered graphs. Graph Drawing 96, LNCS 1190 (1997), 113-128.
9. P. Eades, J. Marks and S. North: Graph-Drawing Contest Report. Graph Drawing 96, LNCS 1190 (1997), 129-138.
10. M.R. Garey and D.S. Johnson: Computers and Intractability: A Guide to the Theory of NP-Completeness. Freeman, San Fransisco (1979).
11. D.J. Gschwindt and T.P. Murthagh: A recursive algorithm for drawing hierarchical graphs. Technical Report CS -89-02, Williams College, Williamstown (1989).
12. D. Harel: On visual formalisms. Comm. ACM **31** (1988), 514-530.
13. P. Hell and J. Nešetřil: On the complexity of H-coloring. J. of Combinatorial Theory, Series B **48** (1990), 92-110.
14. M. Himsolt: Konzeption und Implementierung von Grapheditoren. Shaker Verlag, Aachen (1993).
15. I. Holyer: The NP-completeness of some edge partition problems. SIAM J. Comput., **10** (1981), 713-717.
16. J. Kratochvíl: String Graphs. II. Recognizing string graphs is NP-Hard. J. of Combinatorial Theory, Series B **52** (1991), 67-78.
17. J. Kratochvíl, M. Goljan and P. Kučera: String Graphs. Academia, Prague (1986).
18. T. Lengauer: Combinatorial Algorithms for Integrated Circuit Layout. Wiley-Teubner Series (1990).
19. U. Manber: Introduction to Algorithms a Creative Approach. Addison Wesley, Reading (1989).
20. E.B. Messinger, L.A. Rowe and R.R. Henry: A divide-and conquer algorithm for the automatic layout of large directed graphs. IEEE Trans. Systems Man Cybernetics **21** (1991), 1-12.
21. R. Sablowski and A. Frick: Automatic graph clustering. Graph Drawing 96, LNCS 1190 (1997), 396-400.
22. F.-S. Shieh and C.L. McCreary: Directed graphs drawing by clan-based decomposition. Graph Drawing 95, LNCS 1027 (1996), 472-482.
23. K. Sugiyama and K. Misue: Visualization of structural information: automatic drawing of compound graphs. IEEE Trans. Systems Man Cybernetics **21** (1991), 876-892.
24. R.E. Tarjan: Decomposition by clique separators. Discrete Math. **55** (1985), 221-232.
25. S.H. Whitesides: An algorithm for finding clique cut-sets. Inf. Proc. Letters **12** (1981), 31-32.

An Algorithm for Labeling Edges of Hierarchical Drawings *

Konstantinos G. Kakoulis   and   Ioannis G. Tollis

Department of Computer Science
The University of Texas at Dallas
Richardson, TX 75083-0688
email: kostant@utdallas.edu, tollis@utdallas.edu

Abstract. Let $G(V, E)$ be a graph, and let Γ be the drawing of G on the plane. We consider the problem of assigning text labels to every edge of G such that the quality of the label assignment is optimal. This problem has been first encountered in automated cartography. Even though much effort has been devoted over the last 15 years in the area of automated drawing of maps, the Edge Label Placement (ELP) problem remains essentially unsolved. In this paper we investigate the ELP problem. We present an algorithm for the ELP problem more suitable for hierarchical drawings of graphs, but it can be adopted to many different drawing styles and still remain effective. Also, we present experimental results of our algorithm that indicate its effectiveness.

1   Introduction

The area of graph drawing has grown significantly in the recent years motivated mostly by applications in information visualization [4, 17]. When visualizing information, it is essential to display not only the structure of the objects and their relationships, but also important information about them. In other words, it is important to associate text labels with graphical features. This problem is called *automatic label placement* and has applications in many areas including graph drawing and cartography. Also, the automatic label placement problem has been recognized as an important area of research by the ACM Computational Geometry Task Force [7].

Most of the research addressing the labeling problem has been focused on labeling features of geographical and technical maps. Christensen, Marks and Shieber present a comprehensive survey of algorithms for the labeling problem [3]. Significant progress has been made in solving the *Node (point) Label Placement* (NLP) problem which consists of labeling points in the plane so that no labels overlap points or other labels (see [1, 3, 6, 8, 10, 18, 20]). However, the *Edge Label Placement* (ELP) problem remains essentially unsolved [14, 20]. It is worth noting that both the NLP [8, 13, 15] and ELP [12] problems are NP-Hard.

---

* Research supported in part by NIST, Advanced Technology Program grant number 70NANB5H1162. A patent on these and related results is pending.

It has been suggested [3, 14] that the techniques used to solve the NLP problem can be used to solve the ELP problem by positioning a number of points on each edge, and then solving the NLP problem on those points. However, choosing the correct points is almost as hard as solving the original problem. Hence, no efficient techniques for the ELP problem are known to take advantage of the algorithms that solve the NLP problem.

Because the ELP problem is NP-Hard [12], any effective approach to solve the ELP problem must be directed towards devising heuristics. In this paper we first define the ELP problem in a formal way in Section 2. In Section 3 we present the main ideas of our algorithm. In Section 4 we present some improvements that can make the main algorithm more efficient. In Section 5 we present experimental results of an implementation of our algorithm, and we conclude in Section 6.

2 The ELP problem

A *drawing* of a graph $G(V, E)$ is a representation of G in the plane, where nodes are represented by symbols such as circles or boxes, and each edge (u, v) is represented by a simple open curve between the symbols associated with the nodes u and v. A *straight − line* drawing maps each edge into a straight-line segment. A *polyline* drawing maps each edge into a polygonal chain. A *hierarchical* drawing is a polyline drawing where the nodes and bends are constrained to lie on a set of equally spaced horizontal (vertical) lines, called layers. Next, we introduce some terminology specific to the ELP problem: (*i*) Λ_e is the set of all label positions for edge e; (*ii*) Λ is the set of all label positions for all edges; (*iii*) $\lambda : E \rightarrow \Lambda$ is a function that assigns to edge e in E one label position from Λ, and $\lambda(e) = \lambda_e \in \Lambda_e$. Next, we define the ELP problem for a general drawing of a graph.

ELP is the problem of assigning text labels to each edge of a given drawing of a graph. The goal is to communicate the information for each edge via text labels in the best possible way, by positioning the labels in the most appropriate place. Here we consider the problem of assigning one label to each edge of the graph.

Cartographers Imhof [11] and Yoeli [19], have devised general rules that measure the quality of a label assignment.

Basic rules for labeling quality [11, 19]:

1. No overlaps of a label with other labels or other graphical features of the layout are allowed.
2. Each label can be easily identified with exactly one graphical feature of the layout (i.e., the assignment is unambiguous).
3. Each label must be placed in the best possible position (among all acceptable positions).

The first 2 rules evaluate how clear the association of the labels must be when compared with their corresponding edges. A label respects the first rule if it does not overlap any graphical feature, even though it is allowed to touch the edge that it belongs to. Also, a label respects the second rule if it is placed very close or touches the edge that it belongs to. The third rule defines a ranking among all label positions that typically captures the aesthetic preference for specific labels. This is an essential criterion for the quality of labeling of geographical and to some extend technical maps. It also allows to introduce problem specific constraints (i.e., the label of an edge must be closer to the source or destination node).

Each label position that is part of a final label assignment is associated with a cost. $COST : \Lambda \to \mathcal{N}$ is a function that gives us the cost of assigning label λ_e to edge e in the final label assignment. The $COST(\lambda_e)$ is a linear combination of: (i) The cost with respect to the ranking of label λ_e; (ii) The cost which reflects the severity of the violation of the first two basic rules for label λ_e.

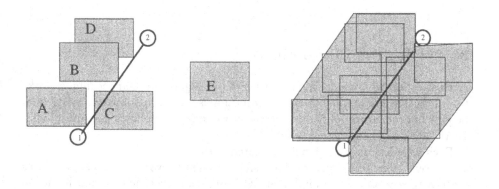

Fig. 1. (a) How to position labels for an edge. (b) Labeling space of an edge.

In $Fig.$ 1(a) labels like A, B or D are preferable but certainly labels like C, which overlaps its associated edge, should be avoided. However, they can be acceptable with some appropriate cost assigned to them. If a drawing has more than one edges, then labels like E (that float between edges) violate the second rule, and must be avoided.

We define the set of all label positions for a given edge either by explicitly defining a set of label positions (*discrete* labeling space) or by defining an area where labels can be placed (*continuous* labeling space). In the *discrete* labeling space, the set of all label positions is finite and each label is identified by its position in the drawing, as shown in $Fig.$ 1(a). In the *continuous* labeling space, the set of all label positions is infinite, and for each edge we define a region which is bounded by a closed line, such that each label position for that edge must be placed inside that region, as shown in $Fig.$ 1(b).

The ELP problem is an optimization problem since the objective is a label assignment of minimum cost. The objective is to find a set of labels, one for each edge, that yields minimum total cost.

The ELP Problem : Let $G(V, E)$ be a graph and let Γ be a drawing of G. **Question** : Find a label assignment that minimizes the following function:

$$\sum_{i \in E} \sum_{j \in \Lambda_i} COST(\lambda(i)) P(i, j)$$

where:

$$P(i, j) = \begin{cases} 1, & if\ \lambda(i) = j, \\ 0, & otherwise \end{cases}$$

and

$$\sum_{i \in E} \sum_{j \in \Lambda_i} P(i, j) = |E|$$

where:

$$\sum_{j \in \Lambda_i} P(i, j) = 1, \qquad i \in E.$$

□

3 Solving the ELP problem

In this section we present the main idea of our algorithm. Our technique works for labels that are parallel to the horizontal axis, and have approximately equal height. In order to simplify our discussion we make the following assumptions:
1. All labels have the same size.
2. Each edge has only one label associated with it.

However, our technique can be easily modified to work for labels of various sizes. Any acceptable solution for the ELP problem must guarantee at least two conditions:
1. Any label must be free of overlaps (except that it may overlap its associated edge).
2. Any label must be very close to its associated edge.

The goal of our algorithm is to assign to each edge a label that is free of overlaps and touches only its associated edge. The main idea of our algorithm is the following:

We divide the input drawing into consecutive horizontal strips of equal height. The height of each strip is equal to the height of the labels. Next we find the set of label positions Λ_e for each edge e. Each label position must be inside a horizontal strip. We slide labels inside each horizontal strip, if a label touches an edge e, and it does not overlap any other edge and/or node, then we assign it as a label position of edge e, as shown in $Fig.$ 2(a). In $Fig.$ 2(b) the drawing is divided into horizontal strips, and the maximum number of label positions that follow the rules of our algorithm have been assigned to the edges of the drawing. As it can be observed from that figure, labels that overlap nodes or edges of the layout are not considered. Also labels are not allowed to intersect their associated edges. Labels must lie entirely into a horizontal strip. Thus, labels can only overlap other labels that belong to the same horizontal strip. Hence, the following lemma is true:

Lemma 1. *Each label position overlaps at most one other label position.*

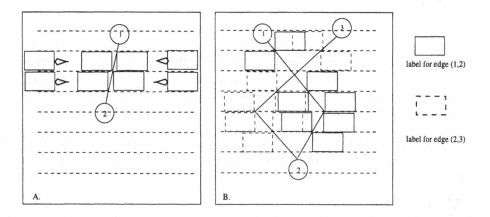

Fig. 2. Assigning label positions to edges of a drawing.

We transform the ELP problem into a matching problem where each edge e is matched with one of its label positions in Λ_e. First, we group label positions together. If two label positions overlap then they belong to the same group. If a label position is free of overlaps then it belongs to a single member group. When we match edges to label positions, only one label position of each group can be part of a label assignment. This restriction guaranties a label assignment where labels do not overlap each other. Let $\mathcal{R}$ be the set of all such groups.

To further clarify the main idea of our algorithm we introduce the matching graph.

Definition 2. Given a drawing Γ of a graph $G(V, E)$, and the set Λ of label positions for Γ (where each label position in Λ overlaps at most one other label position), then a *matching* graph $G_m(V_e, V_g, E_m)$, is defined as follows:

- Each node e in V_e corresponds to an edge e in G.
- Each node r in V_g corresponds to a group r of label positions in $\mathcal{R}$.
- Each edge (e, r) in E_m connects a node e in V_e, to a node r in V_g, if and only if edge e in G has a label position that belongs to group r of $\mathcal{R}$.

By definition graph G_m is bipartite. In addition, the cost of assigning label λ_e to edge e may be the weight of edge (e, r) in G_m, where $\lambda_e \in r$. Then, a maximum cardinality, minimum weight matching for graph G_m will give us a label assignment of minimum cost (among all assignments) such that at most one label position from each group in $\mathcal{R}$ is part of that assignment. Hence, we have the following theorem:

Theorem 3. *Let Λ be the set of label positions for all edges of a drawing Γ of graph $G(V, E)$. If every label position in Λ overlaps at most one other label position in Λ, then a maximum cardinality minimum weight matching of the corresponding matching graph G_m produces an optimal solution to the ELP problem with no overlaps.*

The result of Theorem 3 gives rise to a powerful technique for the solution of the ELP problem, given specific conditions on the set of label positions. In practice we can satisfy these conditions heuristically. The main problems that need to be solved heuristically are: (i) Defining the best set Λ; (ii) Partitioning Λ into groups in $\mathcal{R}$. In our technique, we define the set Λ of label positions by splitting the input drawing into horizontal strips, and assigning label positions to edges as we have explained at the beginning of this section. Also, according to Lemma 1, each label position overlaps at most one other label. Thus, it is trivial to construct the set of groups $\mathcal{R}$. Next, we give a description of the ELP algorithm.

ELP_Algorithm
INPUT: A drawing Γ of graph $G(V, E)$.
OUTPUT: A label assignment.
1. Split Γ into horizontal strips.
2. Find all label positions for each edge and construct the groups in $\mathcal{R}$.
3. Create the matching graph G_m for Γ.
4. Match edges to label positions, by finding a matching in G_m.

The last step in the algorithm solves the maximum cardinality minimum weight matching problem for the matching graph. By Theorem 3, the maximum cardinality minimum weight matching of the matching graph will give an optimal label assignment with respect to the set of labels Λ that satisfy the sufficient conditions of Theorem 3. We can use any known algorithm to find a maximum cardinality minimum weight matching of the matching graph G_m. Notice that the best algorithms take more than quadratic time with respect to the size of G_m [9, 16].

The size of the *matching* graph depends on the size of set Λ of label positions. Unfortunately the size of Λ can be very large with respect to the size of the original graph G. This implies that a typical matching algorithm might take a long time. Additionally, in our application graph G_m has a rather simple structure. In order to reduce the time complexity of the matching we present in the next section a heuristic that finds a maximum cardinality matching with low total weight in linear time with respect to the size of G_m. Some important additional advantages of this heuristic are: reduced memory requirements, simplicity, and that it can be easily embedded into different systems.

3.1 A fast matching heuristic.

Here we present a fast heuristic that solves the maximum cardinality matching problem for the matching graph. The size of the matching graph depends on the size and the total edge length of the drawing, and not only on the size of the input graph. From the definition of the matching graph, and Lemma 1, it is clear that the matching graph is a bipartite graph such that each node in V_g has degree at most 2.

The following algorithm finds a maximum cardinality matching for G_m.

Fast_Matching Heuristic

INPUT: Matching graph G_m.

OUTPUT: A maximum cardinality matching for G_m with low total weight.

1. *If* the minimum weight incident edge of a node in V_e
    connects this node to a node in V_g of degree 1 *then*:
    > Assign this edge as a matched edge.
    > Update G_m.
2. *If* a node in V_e has degree one *then*:
    > Assign its incident edge as a matched edge.
    > Update graph G_m.
3. Repeat Steps 1 and 2 until no new edge can be matched.
4. Delete all nodes of degree 0 from G_m.
    For each node e in V_e do:
    > Remove all but the two incident edges of e with the least weight.
5. The remaining graph consists of simple cycles and/or paths.
    Find the only two maximum cardinality matchings for each component.
    Choose the matching of minimum weight.

Note: The *Update G_m* operation removes the two nodes incident to a new matched edge and stores that edge and its incident nodes as part of the matching. Also removes all incident edges from the two nodes.

In Step 1 we find matched edges that are part of any optimal solution. In Step 2 we find matched edges to those nodes in V_e that are of degree one. If two nodes of degree one in V_e are connected to the same node in V_g, we choose as matched edge the edge with minimum weight. This implies that one of the edges will have no label. Next, in Step 4, for each node in V_e of degree more than two, we keep only its two incident edges of least weight and remove the rest of the edges. Now each node in V_e has degree two and each node in V_g has degree at most two. The remaining bipartite graph has a simple structure: It consists of simple cycles and/or simple paths. Each path or cycle has exactly two maximum cardinality matchings. It is trivial to find both of them by simply traversing the cycle or path and picking as part of the matching only the even or odd numbered edges.

Next, we show that this fast matching heuristic produces a maximum cardinality matching.

Theorem 4. *The matching produced by the Fast_Matching Heuristic is a maximum cardinality matching.*

Sketch of proof: Let us assume that the matching produced by the matching heuristic is not a maximum cardinality matching. Then, there exists an augmenting path according to Berge's theorem [2], which states that: A matching M of a graph G is a maximum matching if and only if G contains no augmenting paths.

Any augmenting path is of odd length, which implies that for graph G_m any such path has one end in V_e and the other end in V_g. So, without loss of generality, we assume that any augmenting path starts from a node e in V_e and ends in a node g in V_g.

It is clear that e has been deleted before Step 5 of the algorithm. Also, g is connected only to nodes in V_e that are matched in the last step of the algorithm. Thus, there exists at least one edge in the augmenting path that connects a node u deleted before Step 3 to a node v not deleted before Step 5. In addition, that edge can only be a non-matched edge, otherwise both nodes would have been deleted (not deleted) before Step 5. A node w in V_e is deleted before Step 5 if it has degree 1 or 0 during the execution of the first three steps. Thus, node w is always connected only to nodes deleted at the same step as w. Consequently, u must be a node in V_g. We have arrived in a contradiction, because edge (u, v) is an even numbered edge in the augmented path which implies that (u, v) must be a matched edge. Thus, we have proved our claim. □

The fast matching heuristic finds a maximum cardinality matching with low total weight because in the last step it considers only the two incident edges of nodes in V_e with the lowest weight. In addition, if each edge in the input drawing has at least two label positions, the fast matching heuristic produces a label assignment that assigns to at least half of the edges their best label position, and to the rest of the edges their second best label position.

4 Improvements and Extensions.

One of the weaknesses of the algorithm presented in this paper is that it considers only a subset of all label positions. Specifically, label positions are defined by the horizontal strips that divide the drawing. Hierarchical drawings are particularly suitable for this algorithm since edges are usually long and almost vertical. If a hierarchical drawing is drawn from left to right then we rotate the drawing by 90 degrees and apply the algorithm.

In order to improve the effectiveness of our algorithm a postprocessing step is often necessary. The postprocessing step explores the solution space in three ways:

1. It locally shifts assigned labels to create space for a new label to be assigned.
2. It searches the solution space to find if there is enough space to assign a label with or without repositioning already assigned labels.
3. It searches the solution space after relaxing the restrictions on the quality of the label assignments by allowing labels to overlap their associated edges.

In the presentation of the algorithm we assumed that labels of different edges are of the same size. In reality, we only need to have labels that are of approximately equal height. Because in order to find the set of label positions for the input drawing we must divide the drawing into equally spaced horizontal strips. If labels have different heights, then if we assume that each label has height equal to the maximum height among all labels we can apply the present algorithm to solve the ELP problem.

One important characteristic of the ELP algorithm is that it can be tailored to take into account the user's preferred positions when it assigns labels to edges. The user can specify preferred positions to place labels (i.e., close to the source or target node), and by ranking the label positions according to her/his specifications, the algorithm will produce a label assignment that respects the user's preferred positions.

After extensive experimentation, we noticed that this algorithm performs very well not only for hierarchical drawings, but also for other straight-line drawings, such as drawings produced by force-directed and circular techniques. One weakness of the algorithm is that it ignores horizontal edges or edge segments. Thus, it is not suitable for orthogonal drawings. One can divide an orthogonal drawing into horizontal and vertical strips in order to find a suitable set of label positions and then use the rest of the algorithm to assign labels to edges.

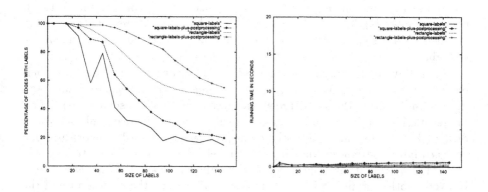

Fig. 3. Experimental results for graphs of 30 nodes and on the average 36 edges.

5 Experimental results.

We have implemented the ELP algorithm on top of the Graph Layout Toolkit (GLT [2]) in C++. We ran the experiments on a SPARC 5 unix based machine. We used a subset of graphs used in the experimental analysis of [5]. Actually, we used all graphs of 30, 60, 80, 90, and 99 nodes in [5].

The goal of our experiments was to give us an indication of the practical effectiveness of the algorithm presented here. It was not our goal to produce

---

[2] GLT is a product of Tom Sawyer Software (http://www.tomsawyer.com).

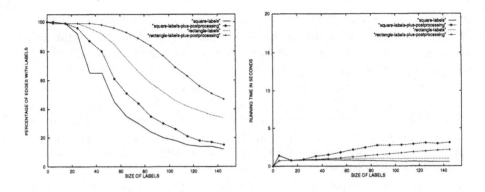

Fig. 4. Experimental results for graphs of 60 nodes and on the average 79 edges.

a comprehensive evaluation of the ELP algorithm, because it is very hard to quantify the complexity of a given drawing with respect to a label assignment. In our case, we drew the graphs using the hierarchical style algorithm implemented in GLT. We drew all graphs using the same tailoring options (i.e., minimum allowed separation of nodes in the same layer, size of nodes and distance between consecutive layers). We were more interested in how the success rate (percentage of edges with assigned labels) was decreasing as we were increasing the size of the labels.

After drawing the graphs, we ran the ELP algorithm with different size labels on the same drawings. We ran 2 sets of experiments; (i) Labels are square, and in successive runs of the algorithm we increase the size of the square; (ii) Labels are rectangles, and in successive runs of the algorithm we increase only the width of the label. In the final label assignment produced by the ELP algorithm each assigned label does not overlap any other label, edge, or node (even though it is allowed to overlap its associated edge). If we allow labels in the final assignment to overlap other graphical features in the drawing, then the success rate of the ELP algorithm is much greater. However, there will be a significant decrease in the quality of the label assignment.

In Figures 3 and 4 we present some results of our experiments. The plain lines represent the results after running only the main algorithm. The marked lines represent the results after running the algorithm with the postprocessing step. In each figure the left plot shows the success rate of the algorithm (the size of the labels is represented in the x-direction, and the percentage of edges with assigned labels is represented in the y-direction). The right plot shows the running time of the same set of experiments that are represented in the left plot.

All drawings are grid drawings, that is nodes and bends have integer coordinates. The label size is with relation to the grid-size. For example, a label with width 20 is a label with width equal to 20 grid units.

A typical example of a hierarchical drawing with labels is shown in $Fig. 5(a)$, and a straight-line drawing (circular style) with labels is shown in $Fig. 5(b)$. In those figures the dark rectangles are the labels.

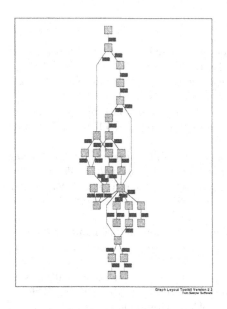 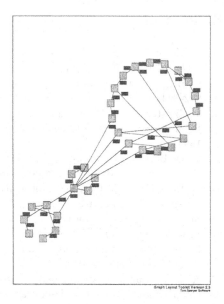

Fig. 5. Sample drawings with edge labels.

6 Conclusions

We have presented an algorithm for the ELP problem. Our experiments have shown that the approach to solve the ELP problem described in this paper is very successful, especially for hierarchical drawings. As we mentioned in the introduction, the ELP and NLP problems have been first proposed in the framework of automated cartography, where fixed geometry is one element that cannot be compromised. When labeling graphs, where the underlying geometry of the drawing is a result of a layout algorithm used to draw the graph, the labeling problem is and can be more flexible. It is an interesting problem to investigate how one can change the drawing of a graph to free up space that can be used to assign labels without compromising the quality of the drawing.

7 Acknowledgements.

We would like to thank Tom Sawyer Software, and especially its President and CEO Brendan Madden, for helpful discussions, and for making their software available to implement our labeling algorithms.

References

1. J. Ahn and H. Freeman. A program for automatic name placement. *Cartographica*, 21(5 & 3):101–109, Summer & Autumn 1984.
2. C. Berge. Two theorems in graph theory. In *Proc. Natl. Acad. Sci.*, 43, pages 842–844, 1957.
3. J. Christensen, J. Marks, and S. Shieber. An empirical study of algorithms for Point Feature Label Placement. *ACM Trans. on Graphics*, 14(3):203–232, July 1995.
4. G. Di Battista, P. Eades, R. Tamassia, and I. G. Tollis. Algorithms for drawing graphs: an annotated bibliography. *Comput. Geom. Theory Appl.*, 4:235–282, 1994.
5. G. Di Battista, A. Garg, G. Liotta, R. Tamassia, E. Tassinari, and F. Vargiu. An experimental comparison of four graph drawing algorithms. *Comput. Geom. Theory Appl.*, 1996. to appear.
6. L. R. Ebinger and A. M. Goulete. Noninteractive automated names placement for the 1990 decennial census. *Cartography and Geographic Informaton Systems*, 17(1):69–78, January 1990.
7. ACM Computational Geometry Impact Task Force. Application challenges to computational geometry. Technical Report TR-521-96, Princeton Univ., 1996.
8. M. Formann and F. Wagner. A packing problem with applications to lettering of maps. In *Proc. 7th Annu. ACM Sympos. Comput. Geom.*, pages 281–288, 1991.
9. A. V. Goldberg and R. Kennedy. An Efficient Cost Scaling Algorithm for the Assignment Problem. *Mathematical Programming*, 71:153–178, December 1995.
10. S. A. Hirsch. An algorithm for automatic name placement around point data. *The American Cartographer*, 9(1):5–17, 1982.
11. E. Imhof. Positioning names on maps. *The American Cartographer*, 2(2):128–144, 1975.
12. K. G. Kakoulis and I. G. Tollis. On the Edge Label Placement Problem. In S. North, editor, *Graph Drawing (Proc. GD '96)*, volume 1190 of *Lecture Notes in Computer Science*, pages 241–256. Springer-Verlag, 1997.
13. T. Kato and H. Imai. The NP-completeness of the character placement problem of 2 or 3 degrees of freedom. In *Record of Joint Conference of Electrical and Electronic Engineers in Kyushu*, pages 11–18, 1988. In Japanese.
14. J. Marks. Personal communication, 1996.
15. J. Marks and S. Shieber. The computational complexity of cartographic label placement. Technical Report 05-91, Harvard University, 1991.
16. R. E. Tarjan. *Data Structures and Network Algorithms*, volume 44 of *CBMS-NSF Regional Conference Series in Applied Mathematics*. Society for Industrial and Applied Mathematics, 1983.
17. I. G. Tollis. Graph Drawing and Information Visualization. *ACM Computing Surveys*, 28A(4), 1996.
18. F. Wagner and A. Wolff. Map labeling heuristics: Provably good and practically useful. In *Proc. 11th Annu. ACM Sympos. Comput. Geom.*, pages 109–118, 1995.
19. P. Yoeli. The logic of automated map lettering. *The Cartographic Journal*, 9(2):99–108, 12 1972.
20. S. Zoraster. The solution of large 0-1 integer programming problems encountered in automated cartography. *Operation Research*, 38(5):752–759, September-October 1990.

Elastic Labels: The Two-Axis Case

Claudia Iturriaga*
Anna Lubiw **

Dept. of Computer Science, University of Waterloo

Abstract. One of the most challenging tasks of cartographic map lettering is the optimal placement of region information on a map. We propose
as an approach to this task the *elastic labeling problem*, in which we are
given a set of *elastic rectangles* as labels, each associated with a point
in the plane. An elastic rectangle has a specified area but its width and
height may vary. The problem then is to choose the height and width of
each label, and the corner of the label to place at the associated point,
so that no two labels overlap.

This problem is known to be NP-hard even when there is no elasticity
(just because of the choice of the corners). We show that the problem
remains NP-hard when we have elasticity but no choice about which
corner of the label to use—we call this the *one-corner elastic labeling
problem*. We give a polynomial time algorithm for the special case of the
one-corner elastic labeling problem in which the points lie on the positive
x and y axes and the labels lie in the first quadrant. We call this the
two-axis labeling problem.

1 Introduction

An essential part of communication in a map or a diagram is the written information included on it ([DFr92]). In the case of cartography, work has been done
to determine a good set of rules for how a region should be labeled to agree with
its geographical characteristics ([Imh75]). In this paper we focus on the actual
placement of information on a map.

A special case is when we must label points. A problem arises when the map
contains thousands of points that are too close together. In this case, it is difficult
to place the labels so that they do not overlap and are easy to read without mag-
nification. Algorithms have been developed to automate the process of placing
labels on a map ([Yoe72], [CMS95], [KRa92], [FWa91]). One formulation of the
label placement problem is the *point-feature-label placement problem* of [FWa91],
[KIm88], [KRa92], and [MSh91], where we are given a set of points in the plane
and with each point an associated axis-parallel rectangular label. The problem
is to place the labels so that they do not overlap, and so that each label has one
of its corners at its associated point. This problem is known to be NP-complete
[FWa91], [KIm88], [KRa92], and [MSh91] and we can find heuristics in [CMS95],

* email: cciturri@barrow.uwaterloo.ca
** email: alubiw@daisy.uwaterloo.ca. Research partially supported by NSERC

[FWa91] and [WWo97]. For the special case where only two corners of each label can be placed at the point, there is an $O(n \log^2 n)$ algorithm [FWa91].

In this paper we introduce the *elastic labeling problem* which generalizes the above point-feature-label placement problem in that, in addition to choosing the corner of the label to place at the point, we are free to choose the height and width as long as we achieve the specified area.

This problem is useful when the goal of placing a label at a given point is to associate some text (more than one word) with the point. In this case we are able to write the specified text inside the label by using one, two, or more rows, as long as the label is placed at the specified point.

It is clear that the elastic labeling problem is NP-hard since it generalizes the point-feature-label placement problem. We show that even the *one-corner elastic labeling problem*—in which we fix the corner of each label that must be placed at the point—is NP-hard.

Our main result is a polynomial time algorithm for the *two-axis labeling problem*. This is the special case of the one-corner elastic labeling problem in which the points lie on the positive x and y axes, and the labels lie in the first quadrant. The two-axis labeling problem is the main step in solving the more general and more natural problem when the labels lie on four sides of an enclosing rectangle. This "rectangle perimeter labeling problem" arises when the perimeter of a map is labeled with information about things that lie beyond the boundary of the map, e.g. where the roads lead to, etc. This problem is likely to be relevant in GIS as maps are displayed dynamically on a computer screen using clipping, panning, and zooming.

The paper is structured as follows. In section 2, we define elastic rectangles, the elastic labeling problem and the two-axis labeling problem. Section 3 contains an algorithm for the two-axis labeling problem. In section 4 we present the NP-hardness proof for the one-corner elastic labeling problem.

2 Definitions

2.1 Elastic Rectangles

We define an orthogonal rectangle R by a quadruple (x, y, x', y') where the pair (x, y) sets the coordinates of the bottom left corner and the pair (x', y') sets the coordinates of the upper right corner. For a rectangle R we define $left(R) = x$, $bottom(R) = y$, $right(R) = x'$, $top(R) = y'$, $height(R) = y' - y$, and $width(R) = x' - x$. We consider rectangles to be topologically open.

Definition 1. An *elastic rectangle* $\mathcal{E}$ is a family of rectangles specified by a quintuplet (p, α, H, W, Q) where p is a point that is a corner of any rectangle in $\mathcal{E}$, α is the area of any rectangle in $\mathcal{E}$, H is the range of the height of the rectangles $[h^{\min}, h^{\max}]$, W is the range of the width $[w^{\min}, w^{\max}]$, and $Q \subseteq \{1, 2, 3, 4\}$ is a set of possible positions of p allowed in the family. The value of the position is 1 when p is a bottom left corner, 2 when p is a top left corner, 3 when p is a top right corner, and 4 when p is a bottom right corner.

We use the notation $p(\mathcal{E})$, $\alpha(\mathcal{E})$, $H(\mathcal{E})$, $W(\mathcal{E})$, and $Q(\mathcal{E})$ for the parameters of an elastic rectangle $\mathcal{E}$. The point $p(\mathcal{E})$ will be called the *anchor* of $\mathcal{E}$.

When Q is a singleton, the family of rectangles $\mathcal{E}$ is described by a hyperbola segment tracing out the locus of the corner of the elastic rectangle opposite p. Figure 1 shows an elastic rectangle with $Q(\mathcal{E}) = \{1\}$, with the hyperbola as a dashed curve.

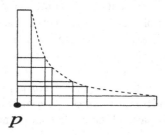

p

Fig. 1. An elastic rectangle $\mathcal{E}$.

An elastic rectangle $\mathcal{E}$ is *x-base* if $Q(\mathcal{E}) = \{1\}$ or $Q(\mathcal{E}) = \{4\}$ and the bottom side of its rectangles lies on the positive x-axis. Similarly $\mathcal{E}$ is *y-base* if $Q(\mathcal{E}) = \{1\}$ or $Q(\mathcal{E}) = \{2\}$ and the left side of its rectangles lies on the positive y-axis.

2.2 Elastic Labeling Problems

In this subsection we define the elastic labeling problem and the special case of the one-corner labeling problem.

A *realization* of an elastic rectangle $\mathcal{E}$, denoted E, is a single rectangle from the family—i.e. we must choose a valid height, width, and corner to place at p. A realization of a set of elastic rectangles is called a *good realization* if the rectangles do not intersect pairwise (Figure 2).

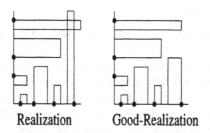

Realization   Good-Realization

Fig. 2. Realizations.

The *elastic labeling problem* is: Given a set of elastic rectangles, find a good realization.

The special case when $|Q| = 1$ for each point is called the *one-corner elastic labeling problem*. This problem is shown to be NP-hard in section 4. In the following subsection we describe a subproblem that we are able to solve in polynomial time.

2.3 The Two-axis Labeling Problem

This section contains the definition and preliminary discussion of the *two-axis labeling problem*: given a set of x-base elastic rectangles, and a set of y-base elastic rectangles, find a good realization for the union of the two sets. See Figure 2.

Let $B = \{\mathcal{B}_1 \ldots \mathcal{B}_n\}$ be a set of x-base elastic rectangles with $p_i = p(\mathcal{B}_i)$ and $x(p_i) \leq x(p_{i+1})$ for $i = 1, \ldots, n - 1$. Let $L = \{\mathcal{L}_1 \ldots \mathcal{L}_m\}$ be a set of y-base elastic rectangles with $q_i = p(\mathcal{L}_i)$ and $y(q_i) \leq y(q_{i+1})$ for $i = 1, \ldots, m - 1$.

We need one more definition: the *height* [*width*] of a realization is the height [width] of the smallest axis-parallel rectangle containing all the rectangles of the realization.

Our algorithm for the two-axis labeling problem will add points one by one. To gain some intuition about where the difficulty of the problem lies, observe that when two consecutive points along the x-axis have their rectangles facing each other (as in Figure 3), we can trade-off the height of one against the height of the other, and which solution is best depends on the particular y-base rectangles we have.

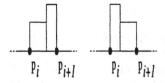

Fig. 3. Elastic rectangles facing each other.

3 Algorithm for the Two-axis Labeling Problem

Our algorithm for the two-axis labeling problem is a dynamic programming algorithm in which we will add points on by one along the x and y axes. We will begin in section 3.1 by describing the global structure of a solution. This allows us to identify the subproblems to be solved in the dynamic programming algorithm. Section 3.2 contains the complete algorithm.

3.1 Structure of a Good Realization

Our dynamic programming algorithm hinges on the following.

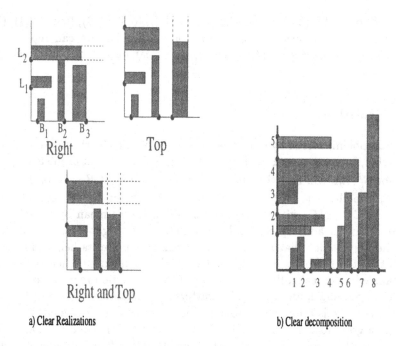

a) Clear Realizations   b) Clear decomposition

Fig. 4. Good clear realizations.

Observation 1. In a good realization of $B_1, \ldots, B_n$ and $\mathcal{L}_1, \ldots, \mathcal{L}_m$, using rectangles $B_1, \ldots B_n$ and $L_1, \ldots, L_m$, respectively, either B_n is *top-clear*—i.e. all the other rectangles lie to the left of the x-coordinate $left(B_n)$—or L_m is *right-clear*—i.e. all the other rectangles lie below the y-coordinate $bottom(L_m)$. See Figure 4.a.

We will call a realization *top-clear* if its last x-base rectangle is top-clear; we will call a realization *right-clear* if its last y-base rectangle is right-clear. Note that a realization can be both top-clear and right-clear.

The above observation tells us that we can always "peel away" the rectangles of a good realization, always removing a top-clear or right-clear one. Such an ordering (actually, its reverse) will drive our dynamic algorithm. More formally, let $\mathcal{S}_{i,j}$ be the subproblem of finding a good realization for $B_1, \ldots, B_i, \mathcal{L}_1, \ldots, \mathcal{L}_j$. Let $\mathcal{S}_{0,0}$ be the null subproblem.

Theorem 2. *Suppose* $B_1, \ldots, B_n, L_1, \ldots, L_m$ *is a good realization of an instance of the two-axis labeling problem, and let* $S_{i,j}$ *be the good realization induced for the subproblem* $\mathcal{S}_{i,j}$. *Then there is a sequence* $(0,0), \ldots, (n,m)$ *such that, if* (i,j) *appears in the sequence then the next element of the sequence is either* $(i+1, j)$ *and the good realization* $S_{i+1,j}$ *is top-clear, or* $(i, j+1)$ *and the good realization* $S_{i,j+1}$ *is right-clear.*

The proof of this theorem follows from the above observation. We call such a sequence a *clear decomposition*. For example, in Figure 4.b a clear decomposition is (0,0), (1,0), (2,0), (2,1), (3,1), (3,2), (4,2) (4,3), (5,3), (6,3), (6,4), (7,4), (7,5), (8,5). Note that the clear decomposition is not unique, we can replace (6,4) by (7,3) in this sequence and obtain another clear decomposition of the same realization.

3.2 Algorithm

The subproblems solved by our dynamic programming algorithm are as follows. For each $i, j, 1 \leq i \leq n, 1 \leq j \leq m$ we will solve two subproblems $S_{i,j}^T$ and $S_{i,j}^R$. For $S_{i,j}^T$ we will find the "best" good realizations of $\mathcal{B}_1, \ldots, \mathcal{B}_i, \mathcal{L}_1, \ldots, \mathcal{L}_j$ with $\mathcal{B}_i$ top-clear. For $S_{i,j}^R$ we will find all the "best" good realizations for $\mathcal{B}_1, \ldots, \mathcal{B}_i, \mathcal{L}_1, \ldots, \mathcal{L}_j$ with $\mathcal{L}_j$ right-clear. What do we mean by "best"? Since we will only ever add top-clear x-base rectangles and right-clear y-base rectangles, the only things that matter about a realization are its height and width. Recall that the height [width] of a realization is the maximum height [width] of a rectangle in the realization. The height of the realization matters for adding a right-clear rectangle, and the width matters for adding a top-clear rectangle.

In case there is a unique minimum height and width for a good realization, we want a good realization of that height and width. The good realization itself will in general not be unique, but that does not matter. The left pane of Figure 5 shows a case where there is a unique minimum height and width top-clear good realization.

Often there will be no unique minimum height and width—height and width may be traded off as they are in a single elastic rectangle. For example in the right pane of Figure 5, if we take the highest rectangle in $\mathcal{B}_2$ we have a top-clear good realization with minimum width, but if we take the widest rectangle then we have a top-clear good realization with minimum height. In this case we will preserve all possible height-width combinations. As we shall see, we can fix the realizations of all rectangles except the last one (the clear one), which must remain elastic.

To be more specific, in solving $S_{i+1,j}^T$ we are adding $\mathcal{B}_{i+1}$ as a top-clear rectangle. If $\mathcal{B}_{i+1}$'s anchor is to its right, we will find one good realization of $\mathcal{B}_1, \ldots, \mathcal{B}_{i+1}, \mathcal{L}_1, \ldots, \mathcal{L}_j$ of minimum height and width, and with $\mathcal{B}_{i+1}$ top-clear. To justify this, note that since the width of any good realization is fixed (namely at $p(\mathcal{B}_{i+1})$), we can simply minimize height, and fix one good realization. On the other hand, if $\mathcal{B}_{i+1}$'s anchor is to its left then we will find one good realization of $\mathcal{B}_1, \ldots, \mathcal{B}_i, \mathcal{L}_1, \ldots, \mathcal{L}_j$ of width no more than $p(\mathcal{B}_{i+1})$ and then of minimum height, together with a restricted range for the elastic rectangle $\mathcal{B}_{i+1}$. It makes sense to leave $\mathcal{B}_{i+1}$ elastic, since, as noted before, the realization we choose for $\mathcal{B}_{i+1}$ will affect the realizations we can choose for the next x-base or y-base rectangle we add. And on the other hand, it makes sense to fix realizations of all prior rectangles, since we now have a bound on the width, and can simply minimize height.

It is easy to make these justifications more formal, and prove by induction as the algorithm progresses:

Lemma 3. *The solution to subproblem $S^T_{i+1,j}$ captures all minimal height-width combinations for good realizations of $\mathcal{B}_1, \ldots, \mathcal{B}_{i+1}, \mathcal{L}_1, \ldots, \mathcal{L}_j$ with $\mathcal{B}_{i+1}$ top-clear. (By minimal, we mean minimal in the ordering $(h, w) \leq (h', w')$ if $h \leq h'$ and $w \leq w'$.)*

Algorithm. Assume we have solved subproblems $S^T_{i,j}$ and $S^R_{i,j}$. This is how we solve subproblem $S^T_{i+1,j}$. (We use a symmetric solution for $S^R_{i,j+1}$.)

We need to add $\mathcal{B}_{i+1}$, top-clear. We will divide into two cases: either $p(\mathcal{B}_{i+1})$ is on the left of the elastic rectangle or on its right. In the first case the new rectangle will remain elastic in our solution; in the second case, because the width of the solution is fixed, we can simply minimize the height, and will fix $\mathcal{B}_{i+1}$. See Figure 5.

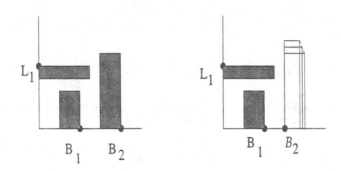

Fig. 5. Adding elastic rectangle $\mathcal{B}_2$ to a right-clear realization.

Case I. $p = p(\mathcal{B}_{i+1})$ is to the left. We may build on $S^T_{i,j}$ or on $S^R_{i,j}$. We will compare to see which is best. We need a solution to $S_{i,j}$ with width no more than $x(p)$. From the stored solutions to each of $S^T_{i,j}$ and $S^R_{i,j}$, take a good realization with width less than or equal to $x(p)$ and with minimum height. Of these two solutions, keep the one of minimum height h. We now add $\mathcal{B}_{i+1}$, keeping it as an elastic rectangle, though with a height and width range that may be restricted. Since the height of any good realization must be at least h, we want the minimum width solution of height h, and all the possible trade-offs for solutions of height greater than h. Thus, if the maximum height of $\mathcal{B}_{i+1}$ is less than h, we fix $\mathcal{B}_{i+1}$ to the maximum possible height, and otherwise, we restrict the height range of $\mathcal{B}_{i+1}$ from h on up and take the corresponding width range.

Case II. $p(\mathcal{B}_{i+1})$ is to the right. The width of our solution is then fixed (namely at $p(\mathcal{B}_{i+1})$) and we are looking for the solution of minimum height. We will fix $\mathcal{B}_{i+1}$. We may build on $S^T_{i,j}$ or on $S^R_{i,j}$. We will compare to see which provides us with the best—i.e. minimum height—solution.

Case 1. Building on $S^T_{i,j}$. If we have a unique realization stored for this subproblem, then we simply add the minimum height realization for $\mathcal{B}_{i+1}$ that fits, i.e. whose left side is greater than or equal to the width of the subproblem's realization (Figure 6.a). Otherwise, $\mathcal{B}_i$ is still elastic in the solution to the sub-problem. In this case we must find the minimum height good realization for two x-base elastic rectangles $\mathcal{B}_i$ and $\mathcal{B}_{i+1}$ with their anchors on the left and right respectively. See Figure 6.b. This involves intersecting two hyperbolas. The details of the computation can be found in the following section.

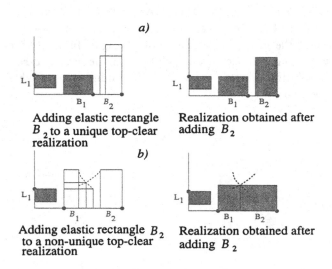

a)

Adding elastic rectangle B_2 to a unique top-clear realization

Realization obtained after adding B_2

b)

Adding elastic rectangle B_2 to a non-unique top-clear realization

Realization obtained after adding B_2

Fig. 6. Case I.1.

Case 2. Building on $\mathcal{S}_{i,j}^R$. If we have a unique realization stored for this sub-problem, then we simply add the minimum height realization that fits, i.e. whose left side is greater than or equal to the width of the subproblem's realization (Figure 7.a). Otherwise, $\mathcal{L}_j$ is still elastic in the solution to the subproblem. In this case we must find the minimum height good realization for two elastic rectangles $\mathcal{L}_j$ and $\mathcal{B}_{i+1}$ with their anchors at the bottom left and bottom right respectively. See Figure 7.b. This is very similar to the problem we needed to solve in Case 1—the only difference is that the first elastic rectangle does not lie on the x-axis. The details of the computation can be found in the following section.

This completes the description of the the main step of the algorithm. We can start the algorithm off with the solutions to the subproblems $\mathcal{S}_{0,0}^T$ and $\mathcal{S}_{0,0}^R$ as rectangles of 0 width and height at the origin. When the algorithm has solved $\mathcal{S}_{n,m}^T$ and $\mathcal{S}_{n,m}^R$ we learn whether there is a solution to the whole problem.

It is easy to see that the algorithm takes $O(nm)$ time.

The correctness of the algorithm is fairly obvious based on two results: Theorem 2 which promises a clear decomposition, and justifies solving subproblem $\mathcal{S}_{i+1,j}^T$ by only looking at solutions to $\mathcal{S}_{i,j}^T$ and $\mathcal{S}_{i,j}^R$; and Lemma 3, which tells us about the form of the solution to subproblem $\mathcal{S}_{i+1,j}^T$.

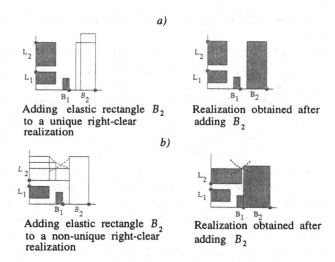

a)

Adding elastic rectangle B_2 to a unique right-clear realization

Realization obtained after adding B_2

b)

Adding elastic rectangle B_2 to a non-unique right-clear realization

Realization obtained after adding B_2

Fig. 7. Case I.2.

3.3 Hyperbolas

In this subsection we will explain the one missing detail of the algorithm: how to compute the minimum height good realization for two x-base elastic rectangles $\mathcal{B}_i$ and $\mathcal{B}_{i+1}$ with their anchors on the left and right respectively. This problem arises in case II, subcase 1 (Figure 6.b).

Consider the hyperbola segments $\Psi(\mathcal{B}_i)$ and $\Psi(\mathcal{B}_{i+1})$ described by the elastic rectangles. If the hyperbola segments intersect in a point c, then we take the rectangles B_i and B_{i+1} that have height equal to $y(c)$. (This is easy on a real-RAM model of computation.) The rectangles will touch each other. Otherwise the minimum height good realization—if it exists— is obtained at a "boundary value" using a rectangle from $\mathcal{B}_i$ or $\mathcal{B}_{i+1}$ (or both) of maximum or minimum possible height. We need to compare the 4 extreme solutions.

The slightly more general problem that arises in case II, subcase 2, where the right-hand elastic rectangle no longer lies on the x-axis, can be solved in exactly the same way.

4 NP-hardness of the One-corner Elastic Labeling Problem

The elastic labeling problem is NP-hard because the special case when we fix the height and width of each elastic rectangular label is precisely the point-feature-label placement problem which was shown to be NP-complete by [FWa91], [KIm88], [KRa92], [MSh91].

In this section we will prove the NP-hardness of the one-corner elastic labeling problem. Thus, either elasticity, or the choice of which corner of the label to use, results in an NP-hard problem. Our proof follows the idea of the proof given by

Formann and Wagner in [FWa91]. The 3-SAT problem is reduced to the labeling problem as follows.

For an instance of the 3-SAT problem we build a configuration of points with elastic rectangles as labels. For each variable we place a special point whose label is forced by the other points to have height in one of two subranges. The subrange that the height of this label takes indicates the setting of the variable. Let $[h^{\min}, h^{\max}]$ be the range of the height for such a label. We will choose an $h' \in [h^{\min}, h^{\max}]$ so that if the label is forced to have height in subrange $[h^{\min}, h']$ the variable is set to true and otherwise it is set to false.

We will construct *variable gadgets*, *clause gadgets* and *pipes* connecting the variable gadget for x to the clause gadgets for clauses containing x or $\overline{x}$. Each gadget and each pipe is an arrangement of points each with an associated elastic label. The pipes propagate the truth-value chosen for the variable. When the variable is set to true [false] its value will be propagated by the pipes to the bottom [top] pipe connected to the variable gadget. The pipes that propagate the true [false] value are connected to the negated [unnegated] variables in the clause gadget. The idea is that the pipes transmit "flow" to the clauses. For each clause we construct a gadget that can support flow from at most two pipes. Since flow in the pipes corresponds to values that do not satisfy the clause, thus a clause admits flow from at most two pipes. If a clause gadget receives flow from all three pipes then the corresponding clause is not satisfied. So, there is a solution for a given instance of 3-SAT iff the constructed instance of the one-corner labeling problem has a solution.

Figures 8, 9 and 10 show the gadgets needed. A *crossing gadget* is a set of points that allows two pipes to cross, with flow properly transmitted by each pipe. Details of the construction and the proof that it works can be found in [ILu97].

5 Discussion

We are currently working on heuristics for the elastic labeling problem [Itu]. A main motivation for the elastic labeling problem is when labels may consist of multiple lines of text. This is formulated most naturally as a *discretized* version of the problem; in particular, the height of each rectangle should be an integer. Both our NP-hardness result for the one-corner problem, and the good algorithm for the two-axis problem carry over to the discretized versions [Itu].

As mentioned in the introduction, the solution to the two-axis labeling problem is the main ingredient needed in solving the more general and useful rectangle perimeter labeling problem. A paper on this is in progress.

We mention one open problem: find a good algorithm for the generalization of the two-axis labeling problem where, in addition to elasticity, we may choose which corner of the label to place at the anchor point, so long as the labels stay in the first quadrant.

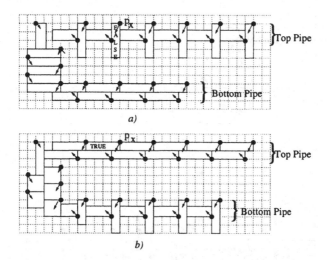

Fig. 8. a) Variable gadget with x set to true and flow transmitted through the top pipe. b) Variable gadget with x set to false and flow transmitted through the bottom pipe.

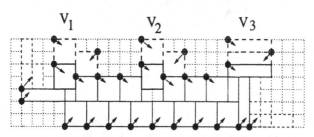

Fig. 9. Clause gadget with flow shown entering on pipes V_1 and V_2.

References

[CMS95] J. Christensen, J. Marks, and S. Shieber. An empirical study of algorithms for point feature label placement . *ACM Transactions on Graphics.* (1995) 14 (3), 203-232.

[DFr92] J.S. Doerschler and H. Freeman. A rule-based system for dense-map name placement. *Communications of the Association of Computing Machinery.* (1992) 35(1), 68-79.

[FWa91] M. Formann and F. Wagner. A packing problem with applications in lettering of maps. *Proceedings of the 7th ACM Symposium on Computational Geometry.* (1991) 281-288.

[Imh75] E. Imhof. Positioning names on maps. *The American Cartographer.* (1975) 2, 128-144.

[ILu97] C. Iturriaga and A. Lubiw. NP-hardness of some map labeling problems. *Technical Report CS-97-18.* University of Waterloo (1997).

[Itu] C. Iturriaga. Ph.D. Thesis, in progress. University of Waterloo.

[KIm88] T. Kato and H. Imai. The NP-completeness of the character placement prob-

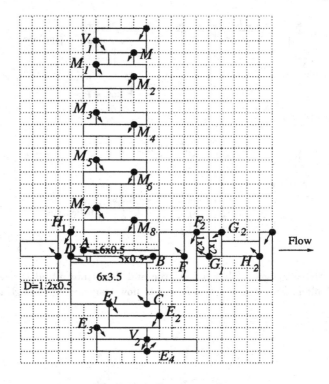

Fig. 10. A Crossing gadget.

lem of 2 or 3 degrees of freedom. *Record of Joint Conference of Electrical and Electronic engineers in Kyushu.* (1988), 1138. In Japanese.

[KRa92] D. Knuth and A. Raghunathan. The problem of compatible representatives. *SIAM Disc. Math.* (1992) 5 (3), 422-427.

[MSh91] J. Marks and S. Shieber. The computational complexity of cartographic label placement. *Technical Report CRCT-05-91.* Harvard University (1991).

[WWo97] F. Wagner and A. Wolff. A practical map labeling algorithm. *Computational Geometry: Theory and Applications.* (1997) 387-404.

[Yoe72] P. Yoeli. The logic of automated map lettering. *The Cartographic Journal.* (1972) 9, 99-108.

Pitfalls of Using PQ-Trees in Automatic Graph Drawing*

Michael Jünger[1], Sebastian Leipert[2] and Petra Mutzel[3]

[1] Institut für Informatik, Universität zu Köln, mjuenger@informatik.uni-koeln.de
[2] Institut für Informatik, Universität zu Köln, leipert@informatik.uni-koeln.de
[3] Max-Planck-Institut für Informatik, Saarbrücken, mutzel@mpi-sb.mpg.de

Abstract. A number of erroneous attempts involving PQ-trees in the context of automatic graph drawing algorithms have been presented in the literature in recent years. In order to prevent future research from constructing algorithms with similar errors we point out some of the major mistakes.

In particular, we examine erroneous usage of the PQ-tree data structure in algorithms for computing maximal planar subgraphs and an algorithm for testing leveled planarity of leveled directed acyclic graphs with several sources and sinks.

1 Introduction

A PQ-tree is a powerful data structure that represents the permutations of a finite set in which the members of specified subsets occur consecutively, and in which updates require linear time. This data structure has been introduced by Booth and Lueker (1976) to solve the problem of testing for the consecutive ones property. The most well known applications of PQ-trees in automatic graph drawing are planarity testing (see Lempel *et al.* (1967), Booth and Lueker (1976)) and embedding (see Chiba *et al.* (1985)). Therefore PQ-trees have become standard tools in automatic graph drawing systems.

Other attempts to use algorithms based on PQ-trees for automatic graph drawing problems have not been successful. One well known example is the computation of maximal planar subgraphs. Given a simple, connected graph $G = (V, E)$ with n vertices and m edges, a planar subgraph G' of G is a *maximal planar* subgraph, if for all edges $e \in G - G'$ the addition of e to G' destroys planarity. Several efforts have been made in the literature to solve the problem with PQ-trees following a certain strategy that was first presented by Ozawa and Takahashi (1981). They describe an O(nm) algorithm using PQ-tree techniques based on the vertex addition algorithm of Lempel *et al.* (1967). Jayakumar *et al.* (1986) show that in general this algorithm does not determine a maximal planar subgraph. Moreover, the resulting planar subgraph may not

---

* Partially supported by DFG-Grant Ju204/7-2, Forschungsschwerpunkt "Effiziente Algorithmen für diskrete Probleme und ihre Anwendungen" and ESPRIT Long Term Research Project Nr. 20244 (ALCOM-IT)

even contain all vertices. Jayakumar *et al.* (1989) present an algorithm called PLANARIZE that computes a spanning planar subgraph G_p of G in $O(n^2)$ time. Furthermore, they present an algorithm called MAX-PLANARIZE that augments G_p to a subgraph G' of G by adding additional edges in $O(n^2)$ time. They claim that G' is a maximal planar subgraph of G if G_p (the result of phase 1 of the two phase algorithm) turns out to be biconnected. Kant (1992) shows that this algorithm is incorrect, and suggests a modification of the second phase of the algorithm that augments G_p to a maximal planar subgraph of G, even if G_p is not biconnected, maintaining $O(n^2)$ time requirement. In Jünger *et al.* (1996) we show that this modification is not correct either. Here we point out a substantial flaw in both the original and the modified two phase algorithm that was not detected previously. This is the subject of section 2.

PQ-trees have also been proposed by Heath and Pemmaraju (1996a, 1996b) to test leveled planarity of leveled directed acyclic graphs with several sources and sinks. In section 3 we show why this application leads to an incorrect algorithm as well. Since this "algorithm" is the only attempt to prove the polynomial time complexity in the literature, the complexity status of leveled planarity testing is still open. Only in the special case in which there is only one source (or only one sink) the algorithm is correct and implies linear time solvability, but this has already been shown previously by Di Battista and Nardelli (1988).

2 Case study: maximal planar subgraphs

2.1 *PQ*-trees for planarity testing

Let $G = (V, E)$ be a simple graph with n vertices and m edges. A graph is *planar*, if it can be embedded in the plane without any edge crossings. A graph is obviously planar, if and only if its biconnected components are planar. We therefore assume that G is biconnected. The planarity testing algorithm of Lempel *et al.* (1967) first labels the vertices of G as $1, 2, \ldots, n$ using an st-numbering (see Even and Tarjan (1976)). A numbering of the vertices of G by $1, 2, \ldots, n$ is an *st-numbering*, if the vertices "1" and "n" are adjacent and each other vertex j is adjacent to two vertices i and k such that $i < j < k$. The vertex 1 is denoted by s and the vertex n is denoted by t. The st-numbering induces an orientation of the graph, in which every edge is directed from the incident vertex with the higher st-number towards the incident vertex with the lower st-number. From now on we refer to the vertices of G by their st-numbers and call an edge (u, v), with $v < u$, *incoming* edge of v and *outgoing* edge of u.

For $1 \leq k \leq n$, let G_k denote the subgraph of G induced by the vertex set $V_k := \{1, 2, \ldots, k\}$. The graph G'_k arises from G_k as follows: For each edge $e = (u, v)$, where $v \in V_k$ and $u \in V \setminus V_k$, we introduce a virtual vertex u_e with label u and a virtual edge (u_e, v). Let B_k be a planar embedding of G'_k such that all virtual vertices are placed on the outer face. Then, B_k is called a *bush form*. It has been shown by Lempel *et al.* (1967) that G is planar, if and only if for every B_k, $k = 1, 2, \ldots, n-1$, there exists a bush form B'_k isomorphic to B_k, such that all virtual vertices in B'_k labeled $k + 1$ appear consecutively.

The *PQ-tree* T_k corresponding to the bush form B_k is a rooted ordered tree that consists of three types of nodes:

1. Leaves in T_k represent virtual edges in B_k.
2. *P-nodes* in T_k represent cutvertices in B_k.
3. *Q-nodes* represent maximal biconnected components in B_k.

The *frontier* of a *PQ*-tree is the sequence of all leaves of T_k read from left to right. The frontier of a node X, denoted by frontier(X), is the sequence of its descendant leaves read from left to right.

Let E_{k+1} denote the set of leaves in T_k that correspond to the virtual vertices labeled $k + 1$. A node X is called *full*, if all leaves in its frontier are in E_{k+1}. A node X is *empty*, if its frontier does not contain any leaf of E_{k+1}. Otherwise, X is called *partial*. A node is called *pertinent*, if it is full or partial. The *pertinent subtree* is the smallest connected subtree that contains all leaves of E_{k+1} in its frontier. The root of the pertinent subtree is called *pertinent root*. Two *PQ*-trees are *equivalent*, if one can be obtained from the other by one or more of the following operations:

1. Permuting the children of a *P*-node.
2. Reversing the order of the children of a *Q*-node.

These operations are called *equivalence transformations* and describe *equivalence classes* on the set of all *PQ*-trees. An equivalence class of *PQ*-trees corresponds to a class of permutations called the *permissible permutations*.

It has been shown by Booth and Lueker (1976) that B'_k exists if and only if T_k can be converted into an equivalent *PQ*-tree T'_k such that all pertinent leaves appear consecutively in the frontier of T'_k. Booth and Lueker (1976) have defined a set of patterns and replacements called *templates* that can be used to reduce the *PQ*-tree such that the leaves corresponding to edges of the set E_{k+1} appear consecutively in all permissible permutations. To construct T_{k+1} from T_k they first reduce T_k by use of the templates and then replace all leaves corresponding to virtual edges incident to vertices labeled $k + 1$ by a *P*-node, whose children are the leaves corresponding to the incoming edges of the vertex $k + 1$ in G.

The planarity testing algorithm now starts with T_1 and constructs a sequence of *PQ*-trees $T_1, T_2, \ldots$. If the graph is planar, the algorithm terminates after constructing T_{n-1}. Otherwise it terminates after detecting the impossibility of reducing some T_k, $1 \leq k < n$.

2.2 Principle of an approach for planarization

The basic idea of a planarization algorithm using *PQ*-trees presented by Jayakumar *et al.* (1989) is to construct the sequence of *PQ*-trees $T_1, T_2, \ldots, T_{n-1}$ by deleting an appropriate number of pertinent leaves every time the reduction fails, such that the resulting *PQ*-tree becomes reducible. In every step of the algorithm PLANARIZE, a maximal consecutive sequence of pertinent leaves is computed by using a $[w, h, a]$-*numbering* (see Jayakumar *et al.* (1989)). All pertinent leaves that are not adjacent to the maximal pertinent sequence are removed from the PQ-tree in order to make it reducible. Hence the edges corresponding to the leaves are removed from G and the resulting graph G_p is planar.

It has been shown by Jayakumar *et al.* (1989) that the graph G_p computed by PLANARIZE is not necessarily maximal planar. The authors therefore suggest to apply a second phase MAX-PLANARIZE, also based on PQ-trees. Knowing which edges have been removed from G to construct G_p, edges from $G - G_p$ are added back to G_p in the second phase without destroying planarity.

During the reduction of a vertex v, there may exist nonpertinent leaves that are in all permissible permutations of the PQ-tree T_{v-1} between a pertinent leaf l_v and its maximal pertinent sequence. This maximal pertinent sequence has been determined with the help of the $[w, h, a]$-numbering. In order to make the tree T_{v-1} reducible, the leaf l_v is removed from the tree and the corresponding edge is removed from the graph G, guaranteeing that the subgraph G_p will be planar. However, it may occur that the nonpertinent leaves that are positioned between l_v and its maximal pertinent sequence in T_{v-1}, are removed as well from a tree T_k, $v \leq k < n$, in order to obtain reducibility. Therefore, there is no need to remove the edge corresponding to l_v from the graph G.

In order to find leaves such as l_v, Jayakumar *et al.* (1989) use the algorithm MAX-PLANARIZE. In step i, both PLANARIZE as well as MAX-PLANARIZE reduce the same vertex i. The difference between the PQ-trees in the two algorithms is, according to the authors, that all leaves that have been deleted in PLANARIZE are ignored in MAX-PLANARIZE from the moment they are introduced into the tree until they get pertinent. This causes the nonpertinent leaves between the pertinent leaf l_v and its maximal pertinent sequence to be ignored. Hence l_v is adjacent to its maximal pertinent sequence and the corresponding edge can be added back to G_p, while the leaves between l_v and the maximal pertinent sequence are removed from the PQ-tree.

2.3 On the incorrectness of the algorithm

While some incorrect details of the approach of Jayakumar *et al.* have been described in a technical report by Kant (1992), who attempted to correct the algorithm, a major problem has not been detected.

Jayakumar *et al.* assume that the maximal planar subgraph G_p is biconnected for the correct application of the Lempel-Even-Cederbaum algorithm. Furthermore, as they have stated correctly, this is necessary in order to have an st-numbering. Nevertheless, the PQ-trees in MAX-PLANARIZE are constructed according to the st-numbering that was computed for the graph G.

As a matter of fact, the st-numbering of G does not imply an st-numbering of any subgraph G_p even if the subgraph G_p is biconnected. This results in two problems, of which one is crucial and cannot be dealt with even by the ideas described by Kant (1992).

Both problems are based on the fact that during the application of PLA-NARIZE for some vertices of V all incoming edges may be deleted from the graph while the resulting graph G_p stays biconnected. In this abstract, we consider only the crucial problem. The other problem is described in detail by Jünger et al. (1996).

The planarization algorithm of Jayakumar et al. (1989) does not obey an important invariant implied by the following lemma, shown by Even (1979).

Lemma 2.1 Let $G = (V, E)$ be a planar graph with an st-numbering and let $1 \leq k \leq n$. If the edge (t, s) is drawn on the boundary of the outer face in an embedding of G, then all vertices and edges of $G - G_k$ are drawn in the outer face of the plane subgraph G_k of G.

This result allowed Lempel et al. (1967) to transform the problem of planarity testing to the construction of a sequence of bush forms B_k, $1 \leq k \leq n$. For a planar graph G, edges and vertices that have not been introduced into the current subgraph G_k are always embedded into the outer face of G_k.

The approach of Jayakumar et al. (1989) does not obey this invariant in the second phase. There exist edges that have to be embedded into an inner face of some G_k, even if (t, s) is drawn on the outer face. Due to the above lemma, the correction step MAX-PLANARIZE only considers edges for reintroduction into the planar subgraph G_p that are on the outer face of the current graph G_k. Since the numbering that is used to determine the order in which the vertices are reduced does not correspond to an st-numbering of G_p in general, the algorithm of Jayakumar et al. (1989) ignores edges that have to be added into an inner face of the embedding of a current graph G_k. This fact is fatal, as we are about to show now.

In Fig. 1, a part of a bush form B_{k-1}, $1 < k \leq n$ of a graph G is shown. The virtual vertices corresponding to the vertex k are labeled $k_1, k_2, \ldots, k_5$ and all other virtual vertices are left unlabeled. The corresponding part of the PQ-tree is shown in Fig. 2. Obviously, there do not exist any reversions or permutations such that the virtual vertices of k occupy consecutive positions. Hence, the graph G is not planar. Applying the $[w, h, a]$-numbering of Jayakumar et al. (1989) allows us to delete the virtual vertex k_5 and to reduce the other four vertices k_1, k_2, k_3, k_4. The resulting bush form B_k is planar and the relevant part is shown in Fig. 3. Figure 4 shows the corresponding part of the PQ-tree. Assume now that all descendants of k have to be removed from the PQ-tree in a later step. Hence all incoming edges incident on k are removed from the tree. Now assume further that there exists a path $v_1, v_2, \ldots, v_l$ in G_p such that

- for all i, j, $1 \leq i < j \leq l$ the inequality $v_i < v_j$ holds,
- the edge (v_2, v_1) corresponds to one of the virtual edges that are between the leaf k_5 and the maximal pertinent sequence k_1, k_2, k_3, k_4 in all PQ-trees equivalent to T_{k-1},
- $v_l = t$.

This path guarantees that all outgoing edges of the vertex k cannot be embedded into the outer face of the embedding of B_{k-1} without crossing an edge on this path. Hence the edge e_{k_5} corresponding to the leaf k_5 is not considered

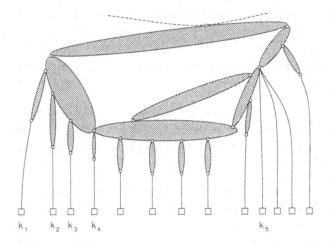

Fig. 1. Part of a bush form B_{k-1}

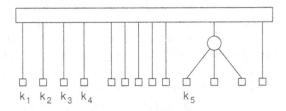

Fig. 2. Part of a PQ-tree corresponding to bush form B_{k-1}

by the algorithm MAX-PLANARIZE as being an edge that does not destroy planarity. Therefore, e_{k_5} is not added back to the planar subgraph G_p.

Nevertheless adding the edge e_{k_5} to G_p may not destroy planarity of G_p as is shown in our example in Fig. 5. Since all incoming edges of the vertex k have been deleted by PLANARIZE and are not added back by MAX-PLANARIZE, it may be possible to swap the vertex k into an inner face of the embedding of B_k such that the virtual vertex k_5 can be identified with k and the edge e_{k_5} is embedded into the bush form B_k without destroying planarity.

Therefore, the strategy of using PQ-trees presented by Jayakumar *et al.* (1989) does not compute a maximal planar subgraph in general. Furthermore, we point out that the same problem holds for the modified version of this algorithm, presented by Kant (1992). This version follows a similar strategy of computing a spanning planar subgraph G_p using PLANARIZE and then adding edges that do not destroy planarity in a second phase. The order of reductions that is used to insert vertices into existing bush forms is the same as the one implied by the st-numbering on G. Hence this approach is not able to compute a maximal planar subgraph for the same reason.

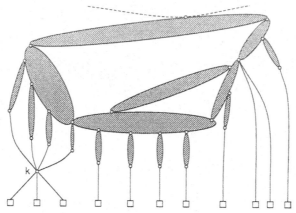

Fig. 3. Part of a bush form B_k

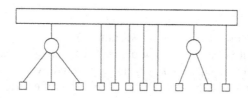

Fig. 4. Part of a PQ-tree corresponding to bush form B_k

Summarizing, we state the following lemma that has been shown in the discussion above.

Lemma 2.2 *Let $G = (V, E)$ be a nonplanar graph. Let $G_p = (V, E_p)$, $E_p \subseteq E$, be a planar subgraph of G, such that G_p was obtained from G by*

1. *computing an st-numbering for all vertices and*
2. *applying the algorithm of Lempel et al. (1967) constructing a sequence of bush forms B_k, $1 \leq k \leq n$, by embedding a maximal number of outgoing edges of a vertex k, $1 < k \leq n$, in the outer face of B_{k-1} without crossings, deleting all other outgoing edges of k.*

Let $G'_p = (V, E'_p)$, be a planar subgraph of G such that

1. *$E_p \subseteq E'_p \subseteq E$,*
2. *the graph G'_p is computed by constructing a sequence of bush forms B'_k, $1 \leq k \leq n$, based on the st-numbering used for determining G_p, and possibly embedding outgoing edges $e \in E \setminus E_p$ of every vertex k, $1 < k \leq n$, without crossings in the outer face of B_{k-1}.*

Then the subgraph G'_p is not necessarily maximal planar.

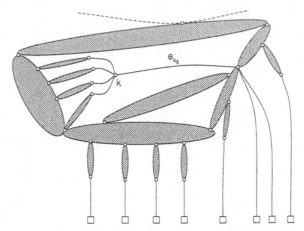

Fig. 5. Part of a bush form B_k with e_{k_5} embedded

Considering a computation of an *st*-numbering for the planar subgraph G_p in order to augment G_p to a maximal planar subgraph of G and then construct a sequence of bush forms B'_k, $1 \le k \le n$, is aggravated by the fact that the graph G_p is not biconnected in general. Furthermore, the difference between the bush forms of the first phase and the second phase may result in the deletion of the edges of G_p as soon as edges of $E \setminus E_p$ are added to G_p.

3 Case study: leveled planarity testing

3.1 Principle of an approach for recognizing leveled planar dags

Let $G = (V, E)$ be a directed acyclic graph. A leveling of G is a function *lev* : $V \to \mathbb{Z}$ mapping the nodes of G to integers such that $lev(v) = lev(u) + 1$ for all $(u, v) \in E$. G is called a *leveled dag* if it has a leveling. If $lev(v) = j$, then v is a *level-j vertex*. Let $V_j = lev^{-1}(j)$ denote the set of level-j vertices. Each V_j is a *level* of G.

For the rest of this section, we consider G to be a leveled dag with $m \in \mathbb{N}$ levels. An embedding of G in the plane is called *leveled* if the vertices of every V_j, $1 \le j \le m$, are placed on a horizontal line $l_j = \{(x, m - j) \mid x \in \mathbb{R}\}$, and every edge $(u, v) \in E$, $u \in V_j$, $v \in V_{j+1}$ is drawn as straight line segment between the lines l_j and l_{j+1}. A leveled embedding of G is called *leveled planar* if no two edges cross except at common endpoints. A leveled dag is leveled planar, if it has a leveled planar embedding. The dag G is obviously leveled planar, if all its components are leveled planar. We therefore assume that G is connected.

A leveled embedding of G determines for every V_j, $1 \le j \le m$, a total order $\le_j$ of the vertices of V_j, given by the left to right order of the nodes on l_j. In order to test whether a leveled embedding of G is leveled planar, it is sufficient to find an order of the vertices of every set V_j, $1 \leq j < m$, such that for every pair of edges $(u_1, v_1), (u_2, v_2) \in E$ with $lev(u_1) = lev(u_2) = j$ and $u_1 \leq_j u_2$ it follows that $v_1 \leq_{j+1} v_2$. Apparently, the ordering $\leq_j$, $1 \leq j \leq m$, describes a permutation of the vertices of V_j. Let G_j denote the subgraph of G, induced by $V_1 \cup V_2 \cup \ldots \cup V_j$. Unlike G, G_j is not necessarily connected.

The basic idea of the leveled planarity testing algorithm presented by Heath and Pemmaraju Heath and Pemmaraju (1996a, 1996b) is to perform a top-down sweep processing the levels in the order $V_1, V_2, \ldots, V_m$ computing for every level V_j, $1 \leq j \leq m$, a set of permutations of the vertices of V_j that appear in some leveled planar embedding of G_j. In case that the set of permutations for G_m is not empty, the graph $G = G_m$ is obviously leveled planar.

As long as the graph G_j is connected for some $j \in \{1, 2, 3, \ldots, m\}$ standard PQ-tree techniques similar to the ones used in the planarity test can be applied in order to determine the required set of permutations (see Di Battista and Nardelli (1988)). In case that G_j, $1 \leq j < m$, consists of more than one connected component, Heath and Pemmaraju suggest to use a PQ-tree for every component and formulate a set of rules of how to merge components F_1 and F_2, respectively their corresponding PQ-trees T_1 and T_2, if F_1 and F_2 both are adjacent to some vertex $v \in V_{j+1}$.

The authors first reduce the pertinent leaves of T_1 and T_2 corresponding to the vertex v. After successfully performing the reduction, the consecutive sequence of pertinent leaves is replaced by a single pertinent representative in both T_1 and T_2. Going up one of the trees T_i, $i \in \{1, 2\}$, from its pertinent representative, an appropriate position is searched, allowing the tree T_j, $j \neq i$ to be placed into T_i. After successfully performing this step the resulting tree T' has two pertinent leaves corresponding to the vertex v, which again are reduced. If any of the steps fails, Heath and Pemmaraju state that the graph G is not leveled planar.

Merging two PQ-trees T_1 and T_2 corresponds to merging the two components F_1 and F_2 and is accomplished using certain informations that are stored at the nodes of the PQ-trees. For any subset S of the set of vertices in V_j, $1 \leq j \leq m$, that belong to a component F, define ML(S) to be the greatest $d \leq j$ such that $V_d, V_{d+1}, \ldots, V_j$ induces a dag in which all nodes of S occur in the same connected component. For a Q-node q in the corresponding PQ-tree T_F with ordered children $r_1, r_2, \ldots, r_t$ maintain in node q integers denoted ML(r_i, r_{i+1}), where $1 \leq i < t$, satisfying ML(r_i, r_{i+1}) = ML(frontier(r_i) $\cup$ frontier(r_{i+1})). For a P-node p maintain in p a single integer denoted ML(p) that satisfies ML(p) = ML(frontier(p)). Furthermore define LL(F) to be the smallest d such that F contains a vertex in V_d and maintain this integer at the root of the corresponding PQ-tree.

Using these LL- and ML-values, Heath and Pemmaraju (1996a, 1996b) describe a set of rules how to connect two PQ-trees claiming that the pertinent leaves of the new tree T' are reducible if and only if the corresponding component F' is leveled planar.

3.2   On the incorrectness of the algorithm

Within the merge phase, pertinent leaves are reduced pairwise in any given order. This includes the pairwise reduction of pertinent leaves of different components as well. Hence, components that have pertinent leaves of the same vertex in their frontier, are merged in an arbitrary order.

Consider four different components F_1, F_2, F_3, F_4 and their corresponding PQ-trees T_1, T_2, T_3, T_4 each having at least one pertinent leaf corresponding to some level-j vertex k. For simplicity, assume that the pertinent leaves of every component appear consecutively in all permutations on one side of their PQ-trees and assume further that the smallest common ancestor of the pertinent leaves and some other leaves is a Q-node. In Fig. 6 such a component F_i, $i \in \{1, 2, 3, 4\}$, and its corresponding PQ-tree T_i, $i \in \{1, 2, 3, 4\}$, is shown. The number c_i, $i \in \{1, 2, 3, 4\}$, depicts the ML-value between the leftmost pertinent leaf and the frontier of its left neighbor. We have marked all pertinent leaves with a k for simplicity.

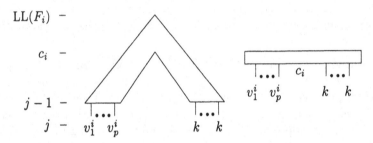

Fig. 6. Component F_i and its corresponding PQ-tree T_i. On the left side of F_i, some levels of F_i are indicated. The value c_i is equal to $\mathrm{ML}(\{v_p^i, k\})$.

Assuming that the following condition,

$$LL(F_1) \leq c_1 < LL(F_2) \leq c_2 < LL(F_3) \leq c_3 < LL(F_4) \leq c_4$$

on the ML- and LL-values of the components holds, it is possible to merge all four components into one component such that the pertinent leaves form a consecutive sequence. Figure 7 shows the four components, indicating how the components can be merged so that a reduction of the pertinent leaves becomes possible.

Consider the following merge operations on the components F_1, F_2, F_3, F_4 and their corresponding PQ-trees:

1. merge F_1 and F_4 into component F',
2. merge F' and F_3 into component F'',
3. merge F'' and F_2 into component F'''.

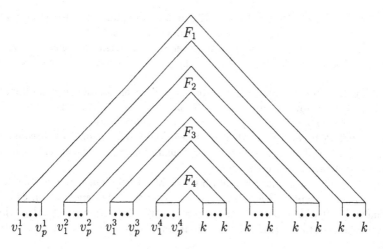

Fig. 7. Possible leveled planar arrangement of the components F_1, F_2, F_3, F_4.

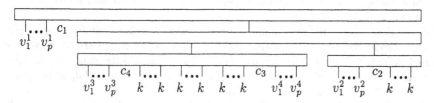

Fig. 8. PQ-tree T''' whose pertinent leaves depicted by k are not reducible.

The resulting PQ-tree T''' corresponding to F''' is shown in Fig. 8. Obviously, the pertinent leaves do not form a consecutive sequence in any permissible permutation of the PQ-tree. Hence the algorithm presented by Heath and Pemmaraju (1996a, 1996b) states leveled nonplanarity although the graph may be leveled planar.

As a matter of fact, the order of merging the components is important for testing a leveled dag. Moreover it is easy to see, that using different orderings while merging three or more components results in different equivalence classes of PQ-trees. So even if every order of merging PQ-trees with pertinent leaves results in a reducible PQ-tree, a PQ-tree may be constructed such that the leaves of some vertex l, $lev(l) > j$ are not reducible, although the graph G is leveled planar. Hence the algorithm presented by Heath and Pemmaraju (1996a, 1996b) may state incorrectly the leveled nonplanarity of a leveled planar graph.

References

Booth, K. and Lueker, G. (1976). Testing for the consecutive ones property, interval graphs, and graph planarity using PQ-tree algorithms. *Journal of Computer and System Sciences*, **13**, 335–379.

Chiba, N., Nishizeki, T., Abe, S., and Ozawa, T. (1985). A linear algorithm for embedding planar graphs using PQ-trees. *Journal of Computer and System Sciences*, **30**, 54–76.

Di Battista, G. and Nardelli, E. (1988). Hierarchies and planarity theory. *IEEE Transactions on systems, man, and cybernetics*, **18**(6), 1035–1046.

Even, S. (1979). *Graph Algorithms*. Computer Science Press, Potomac, Maryland.

Even, S. and Tarjan, R. E. (1976). Computing an st-numbering. *Theoretical Computer Science*, **2**, 339–344.

Heath, L. and Pemmaraju, S. (1996a). Recognizing leveled-planar dags in linear time. In F. J. Brandenburg, editor, *Proc. Graph Drawing '95*, volume 1027 of *Lecture Notes in Computer Science*, pages 300–311. Springer Verlag.

Heath, L. and Pemmaraju, S. (1996b). Stack and queue layouts of directed acyclic graphs: Part II. Technical report, Department of Computer Science, Virginia Polytechnic Institute & State University.

Jayakumar, R., Thulasiraman, K., and Swamy, M. (1986). On maximal planarization of non-planar graphs. *IEEE Transactions on Circuits Systems*, **33**(8), 843–844.

Jayakumar, R., Thulasiraman, K., and Swamy, M. (1989). On $O(n^2)$ algorithms for graph planarization. *IEEE Transactions on Computer-Aided Design*, **8**(3), 257–267.

Jünger, M., Leipert, S., and Mutzel, P. (1996). On computing a maximal planar subgraph using PQ-trees. Technical Report 96.227, Institut für Informatik der Universität zu Köln.

Kant, G. (1992). An $O(n^2)$ maximal planarization algorithm based on PQ-trees. Technical Report RUU-CS-92-03, Department of Computer Science, Utrecht University.

Lempel, A., Even, S., and Cederbaum, I. (1967). An algorithm for planarity testing of graphs. In *Theory of Graphs: International Symposium: Rome, July 1966*, pages 215–232. Gordon and Breach, New York.

Ozawa, T. and Takahashi, H. (1981). A graph-planarization algorithm and its application to random graphs. In *Graph Theory and Algorithms*, volume 108 of *Lecture Notes in Computer Science*, pages 95–107. Springer Verlag.

Graph Drawing with no k Pairwise Crossing Edges[*]

PAVEL VALTR

Department of Applied Mathematics, Charles University
Malostranské nám. 25, 118 00 Praha 1, Czech Republic
e-mail: valtr@kam.ms.mff.cuni.cz

DIMACS Center, Rutgers University
P.O.Box 1179, Piscataway, NJ 08855, U.S.A.

Abstract. A *geometric graph* is a graph $G = (V, E)$ drawn in the plane so that the vertex set V consists of points in general position and the edge set E consists of straight line segments between points of V. It is known that, for any fixed k, any geometric graph G on n vertices with no k pairwise crossing edges contains at most $O(n \log n)$ edges. In this paper we give a new, simpler proof of this bound, and show that the same bound holds also when the edges of G are represented by x-monotone curves (Jordan arcs).

1 Introduction

A *graph drawing* is a drawing of a graph in the plane such that each vertex is represented by a distinct point and each edge is represented by a Jordan arc connecting the corresponding two points (vertices) so that any two arcs meet in at most one point which is either a common endpoint or a common interior point where the two arcs cross each other. A *geometric graph* is a graph drawing in which all arcs are straight line segments. See [8] for a survey of results about geometric graphs.

Most papers on geometric graphs deal with the crossing number or with extremal questions motivated by extremal graph theory and Ramsey theory. One of the extremal results is due to Pach and Törőcsik [10] who proved that any geometric graph on n vertices with no k pairwise disjoint edges contains at most $(k-1)^4 n$ edges. A related ("dual") question is what is the maximum number of edges that can be contained in a geometric graph (or in a graph drawing) on n vertices with no k pairwise crossing edges. Euler's formula shows that this number is $3n - 6$ for $k = 2$ (and $n \geq 3$). Pach [7] proved that any geometric graph with no 3 pairwise crossing edges contains at most $13n^{3/2}$ edges, and that any geometric graph with no k pairwise crossing edges contains at most $O(n^{2-1/(25k^2)})$ edges. This result was improved by Pach, Sharokhi, and Szegedy [9] who proved that any graph drawing on n vertices with no k pairwise

---

[*] Research was supported by DIMACS Center, by Czech Republic Grant GAČR 0194, by Charles University grants No. 193 and 194, and by Czech-U.S. Science and Technology Research Grant no. 94051.

crossing edges contains at most $O(n \log^{2k-4} n)$ edges ($k \geq 3$). This bound was further improved to $O(n \log^{2k-6} n)$ in [2] ($k \geq 3$) (thus, in particular, to $O(n)$ for $k = 3$). In [12] we improved the bound for geometric graphs to $O(n \log n)$ (for $k \geq 4$).

Theorem 1 *Let $k \geq 4$ be a constant. Then any geometric graph on n vertices with no k pairwise crossing edges has at most $O(n \log n)$ edges.*

In this paper we give a simpler proof of Theorem 1, and extend Theorem 1 to graphs drawn in the plane by x-monotone Jordan arcs.

Theorem 2 *Let $k \geq 4$ be a constant, and let G be a graph drawing with n vertices such that all edges are drawn by x-monotone Jordan arcs and no k edges (arcs) are pairwise crossing. Then G has at most $O(n \log n)$ edges.*

In the proof of Theorems 1 and 2 we apply the Erdős-Szekeres theorem, Ramsey's theorem, and results on generalized Davenport–Schinzel sequences. Our paper is organized as follows. In Section 2 we recall generalized Davenport-Schinzel sequences, in Section 3 we give the proof of Theorems 1 and 2 which relies on two lemmas proved in Sections 4 and 5, respectively.

2 Generalized Davenport–Schinzel sequences

For $l \geq 1$, a sequence is called *l-regular*, if any at most l consecutive terms are pairwise different. For $l \geq 2$, a sequence

$$S = s_1, s_2, \ldots, s_{3l-2}$$

of length $3l - 2$ is said to be *of type* up-down-up(l), if the first l terms are pairwise different and, for $i = 1, 2, \ldots, l$,

$$s_i = s_{2l-i} = s_{(2l-2)+i}.$$

Thus, a sequence is of type up-down-up(2) if and only if it is an alternating sequence of length 4. It is well–known [3] (and not difficult to prove) that any 2-regular sequence over an n-element alphabet containing no alternating subsequence of length 4 has length at most $2n - 1$. In the proof of Theorems 1 and 2 we apply the following more general result.

Theorem 3 (Klazar and Valtr [6]) *Let $l \geq 2$ be a constant. Then the length of any l-regular sequence over an n-element alphabet containing no subsequence of type up-down-up(l) has length at most $O(n)$.*

3 Proof of Theorems 1 and 2

Theorem 1 is weaker than Theorem 2. Thus, it suffices to show Theorem 2.

Let $k \geq 2$ be a constant, and let $f_k(n)$ be the maximum number of edges in a graph drawing on n vertices with edges represented by x-monotone Jordan arcs and with no k pairwise crossing edges. Let $G = (V, E)$ be a graph drawing on n vertices with edges represented by x-monotone Jordan arcs and with no k pairwise crossing edges. Introduce Cartesian coordinate system so that the y-axis partitions V into two parts which are as equal as possible, thus the sets

$$V^- = \{v \in V | \text{the } x\text{-coordinate of } v \text{ is negative}\},$$
$$V^+ = \{v \in V | \text{the } x\text{-coordinate of } v \text{ is positive}\}$$

have sizes

$$|V^-| = \lfloor n/2 \rfloor, \qquad |V^+| = \lceil n/2 \rceil.$$

Partition E into three subsets E^+, E^-, E' such that E^+ contains the edges with both endpoints in V^+, E^- contains the edges with both endpoints in V^-, and E' contains the edges crossing the y-axis.

Obviously,

$$|E^-| \leq f_k(\lfloor n/2 \rfloor), \qquad |E^+| \leq f_k(\lceil n/2 \rceil).$$

To obtain Theorem 2, it suffices to show that

$$|E'| = O(n), \tag{1}$$

because this implies

$$f_k(n) \leq f_k(\lfloor n/2 \rfloor) + f_k(\lceil n/2 \rceil) + O(n)$$

and, consequently,

$$f_k(n) = O(n \log n).$$

We may assume without loss of generality that no two edges of E' cross the y-axis in the same point. We label the vertices of V by integers from 1 to n. We do not make any notational distinction between a point and its label. Let

$$e_1, e_2, \ldots, e_r$$

be the edges of E' ordered in the order in which they cross the y-axis (from $-\infty$ to $+\infty$, say). Further, let S_1 and S_2 be the sequences of $r = |E'|$ integers obtained from the sequence

$$e_1, e_2, \ldots, e_r$$

by replacing each edge by (the label of) its left endpoint and by (the label of) its right endpoint, respectively. The proof of (1) is based on a proper handling with the sequences S_1 and S_2. In particular, it relies on the following two lemmas:

Lemma 4 *For each $l \geq 1$, at least one of the sequences S_1, S_2 contains an l-regular subsequence of length at least $|E'|/(4l) = r/(4l)$.*

Lemma 5 *Neither of the sequences S_1 and S_2 contains a subsequence of type up-down-up$((r(k))^2)$, where $r(k)$ is the kth Ramsey number (i.e., $r(k)$ is the minimum number R such that any complete graph on at least R vertices with edges colored by 2 colors contains a monochromatic complete subgraph on k vertices).*

We remark that Lemma 5 holds in a stronger form for geometric graphs (with $(r(k))^2$ replaced by $(k-1)^3 + 1$). This is shown in Section 5. Before proving Lemmas 4 and 5 we complete the proof of (1), which, as argued above, implies Theorem 2.

According to Lemma 4, at least one of the sequences S_1, S_2 contains an $(r(k))^2$-regular subsequence S of length at least $|E|/((4r(k))^2)$. It follows from Lemma 5 that the sequence S contains no subsequence of type up-down-up$((r(k))^2)$. Consequently, Theorem 3 implies that the length of S is at most $O(n)$. Thus,

$$|E|/(4(r(k))^2) \leq \text{length of } S \leq O(n).$$

Consequently,

$$|E| \leq 4(r(k))^2 \cdot O(n) = O(n). \quad \square$$

To complete the proof of Theorems 1 and 2, it remains to prove Lemmas 4 and 5.

4   Proof of Lemma 4

We apply a simple greedy algorithm [1] which, for given integer $l \geq 1$ and finite sequence A, returns an l-regular subsequence $B(A, l)$ of A. At the beginning of the algorithm, an auxiliary sequence B is taken empty. Then the terms of A are considered one by one from left to right, and in each step the considered term is placed at the end of B iff this doesn't violate the l-regularity of B (otherwise B is unchanged). Finally, the obtained l-regular subsequence B of A is taken for $B(A, l)$.

E.g., if $A = 1, 3, 1, 3, 5, 2, 2, 5, 1, 5, 1, 2$ and $l = 3$, then the algorithm returns the sequence $B(A, 3) = 1, 3, 5, 2, 1, 5, 2$. We prove Lemma 4 by showing that, given $l \geq 1$, at least one of the sequences $B(S_1, l)$, $B(S_2, l)$ obtained by the algorithm has length at least $|E'|/(4l) = r/(4l)$.

Let $l \geq 1$ be given. For $i = 1, 2$ and for $1 \leq j_1 \leq j_2 \leq m$, let $S_{i,[j_1,j_2]}$ be the subsequence of S_i obtained from S_i by removing all terms preceding the j_1th term and following the j_2th term. Thus, $S_{i,[j_1,j_2]}$ is a continuous part of S_i of length $j_2 - j_1 + 1$.

Let $|T|$ denote the length of a sequence T, and let $I(T)$ denote the set of integers appearing in T.

Claim 6 *For each $j = 1, 2, \ldots, m$,*

$$\sum_{i=1}^{2} |B(S_{i,[1,j]}, l)| \geq j/(2l).$$

Proof. First, consider two integers j_1, j_2 such that $1 \leq j_1 \leq j_2 \leq m$. Obviously,

$$\{e_{j_1}, e_{j_1+1}, \ldots, e_{j_2}\} \subseteq \{(v_a, v_b) | a \in I(S_{1,[j_1,j_2]}), b \in I(S_{2,[j_1,j_2]})\}.$$

Thus,

$$|\{e_{j_1}, e_{j_1+1}, \ldots, e_{j_2}\}| \leq |\{(v_a, v_b) | a \in I(S_{1,[j_1,j_2]}), b \in I(S_{2,[j_1,j_2]})\}|.$$

Consequently,

$$j_2 - j_1 + 1 \leq |I(S_{1,[j_1,j_2]})| \cdot |I(S_{2,[j_1,j_2]})|.$$

By the inequality between algebraic and geometric means,

$$\frac{\sum_{i=1}^{2} |I(S_{i,[j_1,j_2]})|}{2} \geq \sqrt{j_2 - j_1 + 1}. \tag{2}$$

We now can prove the claim by induction on j. If $j \leq \min\{16l^2, m\}$, then by (2) and by $j \leq 16l^2$

$$\sum_{i=1}^{2} |B(S_{i,[1,j]}, l)| \geq \sum_{i=1}^{2} |I(S_{i,[1,j]})|$$
$$\geq 2\sqrt{j}$$
$$\geq j/(2l).$$

Suppose now that $16l^2 < j_0 \leq m$ and that Claim 6 holds for $j = 1, 2, \ldots, j_0 - 1$. Since for $i = 1, 2$ each integer of $I(S_{i,[j_0-4l^2+1,j_0]})$ not appearing among the last $l - 1$ terms in $B(S_{i,[1,j_0-4l^2]}, l)$ appears more times in $B(S_{i,[1,j_0]}, l)$ than in $B(S_{i,[1,j_0-4l^2]}, l)$, we have

$$|B(S_{i,[1,j_0]}, l)| \geq |B(S_{i,[1,j_0-4l^2]}, l)| + |I(S_{i,[j_0-4l^2+1,j_0]})| - (l-1).$$

Consequently, by the inductive hypothesis and by (2),

$$\sum_{i=1}^{2} |B(S_{i,[1,j_0]}, l)| \geq (j_0 - 4l^2)/(2l) + 2\sqrt{4l^2} - 2(l-1)$$
$$> j_0/(2l). \qquad \square$$

Proof of Lemma 4. Lemma 4 easily follows from Claim 6 (with $j = m$) and from the pigeon-hole principle. $\qquad \square$

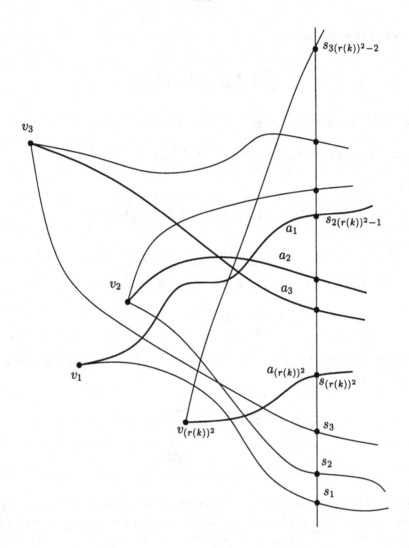

Fig. 1. Arcs a_i.

5 Proof of Lemma 5

Because of symmetry, it suffices to prove Lemma 5 for the sequence S_1. Suppose to the contrary that S_1 contains a subsequence of type up-down-up$((r(k))^2)$. Thus, there is a subsequence

$$S = s_1, s_2, \ldots, s_{3(r(k))^2 - 2}$$

of S_1 such that the integers $s_1, s_2, \ldots, s_{(r(k))^2}$ are pairwise different and that, for $i = 1, 2, \ldots, (r(k))^2$, $s_i = s_{2(r(k))^2 - i} = s_{(2(r(k))^2 - 2) + i}$. For each i, we denote by

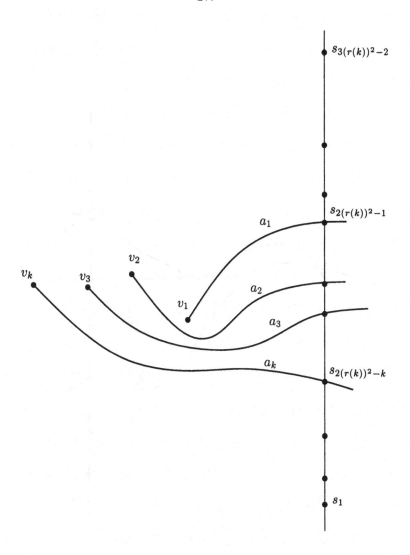

Fig. 2. Case 1.

v_i the vertex of V^- labeled by s_i, and we denote by x_i the x-coordinate of v_i. Further, let a_i be the arc emanating from the vertex v_i and passing through the point on the y-axis corresponding to $s_{2(r(k))^2-i}$ (see Fig. 1). We may assume that the coordinates x_i are pairwise different. By the Erdős-Szekeres theorem [5], the sequence

$$x_1, x_2, \ldots, x_{(r(k))^2}$$

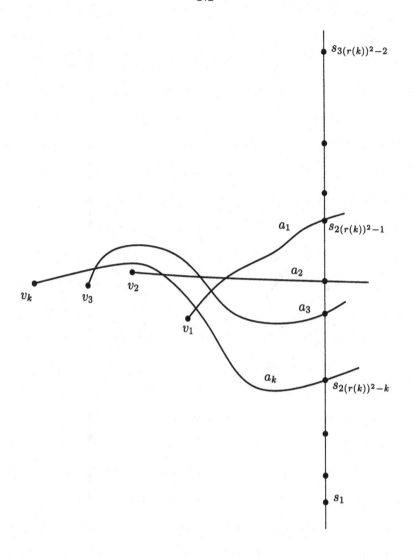

Fig. 3. Case 2.

contains a monotone subsequence of length $r(k)$. For simplicity of notation (and without loss of generality), we assume that the sequence

$$x_1, x_2, \ldots, x_{r(k)}$$

is monotone. Let $T = \{v_1, \ldots, v_{r(k)}\}$. We color the edges of a complete graph on the vertex set T by 2 colors as follows. An edge $v_i v_j$ with $x_i < x_j$ is colored blue if v_j lies below the arc a_i, otherwise its color is red. By Ramsey's theorem [11], our graph contains a monochromatic complete subgraph on k vertices. For simplicity of notation and without loss of generality, we assume that

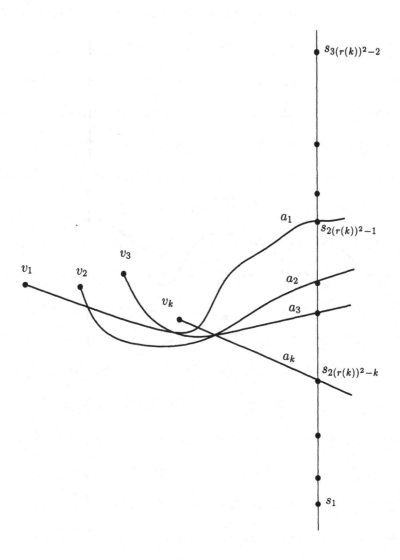

Fig. 4. Case 3.

the complete subgraph on vertices $v_1, v_2, \cdots, v_k$ is monochromatic. Then, one of the four cases in Figures 2–5, respectively, applies. In cases shown in Figs. 3 and 4, the arcs $a_i, i = 1, \ldots, k$, are pairwise crossing. In Fig. 2 the k arcs passing through the points on the y-axis corresponding to $s_1, s_2, \ldots, s_k$ are pairwise crossing (see Fig. 6), in Fig. 5 the k arcs passing through the points on the y-axis corresponding to $s_{2(r(k))^2-1}, s_{2(r(k))^2}, \ldots, s_{2(r(k))^2+k-2}$ are pairwise crossing (see Fig. 7). Thus, we have obtained a contradiction by showing that k of the arcs in our graph drawing are pairwise crossing. □

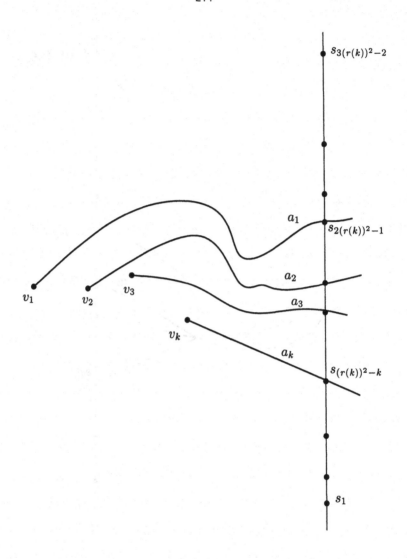

Fig. 5. Case 4.

We now show that Lemma 5 holds in a stronger form for geometric graphs. The strengthening is based on an application of Dilworth's theorem instead of Ramsey's theorem.

Lemma 7 *If G is a geometric graph, then neither of the sequences S_1 and S_2 contains a subsequence of type up-down-up$((k-1)^3 + 1)$.*

Proof. Because of symmetry, it suffices to prove Lemma 7 for the sequence S_1. Suppose to the contrary that S_1 contains a subsequence of type up-down-up(l),

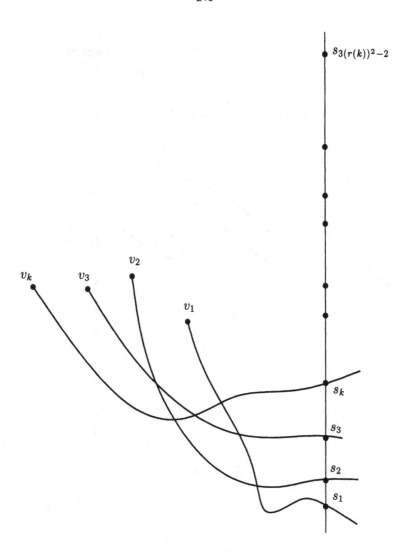

Fig. 6. k pairwise crossing arcs in case 1.

where $l = (k-1)^3 + 1$. Thus, there is a subsequence

$$S = s_1, s_2, \ldots, s_{3l-2}$$

of S_1 such that the integers $s_1, s_2, \ldots, s_l$ are pairwise different and that, for $i = 1, 2, \ldots, l$, $s_i = s_{2l-i} = s_{(2l-2)+i}$. For each i, we denote by v_i the vertex of V^- labeled by s_i, and we denote by x_i the x-coordinate of v_i. We define three relations on the set $L = \{1, 2, \ldots, l\}$ as follows:

$$\prec_1 = \{(i,j) : i < j \text{ and the ray } \overrightarrow{v_i v_j} \text{ intersects the } y\text{-axis below the point}$$

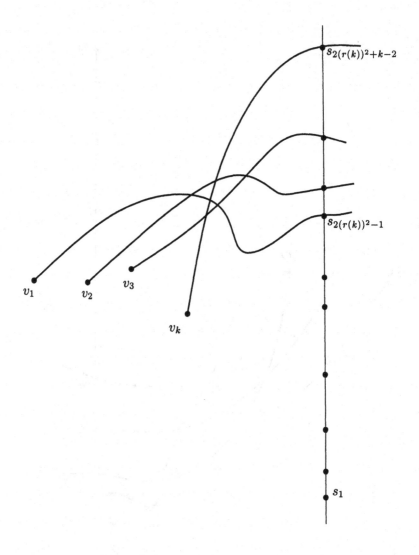

Fig. 7. k pairwise crossing arcs in case 4.

corresponding to s_{2l-1}},

$\prec_2 = \{(i,j) : i < j$ and the ray $\overrightarrow{v_j v_i}$ intersects the y-axis above the point corresponding to s_l},

$\prec_3 = \{(i,j) : i < j$ and the pair (i,j) lies neither in $\prec_1$ nor in $\prec_2$}.

It is not difficult to see that each of the relations $\prec_i$ is a partial ordering. It follows from Dilworth's theorem [4] that either there are two elements in L which are not comparable by any of the partial orders $\prec_i$, or one of the partial orders $\prec_i$ contains a chain of length k. The first case is in contradiction with the definition of the partial orders $\prec_i$, thus one of the partial orders, $\prec_i$, contains a chain $j_1 \prec_i j_2 \prec_i \ldots \prec_i j_k$ of length k. If $i = 1$ then the k segments corresponding to the terms $s_{(2l-2)+j_t}, t = 1, 2, \ldots, k$, are pairwise crossing. If $i = 2$ then the k segments corresponding to the terms $s_{j_t}, t = 1, 2, \ldots, k$, are pairwise crossing. Finally, if $i = 3$ then the k segments corresponding to the terms $s_{2l-j_t}, t = 1, 2, \ldots, k$, are pairwise crossing. Thus, we have obtained a contradiction by showing that k of the edges in our geometric graph are pairwise crossing. □

Concluding remarks

1. Our proof of Theorems 1 and 2 has some similarities with the original proof [12] of Theorem 1. In this paper we obtained Theorems 1 and 2 by analyzing two sequences S_1 and S_2 of endpoints of segments listed in the order in which the segments cross a fixed halving line. In [12] we considered analogous sequences when, roughly speaking, the segments are ordered according to increasing slope. Similarly as in [12], we applied results about generalized Davenport-Schinzel sequences (Theorem 3) and a Ramsey-type argument. In [12], we still needed to use a dual-type projection. Moreover, the proof in [12] doesn't seem to generalize to a proof of Theorem 2.

2. The most interesting question related to our paper is if any geometric graph on n vertices with no k pairwise crossing edges contains at most $c_k n$ edges. Another open problem is how many edges can be contained in a geometric graph on n vertices with no k pairwise disjoint edges. In the introduction we mentioned the upper bound $(k - 1)^4 n$ due to Pach and Törőcsik [10]. This bound was very recently improved to $O(k^3 n)$ by Géza Tóth and the author. On the other hand, no lower bound bigger than $\Theta(kn)$ is known.

References

1. R. Adamec, M. Klazar, and P. Valtr, Generalized Davenport–Schinzel sequences with linear upper bound, *Discrete Math.* 108 (1992), 219-229.
2. P.K. Agarwal, B. Aronov, J. Pach, R. Pollack, and M. Sharir, Quasi-planar graphs have a linear number of edges, Graph Drawing (Passau, 1995), *Lecture Notes in Comput. Sci.* 1027 (1995), 1-7.
3. H. Davenport and A. Schinzel, A combinatorial problem connected with differential equations, *Amer. J. Math.* 87 (1965), 684-694.
4. R. Dilworth, A decomposition theorem for partially ordered sets, *Annals of Mathematics* 51 (1950), 161-166.
5. P. Erdős and G. Szekeres, A combinatorial problem in geometry, *Compositio Math.* 2 (1935), 463-470.
6. M. Klazar and P. Valtr, Generalized Davenport–Schinzel sequences, *Combinatorica* 14 (1994), 463-476.

7. J. Pach, Notes on geometric graph theory, in: *Discrete and Computational Geometry: Papers from DIMACS Special Year*, DIMACS Series, Vol. 6, AMS, Providence, RI, 1991, pp. 273-285.

8. J. Pach and P.K. Agarwal, *Combinatorial Geometry*, Wiley Interscience, New York (1995).

9. J. Pach, F. Sharokhi, and M. Szegedy, Applications of the crossing number, *Algorithmica* 16 (1996), 111-117.

10. J. Pach and J. Törőcsik, Some geometric applications of Dilworth's theorem, *Discrete Comput. Geom.* 12 (1994), 1-7.

11. F. P. Ramsey, On a problem of formal logic, *Proc. Lond. Math. Soc.*, II. Ser., 30 (1930), 264-286.

12. P. Valtr, On geometric graphs with no k pairwise parallel edges, to appear in *Discrete and Computational Geometry*.

Area Requirements for Drawing Hierarchically Planar Graphs

Xuemin Lin[1] and Peter Eades[2]

[1] Department of Computer Science, The University of Western Australia
Nedlands, WA 6907, Australia. e-mail: lxue@cs.uwa.oz.au.
[2] Department of Computer Science, The University of Newcastle
Callaghan, NSW 2308, Australia. e-mail: eades@cs.newcastle.edu.au.

(Extended Abstract)

Abstract. In this paper, we investigate area requirements for drawing s-t hierarchically planar graphs by straight-lines. Two drawing standards will be discussed: 1) each vertex is represented by a point and 2) grid visibility representation (that is, a line segment is allowed to represent a vertex). For the first drawing standard, we show an exponential area lower bound needed for drawing hierarchically planar graphs. The lower bound holds even for hierarchical graphs without transitive arcs, in contrast to the results for upward planar drawing. Applications of some existing algorithms from upward drawing can guarantee the quadratic drawing area for grid visibility representation but do not necessarily guarantee the minimum drawing area. Motivated by this, we will present another grid visibility drawing algorithm which is efficient and guarantees the minimum drawing area.

Keywords: Graph Drawing, Hierarchically Planar Graph, Straight Line Drawing, Visibility Representation, Drawing Area.

1 Introduction

Automatic graph drawing plays an important role in many computer-based applications such as CASE tools, software and information visualization, and VLSI design. The *upward* drawing convention for drawing acyclic directed graphs has received a great deal of attention since last decade. Many results [2, 4, 6, 9, 11] for drawing upward planar graphs have been published. However, acyclic directed graphs are not powerful enough to model all applications. Hierarchical graphs are then introduced, where layering information is also specified in an acyclic directed graph. Naturally, the "hierarchical" drawing convention (to be defined in Section 2) is proposed to display the specified layering information.

Due to the additional layering constraint, hierarchical drawing is different to upward drawing. Results in upward drawing and hierarchical drawing are not always the same. Issues, such as planar, straight-line, convex, and symmetric representations, have been revisited [7, 8, 10, 11] for drawing "hierarchically planar graphs" (to be defined in Section 2). In this paper, we investigate the problem of area requirements for drawing "s-t hierarchically planar graphs" (to be defined in Section 2) with respect to a given *resolution requirement*.

In [6], it has shown that exponential areas are generally necessary for drawing upward planar graphs by the drawing standard of using points only to represent vertices and straight-lines to represent arcs. However, only quadratic drawing areas are required when upward planar graphs are *reduced*, meaning that no "transitive" arcs exist.

In this paper, we show that the results in [6] do not hold for hierarchically planar graphs. Specifically, we show that by the same drawing standard, exponential drawing areas are necessary even for hierarchically planar graphs without transitive arcs. This is the first contribution of the paper.

Secondly, we study the drawing area problem by allowing line segments to represent vertices. Particularly, we study the drawing area problem for "grid visibility representations" (to be defined in Section 2). An application of the algorithm VISIBILITY_DRAW in [4] gives the quadratic area for the grid visibility representation of hierarchically planar graphs. However, this algorithm does not necessary guarantee the minimal drawing area - an example will be given in Section 4. Motivated by this, we present an efficient algorithm for grid visibility representations of s-t hierachically planar graphs which guarantees the minimum drawing area.

The rest of the paper is organized as follows. Section 2 gives the basic terminology and background knowledge. Section 3 shows an exponential lower bound of drawing area. In Section 4, we present a drawing algorithm for producing a grid visibility representation which minimizes the drawing area. This is followed by the conclusions and remarks.

2 Preliminaries

The basic graph theoretic definitions can be found in [1].

A *hierarchical graph* $H = (V, A, \lambda, k)$ consists of a simple and acyclic directed graph (V, A), a positive integer k, and for each vertex u, an integer $\lambda(u) \in \{1, 2, ..., k\}$ with the property that if $u \to v \in A$, then $\lambda(u) > \lambda(v)$. For $1 \leq i \leq k$ the set $\{u : \lambda(u) = i\}$ of vertices is the *ith layer* of H and is denoted by L_i. An arc $u \to v$ in $H = (V, A, \lambda, k)$ is a *transitive* arc if there exists another directed path from u to v with length at least 2. An arc $u \to v$ is *long* if it spans more than two layers, that is, $\lambda(u) - \lambda(v) \geq 2$.

A *sink* of a hierarchical graph H is a vertex which does not have outgoing arcs, and a *source* of H is a vertex which does not have incoming edges. H is *s-t* if it has only one sink and one source.

A hierarchical graph is *proper* if it has no long arcs. Clearly, adding $\lambda(u) - \lambda(v) - 1$ dummy vertices to each long arc $u \to v$ in an unproper hierarchical graph H results in a proper hierarchical graph, denoted by H_p. H_p is called the *proper image* of H. Note that $H_p = H$ if H is proper.

To display the specified hierarchical information in a hierarchical graph, the *hierarchical drawing convention* is proposed, where a vertex in each layer L_i is separately allocated on the horizontal line $y = i$ and arcs are represented as curves monotonic in y direction; see Figures 1 (a)-(c). A hierarchical drawing is *planar* if no pair of noincident arcs intersect. A hierarchical graph is *hierarchically planar* if it has a planar drawing admitting the hierarchical drawing convention. In this paper, we will discuss only hierarchical drawing convention.

An *embedding* of a proper hierarchical graph H consists of an ordering of the vertices in each layer, and is denoted by E_H. An *embedding* of an unproper hierarchical graph H means an embedding of the proper image H_p of H, and is also denoted by E_H.

A hierarchical drawing α of H *respects* E_H if α gives the same vertex ordering in each layer in H_p as E_H does. An embedding E_H is *planar* if any straight-line drawing of H_p respecting E_H is planar.

Various drawing standards exist for drawing hierarchically planar graphs by retaining planarity. In a *straight* line drawing α, each vertex v is represented as a point $\alpha(u)$ and each arc $u \rightarrow v$ is represented as a line segment connecting $\alpha(u)$ and $\alpha(v)$; see Figure 1 (a). In a *polyline* drawing, each long arc is allowed to be represented as a polygonal chain with bends allocated on some of the k horizontal lines $y = i$ for $1 \leq i \leq k$; see Figure 1 (c). In a *visibility representation* β, each vertex u is represented as a horizontal line segment $\beta(u)$ on $y = \lambda(u)$ and each arc $u \rightarrow v$ as a vertical line segment connecting $\beta(u)$ and $\beta(v)$, such that:

- $\beta(u)$ and $\beta(v)$ are disjoint if $u \neq v$, and
- a vertical line segment and a horizontal line segment do not intersect if the correponding arc and vertex are not incident.

See Figure 1(b), for example. Note that in a visibility representation, a line segment used to represent a vertex may be degenerated into a point.

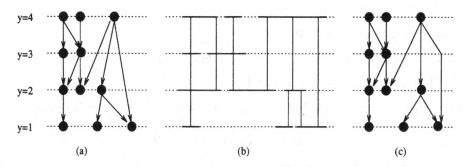

Fig. 1. Various Representations

A straight line drawing is a *grid* drawing if each vertex is at a grid position; and a polyline drawing is a *grid* drawing if vertices and bends are at grid positions. Similarly, in a *grid* visibility representation each horizontal line segment and vertical line segments must use grid points as their ends.

Drawing a hierarchically planar graph H consists of two phases: 1) finding a planar embedding E_H, and 2) finding a hierarchical drawing of H respecting E_H.

A linear time algorithm [10] was proposed for phase 1. In this paper, we concentrated on phase 2 and assume that a planar embedding is given. More restrictly, we study how to draw a planar embedding of an s-t hierarchically planar graph. This does not destroy the generality of the investigation of phase 2, since each planar embedding can be easily extended to a planar embedding of an s-t hierarchically planar graph [7, 8, 11].

3 An Exponential Area Lower Bound

The drawing *area* of a hierarchical drawing α is the minimal rectangle R which contains α and is composed of horizontal and vertical lines.

The *width* of a hierarchical drawing is the horizontal distance between the leftmost vertex and the rightmost vertex, while the *height* is the vertical distance between the top layer and the bottom layer. For a given hierarchical graph H, any hierarchical drawing of H has the fixed height. Consequently, the investigation of drawing area problem is reduced to that of drawing width problem.

In this section, we define a hierarchically planar graph $H_n = (V_n, A_n, \lambda_n, 4n -1)$ with $|V_n| = 10n - 6$ for $n \geq 1$, such that 1) H_n has no transitive arcs, and 2) any planar straight-line drawing of H_n requires exponential drawing area with respect to a given vertex resolution requirement. More specifically, we define H_n by extending H_{n-1} for $n \geq 2$. The extension follows the same topology.

The graph H_1 consists of 4 vertices $\{t_1, c_{1,1}, c_{2,1}, s_1\}$ and three layers $L_3 = \{s_1\}$, $L_2 = \{c_{1,1}, c_{2,1}\}$, and $L_1 = \{t_1\}$. Four arcs connect H_1 in a diamond shape (see Figure 2(a)). To extend H_1 to H_2, ten vertices are added as depicted in Figure 2(b).

Generally, we construct H_{i+1} from H_i by adding the following ten vertices in a way depicted in Figure 2(c):

$$V_{i+1} = V_i \cup \{s_{i+1}, t_{i+1}, a_{1,i+1}, a_{2,i+1}, b_{1,i+1}, b_{2,i+1}, c_{1,i+1}, c_{2,i+1}, d_{1,i+1}, d_{2,i+1}\}.$$

The layering of H_n is described below:

$L_1 = \{t_n\}$, $L_{4n-1} = \{s_n\}$, $L_{2n} = \{c_{1,1}, c_{2,1}\}$; and for $1 \leq i \leq n - 1$, $L_{2n+2i} = \{b_{j,i+1} : 1 \leq j \leq 2\}$, $L_{2n+2i-1} = \{s_i\} \cup \{a_{j,i+1} : 1 \leq j \leq 2\}$, $L_{2i+1} = \{t_{n-i}\} \cup \{c_{j,n-i+1} : 1 \leq j \leq 2\}$, and $L_{2i} = \{d_{j,n-i+1} : 1 \leq j \leq 2\}$.

The arc set A_n of H_n consists of:

$\{s_1 \to c_{j,1}, c_{j,1} \to t_1 : 1 \leq j \leq 2\}$, $\{s_i \to s_{i-1}, s_i \to b_{j,i}, s_i \to c_{j,i} : 1 \leq j \leq 2, 2 \leq i \leq n\}$, $\{b_{j,i} \to a_{j,i}, b_{j,i} \to c_{j,i-1}, a_{j,i} \to t_{i-1}, a_{j,i} \to d_{j,i} : 1 \leq j \leq 2, 2 \leq i \leq n\}$, and $\{c_{j,i} \to d_{j,i}, d_{j,i} \to t_i, t_{i-1} \to t_i : 1 \leq j \leq 2, 2 \leq i \leq n\}$.

The following two lemmas can be immediately verified [12] based on the structure of H_n.

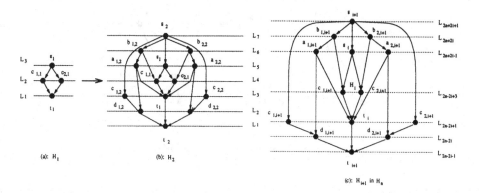

Fig. 2. Construct H_n

Lemma 1. *For $n \geq 1$, H_n is a hierarchically planar graph with no transitive arcs.*

Lemma 2. *For $n \geq 1$, the planar embedding E_{H_n} of H_n is unique up to a complete reversal.*

Here is the main result in this section.

Theorem 3. *For each H_n, suppose that α is a hierarchically planar straight line drawing of H_n, where each pair of vertices in the same layer are at least distance 1 apart. Then α has width at least $\Omega((2n-1)!)$.*

Proof: With respect to α, suppose that for $2 \leq i \leq n$, the distance between $\alpha(c_{1,i})$ and $\alpha(c_{2,i})$ is l_i.

Lemma 2 tells us that the planar embedding given by Figure 2(c) is unique to any hierarchically planar drawing of H_n up to a complete reversal.

Without loss of generality, we may assume that α gives the planar embedding as depicted in Figure 2(c).

Thus, in α, the relationship of the vertices orderings between α restricted to H_{i+1} and α restricted to H_i must be as the one illustrated in Figure 2(c). Consider the two triangles in Figure 3 with respect to α. Since the drawing α is a planar straight line drawing, elementary geometry implies $\frac{l_{i+1}}{l_i} \geq 2i(2i-1)$. Hence, $l_n \geq (2(n-1))!$. Therefore the Theorem holds.
□

Note that any hierarchical drawing of H_n has height $4n-2$. Thus each hierarchically planar straight-line drawing of H_n, in which each pair of vertices in the same layer are at least distance 1 apart, has area at least $\Omega(n(2n-2)!)$.

Note that H_n can be drawn upward planar in quadratic area (with respect to the number of vertices in H_n) by the algorithm in [6], but the layering of H_n is not preserved.

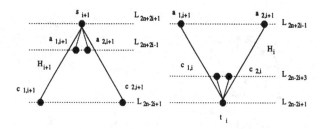

Fig. 3. Relationship among widths

4 Visibility Representation

The algorithm VISIBILITY_DRAW in [4] was developed for grid visibility representations of *s-t* upward planar graphs by using the *dual graph* technique; and it can be immediately applied to s-t hierarchically planar graphs for grid visibility representations. By applying the algorithm VISIBILITY_DRAW to an s-t hierarchically planar graph H, the output visibility representation of H has the width $w - 1$ where w is the length of the longest path from the source to the sink of the dual graph of H. However, the minimum width of a visibility representation of H may be much smaller than $w - 1$; and this is shown by the following examples.

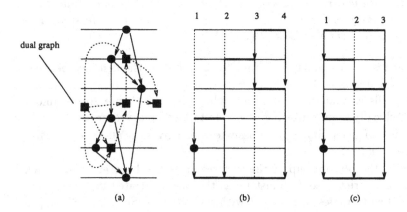

Fig. 4. Example 1

A hierarchical graph H_1 and its dual graph are illustrated in Figure 4(b), where the dual graph is depicted by rectangles and dotted arcs. An application of the algorithm VISIBILITY_DRAW produces the grid visibility representation of H_1 with width 3 as shown in Figure 4(b). However, the minimum width of a grid visibility representation of H is 2 as shown in Figure 4(c). Actually, the drawing in 4(c) is output by our algorithm.

We can generalize the example in Figure 4 to the graph H_2 as shown in Figure 5, where H_1 in Figure 4 is duplicated n times in H_2. It can be immediately verified that the length of the longest path from the source to the sink of the dual graph of H_2 is $4n$. Consequently, the width of the grid visibility representation of H_2 produced by the algorithm VISIBILITY_DRAW is $4n - 1$. However, it easy to show that the minimum width of a grid visibility representation of H_2 is $3n - 1$.

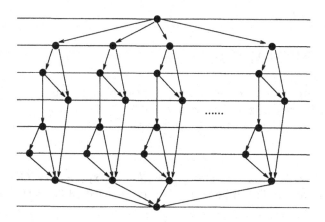

Fig. 5. Example 2

Inspired by the work in [4], in this section we present a new drawing algorithm GVP for the grid visibility representation of an s-t hierarchically planar graph with respect to a given planar embedding. The algorithm guarantees the minimal drawing area; that is, the width is minimized.

Like the algorithm VISIBILITY_DRAW, the algorithm GVP consists of two steps: 1) label each arc a by an integer $l(a)$ and 2) allocate an arc a on the vertical line $x = i$ if $l(a) = i$. However, the labeling technique in our algorithm GVP is different than that in the algorithm VISIBILITY_DRAW and therefore can guarantee the minimum drawing area.

The basic idea of our labeling technique is to push each individual vertical line segment as left as possible. This can be done by the following procedure ARC-LABELING which labels a given arc according to the previous labeling information.

Suppose that an s-t hierarchically planar graph H and its proper image H_p are given. Recall that H_p is a proper hierarchical graph. The label of a long arc $u \rightarrow v$ in H is inherited by the short arcs in H_p of which $u \rightarrow v$ is composed. In our algorithm, the current labeling information of H is kept with respect to H_p to simplify descriptions. For each layer L_i in H_p,

- I_i denotes the maximal integer label used among labeled arcs incoming to L_i,
- O_i denotes the maximal label used among labeled arcs outgoing from L_i, and

– IO_i denotes the maximal label used among arcs incident to the vertices in L_i all of whose outgoing and incoming arcs are labeled.

We also need the following notion for the description of our algorithm. Let E_H be a given planar embedding of H. For two arcs $u \to v$ and $x \to y$ between L_{i+1} and L_i, $u \to v$ is at the *left* of $x \to y$ with respect to E_H if $u \le x$ and $v \le y$ according to E_H.

Procedure ARC-LABELING

INPUT: an arc $u \to v$ to be labeled, a given planar embedding E_H, and the current O_i, I_i and IO_i for each layer L_i in H.
OUTPUT: a label $l(u \to v)$ of $u \to v$ and the updated O_i, I_i and IO_i for each layer L_i.

Suppose that $u \in L_{j+m}$ and $v \in L_j$. Let $L = \max_{1 \le \eta \le m-1}\{IO_{j+\eta}\}$ if $m \ge 2$, and $L = 0$ if $m = 1$. The arc $a = u \to v$ is labeled according to the four different cases:

1. If a is the leftmost outgoing arc from u and the leftmost incoming arc to v, $l(a) = \max\{IO_{j+m}, IO_j, L\} + 1$.

2. If a is the leftmost outgoing arc from u but not the leftmost incoming arc to v, $l(a) = \max\{IO_{j+m}, I_j, L\} + 1$.

3. If a is the leftmost incoming arc to v but not the leftmost outgoing arc from u, $l(a) = \max\{O_{j+m}, IO_j, L\} + 1$.

4. If a is neither leftmost outgoing from u nor leftmost incoming to v, $l(a) = \max\{O_{j+m}, I_j, L\} + 1$.

After labeling a, update $IO_{j+\eta}$ to $l(a)$ for $1 \le \eta \le m-1$. Meanwhile, we modify O_{j+m}, and IO_{j+m}, I_j, and IO_j as follows. $O_{j+m} = l(a)$ and $I_j = l(a)$. If all arcs incident to u are labeled, then $IO_{j+m} = \max\{l(a), I_{j+m}\}$. Similarly, if all arcs incident to v are labeled, then $IO_j = \max\{l(a), O_j\}$. □

To preserve the given planar embedding E_H in the algorithm GVP and then to guarantee the planarity of the drawing, we successively label each arc in H according to the trajectory of a leftmost depth-first search (LDFS) on H_p with respect to E_H. LDFS is a variation of the depth-first search technique [1].

LDFS

Start from the source of H_p. While going down by the depth-first search from vertex u, always first visit the leftmost unvisited outgoing arc $u \rightarrow v$ from u. If all outgoing arcs from v are already visited or v is a sink, then the LDFS procedure continues as follows.

S1: Terminate the current LDFS path. Goto S2.

S2: Along the reverse direction of the current LDFS path, trace back (till reach the source) to the bottom-most vertex v_0 which has unvisited outgoing arcs. If no such v_0 exists then stop the LDFS procedure, otherwise goto S3.

S3 Start the next LDFS path from v_0. □

For example, by applying LDFS to H_2 in Figure 2(b) we successively visit the arcs: $s_2 \rightarrow c_{1,2}$, $c_{1,2} \rightarrow d_{1,2}$, $d_{1,2} \rightarrow t_2$, $s_2 \rightarrow b_{1,2}$, $b_{1,2} \rightarrow a_{1,2}$, $a_{1,2} \rightarrow d_{1,2}$, $a_{1,2} \rightarrow t_1$, $t_1 \rightarrow t_2$, ...

Algorithm GVP

INPUT: an s-t hierarchically planar graph H and its planar embedding E_H.

OUTPUT: a grid visibility representation of H respecting E_H.

Step 1: Labeling. Initially, I_i, O_i and IO_i are set to zero. Label arcs successively as follows according to the ordering given by LDFS till all arcs are labeled. Note that each arc in H_p is a short arc. While an arc $u \rightarrow v$ in H_p is visited in LDFS, there are two cases:

case1: $u \rightarrow v$ is an arc of H. Then call the procedure ARC-LABELING for labeling $u \rightarrow v$.

case2: At least one of u, v is a dummy vertex to H; that is, $u \rightarrow v$ is one part of a long arc $x \rightarrow y$ in H. In this case, we do not label $u \rightarrow v$ separately but give a label to the whole arc $x \rightarrow y$. Thus, call the procedure ARC-LABELING for labeling $x \rightarrow y$. (In LDFS, we should be able to notice this long arc b immediately after x, and be able to reach y by the LDFS path from x.)

Step 2: Drawing. This step follows immediately Step 1 and draws H based on the output of Step 1. It consists the following two phases: drawing vertices and drawing arcs of H.

Drawing vertices. For each vertex $u \in H$, let A_u represent the set of arcs in H which are incident to u. Assume $u \in L_i$. Represent u by the horizontal line segment from $(\min_{a \in A_u}\{l(a)\}, i)$ to $(\max_{a \in A_u}\{l(a)\}, i)$.

Drawing arcs. Represent an arc $a = u \rightarrow v$ with $u \in L_i$ and $v \in L_j$ by the vertical line segment from $(l(a), i)$ to $(l(a), j)$. □

For instance, Figure 6(a) shows the result after applying Step 1 to H_2 in Figure 2(b), and Figure 6(b) illustrates the result after applying Step 2 in the algorithm GVP to the output (Figure 6(a)) of Step 1.

It can be verified [12] that the drawing given by Step 2 respects the given planar embedding; and thus,

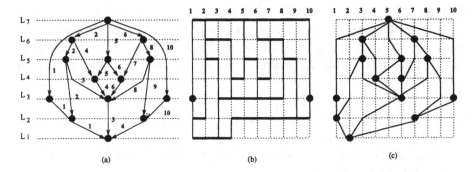

Fig. 6. Apply Algorithm GVP and Algorithm GRID_DRAW to H_2

Lemma 4. *The algorithm GVP gives a grid visibility representation of H respecting a given planar embedding E_H.*

Applying similar arguments as used in [4], we can immediately show that the grid visibility representation given by the algorithm GVP occupies drawing area $O(n^2)$. Furthermore, we can show:

Theorem 5. *Respecting a planar embedding E_H of a hierarchically planar graph H, the grid visibility representation of H produced by the algorithm GVP has the minimum drawing area.*

Sketch of the proof: It can be shown in [12], based on induction, that every arc has been allocated on the "most left-possible" vertical line. The theorem immediately follows. □

It is easy [12] to implement the algorithm GVP in time $O(n + L)$ if the ordering of outgoing arcs from each vertex is pre-specified in a given E_H, where L is the total lengths of long arcs in H. If such ordering has not been provided for each vertex, then the algorithm runs in time $O(n \log n + L)$.

5 Conclusions and Remarks

In this paper, we have shown an exponential area lower bound for planar straight-line drawings of hierarchically planar graphs without transitive arcs in contrast to the result [6] for upward planar drawing. An efficient algorithm has been presented for producing a grid visibility representation with the minimal drawing area.

Finally, we should note that if the algorithm GRID_DRAW is applied to the output of the algorithm GVP, then a grid polyline drawing is obtained, which guarantees the following properties:

- each long arc is represented by a poly line with at most two bends;
- the drawing area is $O(n^2)$.

Figure 4(c) shows the result after applying the algorithm GRID_DRAW to the drawing in Figure 4(b).

Note that our drawing algorithms do not necessarily produce a symmetric drawing when a graph is symmetric. This will be our future study.

References

1. J. A. Bondy and U. S. R. Murty, *Graph Theory with Application*, The Macmilan Press LTD, 1977.
2. G. Di Battista, P. D. Eades, R. Tamassia, and I. Tollis, Algorithms for Automatic Graph Drawing: An Annotated Bibliography, *Computational Geometry: Theory and Application*, 4, 235-282, 1994.
3. G. Di Battista and E. Nardelli, Hierarchies and Planarity Theory, *IEEE Tran. Sys. Man Cybern.* 18(6), 1035-1046, 1988.
4. G. Di Battista and R. Tamassia, Algorithms for Plane Representations of Acyclic Diagraphs, *Theoretical Computer Science*, 61, 175-198, 1988.
5. G. Di Battista and R. Tamassia, On-Line Planarity Testing, *SIAM Journal on Computing*, 25(5), 1996.
6. G. Di Battista, R. Tamassia, and I. G. Tollis, Area Requirement and Symmetry Display of Planar Upward Drawings, *Discrete & Computational Geometry*, 7(381-401), 1992.
7. P. Eades, Q. Feng and X. Lin, Straight-Line Drawing Algorithms for Hierarchical Graphs and Clustered Graphs, *Graph Drawing'96*, LNCS 1190, Springer-Verlag, 113-128, 1996.
8. P. Eades, X. Lin and R. Tamassia, An Algorithm for Drawing a Hierarchical Graph, *International Journal of Computational Geometry and Applications*, 6(2), 145-155, 1996.
9. A. Garg and R. Tamassia, On the Computational Complexity of Upward and Rectilinear Planarity Testing, *Graph Drawing'94*, LNCS, Springer-Verlag, 286-297, 1995.
10. L. S. Heath and S. V. Pemmaraju, Recognizing Leveled-Planar Dags in Linear Time, *Draph Drawing'95*, LNCS 1027, Springer-Verlag, 1995.
11. X. Lin, *Analysis of Algorithms for Drawing Graphs*, PHD thesis, University of Queensland, 1992.
12. X. Lin and P. Eades, Area Requirement for Drawing Hierarchically Planar Graphs, Tec. Rep., The University of Western Australia, 1997.

A Short Proof of a Gauss Problem [*]

H. de Fraysseix and P. Ossona de Mendez

CNRS UMR 0017, EHESS, 54 Boulevard Raspail, 75006, Paris, France.
email : hf@ehess.fr, pom@ehess.fr

Abstract. The traversal of a self crossing closed plane curve, with points of multiplicity at most two, defines a double occurrence sequence.
C.F. Gauss conjectured [2] that such sequences could be characterized by their interlacement properties. This conjecture was proved by P. Rosenstiehl in 1976 [15]. We shall give here a simple self-contained proof of his characterization. This new proof relies on the D-switch operation.

1 Introduction

We first recall and introduce some definitions and notations concerning geometric properties of closed plane curves. For related topics, we refer the reader to the bibliography. P. Rosenstiehl exposed recently a new proof of this theorem, based on patches, that will soon be published.

A *parameterized curve* C is a continuous mapping $C : [0,1] \to \mathbb{R}^2$, such that $C(0) = C(1)$ and such that the *underlying curve* $C([0,1])$ of C has a finite number of multiple points, which all have multiplicity two. Let $P(C)$ denote the set of the points of multiplicity two. To any point $p \in P(C)$, we associate the two parameter values t'_p and t''_p, such that $t'_p < t''_p$ and $C(t'_p) = C(t''_p) = p$. A point $p \in P(C)$ is a *crossing point* if any local deformation of C in a neighborhood of t'_p preserves the existence of a double point. Otherwise, p is a *touching point*. A *touch curve* (resp. a *cross curve*) is a parameterized curve with touching points (resp. crossing points) only.

There are two different types of touching points, depending on the local behavior of the parameterized curve :

Type 1   Type 2

Remark. All the touch points of a touch curve are of type 1.

The sequence of the points of $P(C)$ encountered while the parameter t goes from 0 to 1 (excluded) is the *traversal sequence* of C and is denoted by $S(C)$.

In the following, *sequences* are understood to have two occurrences of each symbol and to be defined up to reversal and cyclic permutation. Given a sequence S, two symbols p, q are *interlaced* in S if exactly one occurrence of q appears in S between the two occurrences of p. We shall denote by $\Lambda(S)$ the *interlacement graph* of S defined by the interlacement relation in S.

A sequence S is *realized* by a parameterized curve C if S is the traversal sequence of C. A sequence is *touch realizable* (resp. *cross realizable*) if it can be realized by a touch curve (resp. a cross curve).

[*] This work was partially supported by the **Esprit LTR Project no 20244-ALCOM IT.**

2 Switches and D-switches

Let us introduce the *switch* operation [4, 8] : Given a point p of $P(C)$, the curve $C' = C \circ p$ is defined by :

$$C'(t) = \begin{cases} C(t), & \text{if } t \notin [t'_p, t''_p] \\ C(t'_p + t''_p - t), & \text{if } t \in [t'_p, t''_p] \end{cases} \tag{1}$$

This curves as the same touching and crossing points as C, with the possible exception of p. The traversal sequence of C' is obtained from the one of C by inverting the order of the points encountered between the two occurrences of p. We shall say that the points that are interlaced with p have been *inverted*. The switch operation on S will be denoted by $S \circ p$, so that : $S(C \circ p) = S(C) \circ p$. Let us remark that these switch operations are involutions : $C \circ p \circ p = C$ and $S \circ p \circ p = S$.

Remark. A switch at a point p of a parameterized curve transforms p in the following way :

- touching point of type 1 $\leftrightarrow$ crossing point,
- touching point of type 2 $\leftrightarrow$ touching point of type 2.

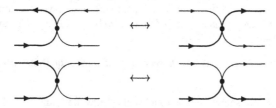

Remark. If q is a touching point of C different from p, then q is a touching point with a type different in C and $C \circ p$ if and only p and q are interlaced (that is if q has been inverted by the switch at p).

The switch of a point p in a sequence S induces a *local complementation* of p in the interlacement graph $\Lambda(S)$: two symbols a, b are adjacent in $\Lambda(S \circ p)$ if and only if

- a or b is not adjacent to p in $\Lambda(S)$ and (a, b) is an edge of $\Lambda(S)$, or
- a and b are both adjacent to p in $\Lambda(S)$ and (a, b) is not an edge of $\Lambda(S)$.

For sake of simplicity, the local complementation of p in $\Lambda(S)$ will be denoted by $\Lambda(S) \circ p$, so that $\Lambda(S \circ p) = \Lambda(S) \circ p$.

Let S be a sequence, a *D-switch* at p consists in a switching at p and in the adding of two occurrences of a new symbol p' (called *twin* of p), one just after the first occurrence of p and one just before the second occurrence of p.

$$S = (\alpha p \beta p \gamma) \mapsto S \infty p = (\alpha p p' \beta^{-1} p' p \gamma)$$

A D-switch of p in S corresponds in $\Lambda(S)$ to a local complementation of p and the addition of a new vertex having the same neighbors as p. This graph operation will be similarly denoted by '∞', so that $\Lambda(S) \infty p = \Lambda(S \infty p)$.

Remark. The sequence obtained from $S \infty p \infty p$ by deleting the two twins of p is equal to S.

3 On realizable sequences

We first state two propositions proved by Dehn, which follow from the remarks of the preceding sections.

Proposition 1. *Consider a cross curve C and any given order $(p_1, \ldots, p_n)$ of the points of C. Then, the parameterized curve $C \circ p_1 \circ \ldots \circ p_n$ obtained from C by switching successively the p_i is a touch curve.* $\qquad\qquad\square$

The converse of this proposition is not true (e.g. the sequence $(abab)$ is not cross realizable).

Remarks. Let us note by $S \xrightarrow{\circ} S'$ the existence of an order $(p_1, \ldots, p_n)$ of the symbols of S, such that $S' = S \circ p_1 \circ \ldots \circ p_n$.

- A cross realizable sequence does not determine the cross curve itself up to an homeomorphism : actually, a cross realizable sequence S can be proved to be realized by $2^{c(\Lambda(S))-1}$, where $c(\Lambda(S))$ is the number of connected components of the interlacement graph of S.
- One may find a cross realizable sequence S_1, a non cross realizable sequence S_2 and a touch realizable sequence S_T, such that $S_1 \xrightarrow{\circ} S_T$ and $S_2 \xrightarrow{\circ} S_T$. Actually, S_1 and S_2 may be proved to have different interlacement graph (using the main theorem).
- Two different cross realizable sequences S_1 and S_2 may have the same interlacement graph (e.g. the sequences $(abcaefdcbefd)$ and $(acbaefdbcefd)$). However, no sequence S_T satisfies $S_1 \xrightarrow{\circ} S_T$ and $S_2 \xrightarrow{\circ} S_T$.

Proposition 2. *A sequence S is touch realizable if and only if its interlacement graph $\Lambda(S)$ is bipartite.*

Proof. The figure bellow shows how a touch curve may be transformed into a bipartite chord diagram with the same interlacement (and conversely). $\qquad\qquad\square$

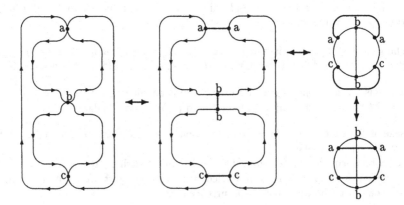

Theorem 3. *Let S be a sequence, and let $(p_1, \ldots, p_n)$ be any order on its symbols. Then, S is cross realizable if and only if the sequence $S_n = S \infty p_1 \infty \ldots \infty p_n$ obtained by successively D-switching the p_i has a bipartite interlacement graph.*

Proof. – Assume S is realized by a cross curve C. As a D-switch of a crossing point of a parameterized curve gives rise to two touching points (that will never become crossing points again), the curve C is iteratively transformed into a touch curve C_n. The traversal sequence S_n of C_n has hence a bipartite interlacement graph.

- Conversely, assume that S_n has a bipartite interlacement graph.

  Let $S_i = S \otimes p_1 \otimes \ldots \otimes p_i$ denote the sequence obtained after the first i D-switches, we shall inductively construct (for i going from n to 0) a parameterized curve C_i, that realizes S_i, and such that the crossings of C_i are the p_j, with $j > i$. Then, the parameterized curve C_0 will be a cross curve realizing S.

  As $\Lambda(S_n)$ is bipartite, there exists a touching curve C_n whose traversal sequence is S_n. If p_i is of type 1, the suppression of p_i' and the switch of p_i transforms p_i into a crossing point and gives rise to a parameterized curve C_{i-1}, having $p_i, \ldots, p_n$ as crossing points and S_{i-1} as traversal sequence. The recursion is then complete if only this case may occur.

  So, we shall prove that p_i is always of type 1 in C_i, that is that p_i has been inverted an even number of times during the D-switch at $p_i, \ldots, p_n$: The symbol p_i and its twin p_i' are not interlaced in S_i, they are alternatively interlaced and not interlaced after each further inversion and, if p_i has been last inverted by a switch at p_j, p_i (resp. p_i') and p_j are interlaced in S_n. As $\Lambda(S_n)$ is bipartite, p_i and p_i' are not interlaced in S_n (else p_i, p_i', p_j would define a triangle of $\Lambda(S_n)$). Hence, the symbol p_i has been inverted an even number of times.

  $\square$

Remark. A cross curve realizing the sequence S could be geometrically derived from a touch curve realizing the sequence S' obtained from S_n by suppressing all twined letters by transforming each touching point into a crossing point.

4 Proof of Rosenstiehl's Theorem

Definition 4. Let G be a graph and let (A, B) be a bipartition of its vertex set.

The property $P(G; A, B)$ is satisfied by a pair $\{u, v\}$ of vertices of G whenever the following equivalence holds :

- the vertices u and v have an odd number of common neighbors,
- the vertices u and v are adjacent and belong to the same class (A or B).

Lemma 5. *Let G be a graph with a vertex bipartition A, B and let p be a vertex of G. Let $G' = G \otimes p$ and let A', B' be the vertex bipartition of G' defined by : $A' = A + N(p), B' = B + N(p)$ and assigning p' to the class of p.*

If G is eulerian and any pair $\{u, v\}$ of vertices of G satisfies $P(G; A, B)$, then G' is eulerian and any pair u, v of vertices of G' satisfies $P(G'; A', B')$.

Proof. We have the following relationship between the neighborhood $N_{G'}$ in G' and the neighborhood N_G in G :

- $N_{G'}(u) = N_G(u)$, if u is not adjacent to p (in G or equivalently in G'),
- $N_{G'}(p') = N_{G'}(p) = N_G(p)$,
- $N_{G'}(u) = N_G(u) + u + N_G(p) + p'$, if u is adjacent to p.

In order to prove that G' is eulerian, we only have to check that the neighbors of p have an even degree : the parity of $N_{G'}(u) = N_G(u) + u + N_G(p) + p'$ is the sum of the parities of $N_G(u)$, $\{u\}$, $N_G(p)$ and $\{p\}$ and hence is even.

Now, we shall prove that any pair $\{u, v\}$ of vertices of G' satisfies $P(G'; A', B')$. If u or v is p', we shall replace it by p as p and p' have the same neighbors, are not adjacent and belong to the same class (A', B').

If two vertices u, v are not adjacent or equal to p, then their adjacencies, their class and their number of common neighbors are the same in G and G'. Thus, the pair u, v satisfies $P(G'; A', B')$.

If u is adjacent to p and v is not adjacent or equal to p, then

$$<N_{G'}(u), N_{G'}(v)> = <N_G(u) + u + N_G(p) + p', N_G(v)>$$
$$= <N_G(u), N_G(v)> + 1$$

As u and v belong to the same class (A', B') if and only if they do not belong to the same class (A, B) and as they are adjacent, the pair u, v satisfies $P(G'; A', B')$.

If u and v are both adjacent to p, then

$$<N_{G'}(u), N_{G'}(v)> = <N_G(u) + u + N_G(p) + p', N_G(u) + v + N_G(p) + p'>$$
$$= <N_G(u), N_G(v)> + <N(u), N(p)> + <N(v), N(p)> + 1$$

As u is adjacent to p, $<N(u), N(p)> = 1$ if and only if u and p belong to the same class (A, B). So, $<N(u), N(p)> + <N(v), N(p)> + 1 = 1$ if and only if u and v belong to the same class (A, B). As $<N_G(u), N_G(v)> = 1$ if and only if u and v are adjacent in G and belong to the same class (A, B), $<N_{G'}(u), N_{G'}(v)> = 1$ if and only if u and v are not adjacent in G and belong to the same class (A, B), that is, if and only if they are adjacent in G' and belong to the same class (A', B'). Thus, the pair u, v satisfies $P(G'; A', B')$. $\square$

Lemma 6. *Let G be a graph with a vertex bipartition A, B and let p be a vertex of G. Let $G' = G \otimes p$ and let A', B' be the vertex bipartition of G' defined by : $A' = A + N(p)$, $B' = B + N(p)$ and assigning p' to the class of p. If G' is eulerian and any pair $\{u, v\}$ of vertices of G' satisfies $P(G'; A', B')$, then G is eulerian and any pair $\{u, v\}$ of vertices of G satisfies $P(G; A, B)$.*

Proof. By Lemma 5, $G'' = G' \otimes p$ has the requested property and this property is still satisfied when deleting the two twins of p. $\square$

Theorem 7 (ROSENSTIEHL)[15]. *A sequence S is cross realizable if and only if its interlacement graph $\Lambda(S)$ satisfies :*

- *the graph is eulerian,*
- *for any non-edge (p, p') of the graph, $N(p) \cap N(p')$ is even,*
- *the set of the of the edges (p, p') of the graph such that $N(p) \cap N(p')$ is even is a cocycle of the graph.*

Proof. The theorem may be restated as follows : A sequence S is cross realizable if and only if its interlacement graph $\Lambda(S)$ is eulerian and if there exists a bipartition A, B of the vertex set of $\Lambda(S)$ such that any pair u, v of vertices of $\Lambda(S)$ satisfies $P(\Lambda(S); A, B)$.

Consider any sequence $S_n = S \otimes p_1 \otimes \ldots \otimes p_n$ obtained by successively D-switching the symbols of S. According to Lemma 5, $\Lambda(S_n)$ is eulerian and has a bipartition A', B' such that any pair of vertices of $\Lambda(S_n)$ satisfies $P(\Lambda(S_n); A', B')$. As all the symbols have been twined and as p and its twin p' have the same neighbors, any two vertices of $\Lambda(S_n)$ have an even number of common vertices. According to property $P(\Lambda(S_n); A', B')$, the graph $\Lambda(S_n)$ is bipartite. Then, from Theorem 3, S is cross realizable.

Conversely, if S is cross realizable, any sequence of D-switches gives rise to a sequence S' having a bipartite interlacement graph. This graph is eulerian (due to the doubling of each symbol) and a bipartition A, B induced by a bicoloration, is such that each pair of vertices satisfies $P(\Lambda(S'); A, B)$. The theorem then follows from Lemma 6. $\square$

5 Matroidal Interpretation

As we wanted to give a short self-contained proof, we did not introduce the usual concepts of binary matroids. In such a context, a proof could be done, relying on the following properties : The graphs which satisfy the conditions given for $\Lambda(S)$ in Rosenstiehl's characterization are exactly the principal interlacement graph of some binary matroid M [14]. Any local complementation of the vertices of such a graph gives rise to a bipartite graph, which is the fundamental interlacement graph of M with respect to some base B of M [3]. The further condition that a principal interlacement graph is an interlacement graph (that is a circle graph) implies that the matroid M is planar and then, the principal interlacement graph corresponds to the interlacement of a left-right path of a planar realization of M [3].

References

1. A. Bouchet. Caractérisation des symboles croisés de genre nul. *C.R. Acad. Sci.*, 274:724–727, 1972. (Paris).
2. H. de Fraysseix. Sur la représentation d'une suite à triples et à doubles occurrences par la suite des points d'intersection d'une courbe fermée du plan. In *Problèmes combinatoires et théorie des graphes*, volume 260 of *Colloques internationaux C.N.R.S.*, pages 161–165. C.N.R.S., 1976.
3. H. de Fraysseix. Local complementation and interlacement graphs. *Discrete Mathematics*, 33:29–35, 1981.
4. M. Dehn. Über kombinatorische topologie. *Acta Math.*, 67:123–168, 1936. (Sweden).
5. H. Fleischner. Cycle decompositions, 2-coverings, removable cycles, and the four-color-disease. *Progress in Graph Theory*, pages 233–246, 1984.
6. G.K. Francis. Null genus realizability criterion for abstract intersection sequences. *J. Combinatorial Theory*, 7:331–341, 1969.
7. C.F. Gauss. *Werke*, pages 272 and 282–286. Teubner Leipzig, 1900.
8. A. Kotzig. Eulerian lines in finite 4-valent graphs and their transformations. In *Proceedings of the Colloquium held at Tihany, Hungary*, pages 219–230, 1969.
9. L. Lovász and M.L. Marx. A forbidden subgraph characterization of gauss codes. *Bull. Am. Math. Soc.*, 82:121–122, 1976.
10. M.L. Marx. The gauss realizability problem. *Trans Am. Math. Soc.*, 134:610–613, 1972.
11. J.V.Sz. Nagy. Über ein topologisches problem von gauss. *Maht. Z.*, 26:579–592, 1927. (Paris).
12. R.C. Read and P. Rosenstiehl. On the gauss crossing problem. In *Colloquia Mathematica Societatis János Bolyai*, pages 843–875, 1976. (Hungary).
13. R.C. Read and P. Rosenstiehl. On the principal edge tripartition of a graph. *Annals of Discrete Maths*, 3:195–226, 1978.
14. P. Rosenstiehl. Les graphes d'entrelacement d'un graphe. In *Problèmes combinatoires et théorie des graphes*, volume 260 of *Colloques internationaux C.N.R.S.*, pages 359–362. C.N.R.S., 1976.
15. P. Rosenstiehl. Solution algébrique du problème de gauss sur la permutation des points d'intersection d'une ou plusieurs courbes fermées du plan. *C.R. Acad. Sci.*, 283 (A):551–553, 1976. (Paris).
16. P. Rosenstiehl. A geometric proof of a Gauss crossingg problem. (to appear).
17. P. Rosenstiehl and R.E. Tarjan. Gauss codes, planar hamiltonian graphs, and stack-sortable permutations. *Jour. of Algorithms*, 5:375–390, 1984.
18. H. Shank. The theory of left-right paths. In *Combinatorial Math.*, volume III of *Lecture Notes in Math.*, pages 42–54. Springer, 1975.
19. W.T. Tutte. On unicursal paths in a network of degree 4. *Amer. Math. Monthly*, 4:233–237, 1941.

A Bayesian Paradigm for Dynamic Graph Layout

Ulrik Brandes and Dorothea Wagner

Fakultät für Mathematik und Informatik
Universität Konstanz, Germany
{Ulrik.Brandes, Dorothea.Wagner}@uni-konstanz.de

Abstract. *Dynamic graph layout* refers to the layout of graphs that change over time. These changes are due to user interaction, algorithms, or other underlying processes determining the graph. Typically, users spend a noteworthy amount of time to get familiar with a layout, i.e. they build a *mental map* [ELMS91]. To retain this map at least partially, consecutive layouts of similar graphs should not differ significantly. Still, each of these layouts should adhere to constraints and criteria that have been specified to improve meaning and readability of a drawing.

In [BW97], we introduced random field models for graph layout. As a major advantage of this formulation, many different layout models can be represented uniformly by random variables. This uniformity enables us to now present a framework for dynamic layout of arbitrary random field models. Our approach is based on Bayesian decision theory and formalizes common sense procedures. Example applications of our framework are dynamic versions of two well-known layout models: Eades' spring embedder [Ead84], and Tamassia's bend-minimum orthogonal layout model for plane graphs [Tam87].

1   Introduction

A sequence of graphs that arises from repeated modification of an initial graph is called a *dynamic graph*. Dynamic graphs occur in many settings, including user interaction, software visualization, animation of graph algorithms, and graph queries. When drawing graphs dynamically, updates shall be economic in the sense that human as well as computer resources are used sparingly. With human resources we mean the effort and time a user spents in order to follow an update and regain familiarity with the drawing. Despite of the large interest in automatic graph drawing, the important extension to dynamic graphs has received little attention, yet. This may partially be due to the lack of a conceptual framework abstracting from issues inherent to specific models. Apparently, most approaches in the literature are tied to classes of admissible graphs [BP90,CDT+92,CDTT95,Nor96,PT96].

A fairly general formulation of the dynamic graph drawing problem is given by North [Nor96]. We use his formalization with minor adjustments. For each graph in the sequence, prescriptions are given for its *representation* and *rendering*, as well as a set of *layout constraints* and *readability criteria*. These define a sequence of static graph drawing problems. Since the computation of a layout and its rendering can typically be separated, we neglect everything related to imaging, and focus on the layout problem. Consecutive layouts are subject to *consistency*, *stability*, and *readability* demands. These can be seen as additional constraints (for consistency) and criteria (for stability and readability of updates), respectively. Moreover, the actual update of a drawing might consist of a number of intermediate drawings (the *physical update*), e.g. to smooth transitions or to add visual update clues. Here, we are only concerned with the *logical update*, i.e. the final layout for each graph in the sequence.

Our approach to dynamic graph drawing relies on the formulation of layout models in terms of random fields [BW97] (see [Guy94] for an introduction to random fields). In a random field model, feasible layouts are assigned probabilities reflecting their conformance to layout goals. A single random variable is sufficient to fully describe a layout model. Random fields are used in many other areas, particularly including image processing [Win95]. In the introduction of [GM85], Geman and McClure explicitly stated their hope to achieve a unification of many image processing problems through a Bayesian framework, which was pioneered in [GG84]. To say the least, random fields and the Bayesian paradigm have influenced a substantial part of the subsequent research [CJ93]. In this paper, we undertake an adaption to dynamic graph layout. We argue that dynamic graphs should be laid out according to an *a-posteriori* model, which forms a compromise between the given static, or *a-priori*, model and some stability criteria.

The main interest of this paper is the derivation of a formal concept for dynamic graph layout. The advantage of an abstract, generic framework is that it provides a common foundation for dynamic versions of arbitrary layout models. Experimental results would be interesting for special cases, but we rather want to emphasize the conceptual elegance and broad applicability of our approach.

This paper is organized as follows. In Section 2 we briefly recall the definition of a random field layout model. The Bayesian approach to dynamic graph layout is developed in Section 3, and applied to a spring model and to bend-minimum orthogonal drawings of embedded planar graphs in the succeeding sections.

2 Preliminaries

A layout of a single graph $G = (V, E)$ is computed by assigning values to certain layout variables. Straight-line embeddings, for example, are completely determined by an assignment of coordinates to each vertex. More general, each element of a set $L = \{l_1, \ldots, l_k\}$ of *layout elements* is assigned a value from a set of allowable *states* $\mathcal{X}_l$, $l \in L$. Clearly, L and $\mathcal{X} = \mathcal{X}^L = \mathcal{X}_{l_1} \times \cdots \times \mathcal{X}_{l_k}$ depend on the desired type of representation. Every vector $x \in \mathcal{X}$ is called a *layout*.

In a random field model [BW97], layouts $x \in \mathcal{X}$ are assigned probabilities reflecting their conformance to layout criteria. These probabilities are based on configurations of subsets of layout elements which mutually affect their states. This interaction of layout elements is modelled by an *interaction graph* $G^\eta = (L, E^\eta)$ that is obtained from a *neighborhood system* $\eta = \bigcup_{l \in L} \eta_l$, where $\eta_l \subseteq$

$L \setminus \{l\}$ is the set of layout elements for which the state assigned to l is relevant in terms of layout quality. Because of physical analogies, we assume that interactions are symmetric. In particular $l_2 \in \eta_{l_1} \Leftrightarrow l_1 \in \eta_{l_2}$ for all $l_1, l_2 \in L$, so G^η is undirected. The set of cliques in G^η is denoted by $\mathcal{C} = \mathcal{C}(\eta)$. Each clique in $\mathcal{C}$ corresponds to a set of pairwise interacting layout elements. By symmetry, the internal interactions are not visible from outside of the clique, but form a potential. We therefore define the *interaction potential* of clique $C \in \mathcal{C}$ to be any function $U_C : \mathcal{X} \to \mathbb{R}$ for which

$$x_C = y_C \quad \Rightarrow \quad U_C(x) = U_C(y)$$

holds for all $x, y \in \mathcal{X}$, where, for $x = (x_l)_{l \in L} \in \mathcal{X}$ and $C \in \mathcal{C}$, $x_C = (x_l)_{l \in C}$. Moreover, x^{y_C} denotes a vector that agrees with $y \in \mathcal{X}$ on $C \in \mathcal{C}$ and with $x \in \mathcal{X}$ on $L \setminus C$. The *energy* $U : \mathcal{X} \to \mathbb{R}$ of a layout equals the sum of all interaction potentials, i.e. $U(x) = \sum_{C \in \mathcal{C}} U_C(x)$. Motivated by results from statistical mechanics and thermodynamics, the probability of a layout x is set to

$$P(X = x) = \frac{1}{Z} e^{-U(x)},$$

where Z is a normalizing constant.[1] Clearly, these probabilities depend on the energy only. For convenience, both the random variable X and its distribution $P(X = x)$ are called a (random field) *layout model* for G. Consequently, a layout of low energy, i.e. little interaction of layout elements, is more likely than a layout of high energy. Many common layout models can be described within this framework [BW97].

3 Dynamic Layout

In this section, we introduce a generic approach to dynamically layout a (finite or infinite) sequence of graphs $G_1, G_2, \ldots$, for which (static) layout models $X_1, X_2, \ldots$ are given. The objective function of a random field layout model is its probability measure. Therefore, we would like to derive models, such that the most desirable sequence of layouts maximizes the joint probability $P(X_1 = x_1, X_2 = x_2, \ldots)$. In case the layout models are assumed to be independent, we have

$$P(X_1 = x_1, X_2 = x_2, \ldots) = \prod_i P(X_i = x_i),$$

which is maximized by maximizing each $P(X_i = x_i)$. Independent random fields correspond to a strategy that layouts each individual graph according to its own model and does not care about the user's mental map (except for, possibly, in the physical update). Since smooth transitions between consecutive layouts are pursued, dependencies among the individual models must be introduced. In the remainder of this section, we develop a Bayesian approach.

---

[1] For readability, we sometimes use $\exp\{x\}$ to denote e^x.

The sequence of graphs to be drawn is typically assumed to be infinite when it is generated over an unknown period of time and the graphs change in an unpredictable manner. Hence, neither the next graph, nor the number of graphs to come is known at any time. Such situations occur in user interaction, network control, phone-call recording, and so on. For a finite sequence $G_1, \ldots, G_t$, the joint probability can be rewritten into conditional (transition) probabilities

$$P(X_1 = x_1, \ldots, X_t = x_t) = \prod_{i=1}^{t} P(X_i = x_i \mid X_{<i} = x_{<i}), \qquad (1)$$

where $X_{<i} = x_{<i}$ is shorthand for $X_1 = x_1, \ldots, X_{i-1} = x_{i-1}$. However, knowledge of all graphs and dependencies is still required to obtain any x_i of a sequence $x_1, \ldots, x_t$ that maximizes (1). Such knowledge is typically not provided.[2]

Throughout this paper we therefore assume that each layout of the sequence has to be computed before anything about the next graph is known, i.e. no look-ahead is available. Thus, the following formalization of the dynamic layout problem is obtained: At time $t > 1$, we are given graphs $G_1, \ldots, G_t$, static layout models $X_1, \ldots, X_t$, and layouts $x_1, \ldots, x_{t-1}$. The goal is to compute a layout $x_t \in \mathcal{X}_t$ that forms a compromise between stability and readability. Obviously, the conditional probability $P(X_t = x_t \mid X_{<t} = x_{<t})$ must reflect this notion of compromise. It is hence called the *dynamic layout model* of G_t.

Suppose, $\lambda : \mathcal{X}_t \times \mathcal{X}_t \to \{0,1\}$ is the (imaginary) zero-one loss function of choosing x_t, when the best choice is x, i.e.

$$\lambda(x_t, x) = \begin{cases} 0 & x_t = x \\ 1 & x_t \neq x. \end{cases}$$

With this loss function, the risk $r(x_t) = \sum_{x \in \mathcal{X}_t} \lambda(x_t, x) P(X_t = x \mid X_{<t} = x_{<t}) = 1 - P(X_t = x_t \mid X_{<t} = x_{<t})$ of selecting x_t equals the average probability of error. It is minimized by choosing an x_t for which

$$P(X_t = x_t \mid X_{<t} = x_{<t})$$

is maximized. Observe that other measures of loss yield other decision rules.

As a uniform means to obtain suitable dynamic layout models, we propose a Bayesian approach which basically provides a formalization of common sense. Note that, by Bayes' rule,

$$\max_{x_t \in \mathcal{X}_t} P(X_t = x_t \mid X_{<t} = x_{<t})$$
$$= \max_{x_t \in \mathcal{X}_t} \frac{P(X_{<t} = x_{<t} \mid X_t = x_t) \cdot P(X_t = x_t)}{P(X_{<t} = x_{<t})}$$
$$\propto \max_{x_t \in \mathcal{X}_t} P(X_{<t} = x_{<t} \mid X_t = x_t) \cdot P(X_t = x_t),$$

where $\propto$ means "proportional to". $P(X_t = x_t)$ is easily recognized to be the static layout model for G_t. It therefore reflects the notion of readability formalized in

---

[2] Observe that there are applications, like animation, where the complete sequence is indeed known in advance.

X_t, and is called the *static model*. $P(X_{<t} = x_{<t} | X_t = x_t)$ is the *likelihood* of a sequence $x_1, \ldots, x_{t-1}$ to result in a (given) layout x_t of G_t. Consequently, it should express our notion of stability, and therefore is called the *stability model*.

In summary, we have argued how dynamic layout models $P(X_t = x_t | X_{<t} = x_{<t})$ can be derived from a sequence of static layout models $P(X_t = x_t)$ by introducing stability models $P(X_{<t} = x_{<t} | X_t = x_t)$. By the very nature of a random field formulation, joint maximization of these two components results in a compromise between those criteria describing readability and those describing stability of layouts. We have thus developed an abstract formulation of a general principle for dynamic graph layout that incorporates given static models.

The following two sections are devoted to examples providing evidence that this formulation results in a general method for uniform integration of stability criteria in a sequence of initially static layout problems.

4 Dynamic Spring Layout

The *spring embedder* [Ead84] is one of the most well-known layout models for straight-line representations. A graph is modelled by a physical system of rings corresponding to the vertices, and springs corresponding to the edges. Rings of adjacent vertices are joined by springs, whereas rings of non-adjacent vertices are repelling. The spring embedder then aims to produce a layout that corresponds to a stable configuration of the system.

The layout elements of a random field formulation of a spring embedder for a graph $G = (V, E)$ simply are the vertices, i.e. $L = V$. Sets of admissible states $\mathcal{X}_v$, $v \in V$, are locations in two or three-dimensional space. In this section, we use attractive and repelling forces between vertices as introduced in the spring embedder variant of [DH96]. Each pair of vertices $u \neq v \in V$ is assigned an interaction potential

$$U_{\{u,v\}}(x) = \begin{cases} \frac{c_1}{d(x_u, x_v)^2} + c_2 \cdot d(x_u, x_v)^2 & \text{if } \{u, v\} \in E \\ \frac{c_1}{d(x_u, x_v)^2} & \text{otherwise} \end{cases}$$

such that the static layout model for G is

$$P(X = x) = \frac{1}{Z} e^{-\sum\limits_{u \neq v \in V} U_{\{u,v\}}(x)}$$

with constant parameters $c_1, c_2 > 0$ and normalizing constant Z. $d(x_u, x_v)$ denotes the Euclidean distance of locations x_u and x_v. While c_1 controls the strength of repelling forces between each pair of vertices, c_2 can be used to define the ideal edge length.[3]

The above spring model is now extended to dynamic graphs. For stability, we use notions of change in vertex locations ("anchoring"), relative locations ("stiffening"), and an accumulated version of the latter. These result in three

---

[3] Potentials $\frac{c_1}{d(x_u, x_v)^2} + c_2 \cdot d(x_u, x_v)^2$ are minimized for $d(x_u, x_v) = \sqrt[4]{\frac{c_1}{c_2}}$.

different dynamic models for G_t. In each case we obtain a new random field model that compromises between readability and stability as proposed in the preceding section.

Anchoring. Let us assume stability is demanded only with respect to consecutive layouts. To simplify notation, we use a number of abbreviations. Let $X = X_t$ be the static spring model for G_t, and $Y = X_{t-1}$ be the static spring model for G_{t-1}, $t > 1$. Furthermore, let V_X, V_Y, and $V_{X,Y}$ denote the vertices of G_t, the vertices of G_{t-1}, and their intersection, respectively. U^X and U^Y denote the energy function with respect to the edge set of G_t and G_{t-1}, respectively, and so on. According to the above dependency assumption, the dynamic model satisfies the Markov property

$$P\left(X_t = x_t \mid X_{<t} = x_{<t}\right) = P\left(X_t = x_t \mid X_{t-1} = x_{t-1}\right)$$
$$= P\left(X = x \mid Y = y\right),$$

such that our formal expression of stability reduces to $P(Y = y \mid X = x)$. Since a straight-line embedding is completely determined by locations assigned to vertices, a very natural criterion of stability is the absence of excessive movement of vertices between consecutive layouts. More formally, let the likelihood of $y \in \mathcal{Y}$ leading to $x \in \mathcal{X}$ be measured by independent two-dimensional Gaussian distributions with mean in the conditioning location x_v of each $v \in V_{X,Y}$:

$$P\left(Y = y \mid X = x\right) = \frac{1}{\sqrt{2\pi\sigma^2}} e^{-\sum\limits_{v \in V_{X,Y}} \frac{\|y_v - x_v\|^2}{2\sigma^2}}$$

where σ is a constant controlling the amplitude of deviation. Since this distribution is symmetric in x and y, the formula can also be read in a more intuitive way: The new location x_v is distributed normally around the current location y_v. Since $\|y_v - x_v\|^2 = d(x_v, y_v)^2$, there is an obvious correspondence to the attracting forces of [DH96].

Combining the models for stability and readability in a dynamic model for G_t yields

$$P\left(X = x \mid Y = y\right) = \frac{1}{Z_{X|Y}} \exp\left\{ - \sum_{v \in V_{X,Y}} \frac{d(y_v, x_v)^2}{2\sigma^2} - \sum_{u \neq v \in V_X} U^X_{\{u,v\}}(x) \right\}$$

with the appropriate normalizing constant $Z_{X|Y}$. Clearly, this is also a random field, and even correspondent to a force directed placement model, yet with additional forces attracting vertices to their previous locations in y. It is interesting to note that by choosing different functions for attracting and repelling forces one obtains the *anchored spring model* introduced by Lyons in the context of (static) graphs with geographic semantics (preferred locations for vertices) [Lyo92].

Stiffening. Assume now, the criterion of stability between two consecutive layouts y and x is relative instead of absolute location, or *structure* instead of *position*. In other words, consecutive layouts should retain pairwise distances rather than single vertex locations.

Let $x \in \mathcal{X}$ be the conditioning layout of the stability model. We consider a layout y of G_{t-1} likely to lead to layout $x \in \mathcal{X}$, if the forces of X are already apparent in y. Therefore, the likelihood of y, given x, is set to

$$P\left(Y = y \mid X = x\right) = \frac{1}{Z_{Y|X}} \exp\left\{ - \sum_{u \neq v \in V_Y} U^Y_{\{u,v\}}(y) - \sum_{\substack{u \neq v: \\ v \in V_{X,Y}, v \in V_Y}} U^Y_{\{u,v\}}(y^{\hat{x}}) \right\}$$

where $\hat{x} = x_{V_{X,Y}}$ and $Z_{Y|X}$ is a normalizing constant. In terms of forces, the stability model states that each vertex contained in both G_{t-1} and G_t contributes twice, once from its location in y, and once from its subsequent location in x. That is, y is a layout with shadow forces of vertices that remain in the graph, excerted from their location in the next layout. Combined with the static layout model X, this notion of stability yields the dynamic model

$$\begin{aligned}
P\,(X = x \mid Y = y) & \\
\propto P\left(Y = y \mid X = x\right) &\cdot P\left(X = x\right) \\
= \frac{1}{Z_{Y|X}} \exp &\left\{ - \sum_{u \neq v \in V_Y} U^Y_{\{u,v\}}(y) - \sum_{\substack{u \neq v: \\ u \in V_{X,Y}, v \in V_Y}} U^Y_{\{u,v\}}(y^{\hat{x}}) \right\} \\
\cdot \frac{1}{Z_X} \exp &\left\{ - \sum_{u \neq v \in V_X} U^X_{\{u,v\}}(x) \right\} \\
= \frac{1}{Z_{X|Y}} \exp &\left\{ - \sum_{\substack{u \neq v: \\ u \in V_{X,Y}, v \in V_Y}} U^Y_{\{u,v\}}(y^{\hat{x}}) - \sum_{u \neq v \in V_X} U^X_{\{u,v\}}(x) \right\}
\end{aligned}$$

with the obvious normalizing constants and $\hat{x} = x_{V_{X,Y}}$. For the last equality, note that the energy function of a random field is unique up to an additive constant. Just like the anchored dynamic model, the above represents a force directed placement model. First note that vertices $u \in V_X$ contribute with their location in $x \in \mathcal{X}$, while vertices $v \in V_Y \setminus V_X$ contribute with their location in $y \in \mathcal{Y}$. Each pair of vertices $u, v \in V_{X,Y}$ contributes twice to the energy of the dynamic model, once according to $U^Y_{\{u,v\}}$, and once according to $U^X_{\{u,v\}}$. Using the spring analogy, the spring connecting u and v is stiffened, i.e. its length remains unchanged, while its strength is increased. Pairs of vertices with exactly one vertex in $V_{X,Y}$ contribute to the energy in the usual way, such that the layout remembers deleted vertices. Unchanged parts of the graph are connected by a stiffer structure than new or altered ones.

Cumulative stiffening. The stiffening stability model is extended to long range dependencies quite easily. Without going into details, we note that pairwise interactions may be accumulated over the sequence of graphs. The longer a relation existed, the less should its displacement be likely to change. Evaluation of the resulting energy function is not quite as complex as it may seem, since time stamps for vertices and edges may be used to store the multiplicity of their contribution.

Most algorithms in force directed placement are designed to find a local minimum of the aesthetic cost function, and hence of the energy function in their corresponding random field model. In dynamic graph layout this is no drawback, but an advantage, since locations in y are suitable initial values for locations of vertices $v \in V_{X,Y}$. If consecutive graphs do not differ significantly, these algorithms should therefore be able to quickly find a satisfactory local optimum close to the initial layout.

5 Dynamic Orthogonal Layout

In this section, the Bayesian paradigm is applied to dynamic orthogonal drawings of 4-planar graphs with fixed planar embbeding. A planar graph is called 4-planar, if no vertex has degree larger than 4, and an embedding is given by a cyclic ordering of the edges incident to each vertex. In an orthogonal drawing, vertices are placed on grid points, while edges are drawn along the grid lines. Edges may overlap at grid points only.

Tamassia describes an algorithm minimizing the number of bends among all planar orthogonal drawings that preserve the embedding [Tam87]. We review briefly his transformation to a minimum cost flow problem, and state it in terms of a random field model. The static model is then combined with a reasonable criterion for stability. Finally, the resulting dynamic model is re-stated in terms of a minimum cost flow problem.

Orthogonal representation. Given a planar graph $G = (V, E)$ with maximum vertex degree 4 and a planar embedding, the drawing is determined from an *orthogonal representation* $H(G)$ of circular lists $H_f = [(e_1, s_1, a_1), \ldots, (e_r, s_r, a_r)]$ for each face f of G. Each tupel (e_i, s_i, a_i) of a list H_f consists of an edge e_i, a string $s_i \in \{0, 1\}^*$, and an integer $a_i \in \{90, 180, 270, 360\}$, such that $e_1, \ldots, e_r$ is a counterclockwise (clockwise, if f is the outer face) traversal of the edges incident to f (note that some edges may appear twice in this traversal), 0's and 1's in s_i represent 90 and 270 degree bends on the right side of e_i, respectively, and a_i is the angle between e_i and its succeeding edge (H_f is cyclic).

Flow network. In [Tam87], a one-to-one correspondence of $H(G)$ to a flow of specified value in some network $N(G) = (U, A, s, t, cap, cost)$ is shown.[4] The vertex set U of $N(G)$ consists of the vertices V of G, an additional vertex for each face of G, and new source and target vertices s and t, respectively. Let $d_G(v)$ denote the number of edges incident to vertex v, and $d_G(f)$ denote the number of edges in the circular list of face f. Arcs are introduced

---

[4] It should always be clear from context, whether t is used to denote an index, or the target vertex of a network, respectively.

- from s to each $v \in V$ (capacity $4 - d_G(v)$ and zero cost),
- from each vertex to its incident faces (infinite capacity, zero cost),
- from s to each internal face with less than 4 edges (capacity $4 - d_G(f)$, zero cost),
- from each face to its neighboring faces (infinite capacity, unit cost),
- from a face to itself, if some edge appears twice in the circular list (infinite capacity, unit cost),
- from each internal face with more than 4 boundary edges to t (capacity $d_G(f) - 4$, zero cost),
- and from the external face to t (capacity $d_G(f) + 4$, zero cost).

Then, each unit of flow represents an angle of 90 degrees, such that the cost of an (s,t)–flow of value $\sum_{(s,u)\in A} cap(s,u)$ equals the number of bends in the corresponding layout.

Random field model. Since a drawing is determined from an orthogonal representation which is obtained from a minimum cost flow in $N(G)$, the layout elements of the above representation are the arcs of $N(G)$, i.e. $L = A$. Admissible values for $a \in L$ are the integers $x_a \in X_a = \{0, \ldots, cap(a)\}$ satisfying the capacity constraint. Since the number of bends is the only criterion of readability, the energy of a layout equals its total cost

$$U(x) = \sum_{a \in A} cost(a) \cdot x_a.$$

Note that the random field model does not incorporate flow constraints. It is therefore a relaxation of the original formulation. The relaxation is used to derive a relaxed dynamic model, on which the flow constraints are then imposed again.

Dynamic model. To simplify presentation, we make a few reasonable assumptions on the dynamic layout problem. Let the layout of G_t be independent of $x_1, \ldots, x_{t-2}$. Moreover, let G_{t-1} and G_t be connected, where G_t is obtained from G_{t-1} by insertion or deletion of a single edge. Since both graphs are connected, insertion or deletion of an edge may also require insertion or deletion of one vertex, respectively.

Now, we are given networks $N(G_{t-1}) = (U_{t-1}, A_{t-1}, s, t, cap_{t-1}, cost)$ and $N(G_t) = (U_t, A_t, s, t, cap_t, cost)$, and a layout x_{t-1} corresponding to a flow of value $\sum_{(s,u)\in A_{t-1}} cap_{t-1}(s,u)$ in $N(G_{t-1})$. Since the criterion of readability is the number of bends, the most natural criterion for stability between consecutive layouts is the difference in the number of bends on edges present in both layouts,

$$\sum_{a \in A_t \cap A_{t-1}} cost(a) \cdot |(x_{t-1})_a - (x_t)_a|.$$

This criterion is taken to be the energy function of a random field model for stability. The dynamic model becomes

$$
\begin{aligned}
P\left(X_t = x_t \mid X_{t-1} = x_{t-1}\right) & \\
\propto\ & P\left(X_{t-1} = x_{t-1} \mid X_t = x_t\right) \cdot P\left(X_t = x_t\right) \\
=\ & \frac{1}{Z_{t-1|t}} \exp\left\{ - \sum_{a \in A_t \cap A_{t-1}} cost(a) \cdot |(x_{t-1})_a - (x_t)_a| \right\} \\
& \cdot \frac{1}{Z_t} \exp\left\{ - \sum_{a \in A_t} cost(a) \cdot (x_t)_a \right\} \\
=\ & \frac{1}{Z_{t|t-1}} \exp\left\{ - \sum_{a \in A_t \cap A_{t-1}} cost(a) \cdot |(x_{t-1})_a - (x_t)_a| - \sum_{a \in A_t} cost(a) \cdot (x_t)_a \right\},
\end{aligned}
$$

such that reducing the number of bends of a remaining edge is just as costly as bends on a new edge, whereas new bends on remaining edges are counted twice.

Implementation. What is probably most appealing about this model is that it can be re-translated into a minimum cost flow problem. Due to space limitations we do not go into details, but note that inserting or deleting an edge in G_{t-1} results in a graph G_t with an associated network $N(G_t)$ that differs from $N(G_{t-1})$ by a small number of vertices and arcs. Let x_{t-1} be a feasible flow of $N(G_{t-1})$ with value $\sum_{(s,u) \in A_{t-1}} cap_{t-1}(s,u)$. Moreover, let x_{t-1} have the property that, for each $(u,v) \in A_t \cap A_{t-1}$, at most one of (u,v) and (v,u) does carry positive flow (which is certainly true for a minimum cost flow). Then, minimization of the energy function of the dynamic model can be performed by solving a new minimum cost flow problem in a modified network $N'(G_t)$. For each arc $(u,v) \in A_t$, for which $(v,u) \notin A_t$, we insert a new arc (v,u) with zero capacity and $cost(v,u) = cost(u,v)$. Now, costs and capacities are modified according to x_{t-1}, where $(x_{t-1})_{(v,u)}$ equals 0, if $(v,u) \notin A_{t-1}$. For each arc (u,v), $\{u,v\} \cap \{s,t\} = \emptyset$, in the symmetric hull of $A_t \cap A_{t-1}$, let $cap'_t(u,v) = cap_t(u,v) - (x_{t-1})_{(u,v)} + (x_{t-1})_{(v,u)}$. Furthermore, let $cost'(u,v) = cost(u,v) + cost(v,u)$. Because of the flow network's special structure, these changes reflect the additional cost of changing flow x_{t-1}. A small number of minimum cost augmenting flow computations in the modified network yields a flow that can be used to transform x_{t-1} into a feasible flow x_t of $N(G_t)$ that is optimal with respect to the dynamic model. Moreover, x_t also satisfies the above property that there are no circulations of length two. It can hence be used in a dynamic model for G_{t+1}. Details will be given in a more general presentation of a Bayesian approach for dynamic orthogonal layouts with few bends, which is in preparation.

6 Discussion

We have presented a general principle for dynamic graph layout that exploits the uniformity of random field modelling. The logical update of a layout is formalized by means of a stochastic estimator composed of the static layout model and an additional stability model. If the stability model is also a random field, it is easy to see that It is possible to bias the inherent compromise between stability and readability by multiplication of a constant factor to the respective energy functions.

Our approach is neither specific to a certain layout model, nor to a set of update operations. Its underlying formalism is based on common sense, yet sound, mature and widely applied. In particular, it cleanly separates the modelling of readability and stability demands. Moreover, it does not imply the use of particular algorithms, even though, in general, algorithms used for static models require only slight modification to apply to the dynamic case as well. Moreover, the flexibility of random fields allows the easy experimentation with different stability models within the same environment. There is no need to implement a new algorithm every time the model is modified. It is still recommended, though, after a suitable model has been identified, since a general optimization procedure is almost certainly inefficient.

For particular layout models, experimental results are needed to identify suitable stability models. It will be interesting to see, how different criteria compare (and perform) in practice. However, the formal value of the framework lies in its generality, simplicity and uniformity. It should not be judged by some good or bad usage.

References

[BP90]   Karl-Friedrich Böhringer and Frances Newbery Paulisch. Using constraints to achieve stability in automatic graph layout algorithms. In *CHI'90 Proceedings*, pages 43–51. ACM, The Association for Computing Machinery, New York, 1990.

[BW97]   Ulrik Brandes and Dorothea Wagner. Random field models for graph layout. Konstanzer Schriften in Mathematik und Informatik 33, Universität Konstanz, 1997.

[CDT+92]   Robert F. Cohen, Giuseppe Di Battista, Roberto Tamassia, Ioannis G. Tollis, and P. Bertolazzi. A Framework for Dynamic Graph Drawing. In *Proc. of 8th Annual Computational Geometry, 6/92, Berlin, Germany*, pages 261–270. ACM, The Association for Computing Machinery, New York, 1992.

[CDTT95]   Robert F. Cohen, Giuseppe Di Battista, Roberto Tamassia, and Ioannis G. Tollis. Dynamic graph drawings: Trees, series-parallel digraphs, and planar *st*-digraphs. *SIAM J. Comput.*, 24(5):970–1001, 1995.

[CJ93]   R. Chellappa and A.K. Jain. *Markov Random Fields: Theory and Applications*. Academic Press, 1993.

[DH96]   Ron Davidson and David Harel. Drawing graphs nicely using simulated annealing. *ACM Transactions on Graphics*, 15(4):301–331, 1996.

[Ead84]   Peter Eades. A heuristic for graph drawing. *Congressus Numerantium*, 42:149–160, 1984.

[ELMS91]   Peter Eades, W. Lai, Kazuo Misue, and Kozo Sugiyama. Preserving the mental map of a diagram. *Proceedings of Compugraphics*, 9:24–33, 1991.

[GG84] Stuart Geman and Donald Geman. Stochastic relaxation, Gibbs distributions, and the Bayesian restoration of images. *IEEE Transactions on Pattern Analysis and Machine Intelligence*, 6(6):721–741, 1984.

[GM85] Stuart Geman and Donald E. McClure. Bayesian image analysis: an application to single photon emission tomography. *Proc. American Statistical Association, Statistical Computing Section*, pages 12–18, 1985.

[Guy94] Xavier Guyon. *Random Fields on a Network*. Springer, 1994.

[Lyo92] Kelly A. Lyons. Cluster busting in anchored graph drawing. In *Proceedings of the '92 CAS Conference/CASCON'92, Toronto, 1992*, pages 7–17, 1992.

[Nor96] Stephen North. Incremental Layout in DynaDAG. *Proceedings of GD'95*, pages 409–418. Springer-Verlag, Lecture Notes in Computer Science, vol. 1027, 1996.

[PT96] Achilleas Papakostas and Ioannis G. Tollis. Issues in Interactive Orthogonal Graph Drawing. *Proceedings of GD'95*, pages 419–430. Springer-Verlag, Lecture Notes in Computer Science, vol. 1027, 1996.

[Tam87] Roberto Tamassia. On embedding a graph in the grid with the minimum number of bends. *SIAM J. Comput.*, 16(3):421–444, 1987.

[Win95] Gerhard Winkler. *Image Analysis, Random Fields and Dynamic Monte Carlo Methods*, volume 27 of *Applications of Mathematics*. Springer, 1995.

Which Aesthetic Has the Greatest Effect on Human Understanding?

Helen Purchase

Department of Computer Science and Electrical Engineering,
The University of Queensland, Australia

Abstract. In the creation of graph drawing algorithms and systems, designers claim that by producing layouts that optimise certain aesthetic qualities, the graphs are easier to understand. Such aesthetics include *maximise symmetry*, *minimise edge crosses* and *minimise bends*.

A previous study aimed to validate these claims with respect to three aesthetics, using paper-based experiments [11]. The study reported here is superior in many ways: five aesthetics are considered, attempts are made to place a priority order on the relative importance of the aesthetics, the experiments are run on-line, and the ease of understanding the drawings is measured in time, as well as in the number of errors. In addition, greater consideration is given to the possible effect of confounding factors in the graph drawings.

The results indicate that reducing the number of edge crosses is by far the most important aesthetic, while minimising the number of bends and maximising symmetry have a lesser effect. The effects of maximising the minimum angle between edges leaving a node and of fixing edges and nodes to an orthogonal grid are not statistically significant.

This work is important since it helps to demonstrate to algorithm and system designers the aesthetic qualities most important for aiding human understanding, the most appropriate compromises to make when there is a conflict in aesthetics, and consequently, how to build more effective systems.

1   Introduction

Automatic graph drawing algorithms produce a diagram which represents an underlying graph structure. The aim of the layout process is to depict relational information in a form that makes it easier to read, understand and use. Designers of such algorithms ensure that certain aesthetics are optimised, and claim that by doing do, the resultant graph drawing helps the human reader to understand and remember the information embodied in the graph. Examples of these aesthetics include: *symmetry* (where possible, a symmetrical view of the graph should be displayed [5, 10]), *minimise edge crosses* (the number of edge crosses in the display should be minimised [6]), and *minimise bends* (the total number of bends in polyline edges should be minimised [13, 15]).

It is important that human experiments be performed on these aesthetics, so that, rather than judging an algorithm by its computational efficiency in conforming to these aesthetics, the aesthetics themselves can be judged with respect to how much they assist human comprehension. Many application domains may make use of automatic graph layout algorithms in order to display relational data in a holistic form: e.g. entity relationship diagrams [1], object oriented design diagrams [4], social networks [3]. If the designers of automatic graph layout algorithms are to claim that their algorithms will illuminate the information embodied therein, it is important that they know that the aesthetic basis for their work is sound.

Many algorithms consider more than one aesthetic in their attempt to create an illuminating graph drawing. For this reason, although the individual aesthetics themselves are important, often it is the *combination* or *prioritisation* of the aesthetics that is most useful. Algorithm designers may need to compromise between more than one aesthetic. For example, in the creation of a particular drawing, minimising the number of crosses may also result in a decrease in symmetry. The knowledge that minimising the number of crosses is of more benefit to understandability than maximising symmetry [11], means that an appropriate compromise can be made.

The previous study performed preliminary paper-based experiments on the human understanding of graph drawings to determine whether three aesthetic criteria (crosses, bends and symmetry) did indeed assist with the understanding of the underlying graph structure. While the hypotheses were confirmed in the case of crosses and bends, there was not enough evidence to either support or reject the symmetry hypothesis.

In this experiment, five aesthetics were considered; there are therefore five primary hypotheses:

– **Bends (b):**
   Increasing the number of edge bends in a graph drawing decreases the understandability of the graph.

– **Crosses (c):**
   Increasing the number of edge crosses in a graph drawing decreases the understandability of the graph.

– **Angles (m):**
   Maximising the minimum angle between edges leaving the nodes in a graph drawing increases the understandability of the graph.

– **Orthogonality (o):**
   Fixing nodes and edges to an orthogonal grid increases the understandability of the graph.

– **Symmetry (s):**
   Increasing the symmetry displayed in a graph increases the understandability of the graph.

Briefly, the experiment entailed subjects answering questions about a number of different drawings of the same graph. Each drawing was drawn such that it varied the aesthetics under consideration in a fixed manner: for example, one drawing had a large number of crosses, while another had less. Measurements were taken of both the number of errors made and the time taken to answer the questions. Using statistical tests, the five primary hypotheses associated with the five different aesthetics under consideration were proved or disproved. In addition, both for the set of "easy" drawings as well as the set of "difficult" drawings, Tukey's WSD pairwise comparison procedure was then used to determine if there were significant understandability priorities between the aesthetics.

Experiments were run online to study these five aesthetics, and the results indicate that crosses is by far the most important aesthetic. Bends and symmetry have a lesser effect, and maximising the minimum angle and maximising orthogonality have no significant effect at all. This paper describes the nature of the on-line system used for the experiments and the experimental methodology (the graph drawings, experiment and the data), and presents and discusses the results.

2 The Experiment

2.1 Definition

There are two ways in which understandability may be measured. A purely *relational* method measures the efficiency and accuracy with which people can read a graph structure and answer questions about it. Such graph-theoretic questions need to be generic and application-independent, and may include questions of the form "What is the shortest path from node A to node B?" A more application-specific method would rather consider a graph *interpretation* task: in this case it is more appropriate that the effectiveness of the graph drawing is measured within the context in which the application-specific graph is usually used. Thus, instead of eliciting answers to specific questions asked about the graph itself, it is more suitable to look at whether the graph has assisted the user in accomplishing a particular application task. Suitable questions for this approach would include (in the area of software engineering) "What object classes would be affected by changing the external interface to class X?"

In this experiment, the *relational* reading of a graph drawing is considered, leaving the *interpretive* consideration of aesthetics for a later study. The questions that are used in this experiment to measure relational understandability are:

- How long is the *shortest path* between two given nodes?

- What is the *minimum number of nodes* that must be removed in order to disconnect two given nodes such that there is no path between them?

- What is the *minimum number of edges* that must be removed in order to disconnect two given nodes such that there is no path between them?

2.2 Scope

A preliminary, more limited, study [11] reported comparable conclusions to those reported here. The study reported here improves on this previous study in a number of important ways, greatly increasing the validity and relevance of the results:

- **Metric definitions:** New metrics for all five aesthetics have been defined [12]. These are all scaled to lie between 0 and 1, where 0 represents an amount of the aesthetic that it is assumed makes the drawing *difficult* to read (e.g. not much orthogonality), while 1 represents an amount of the aesthetic that it is assumed makes the drawing *easy* to read (e.g. not many crosses). A new metric for symmetry has been defined, which more closely represents perceptual symmetry than the one used previously. It takes into account both global and local symmetries, weighting them by their area, and also considers the effects of crosses and bends on perceptual symmetry.

- **Presentation medium:** The experiments are performed online using an experimental system especially designed and implemented for experiments like these. This means that the understandability of the graph drawings is tested using a more valid medium: automatic graph layout algorithms by definition make use of a computer, with the results displayed on a screen, rather than on paper. Experiments where subjects read graph drawings on a screen are therefore more valid than similar paper-based experiments.

- **Dependent variables:** The use of the online system enables two dependent variables to be recorded: the time taken for the subject to answer the question (the "reaction time"), as well as the correctness of the answer. This enables analysis to be performed on two measures of understanding.

- **Confounding factors:** In the drawings that vary a particular aesthetic, it is important that the values of the other four aesthetics are kept constant, to ensure that there is no confounding of variables. It is difficult, and in some cases impossible, to use the extremes of 0 or 1 as the constant value for the other four aesthetics: for example, a metric value of 0 for the bend aesthetic would imply a maximum possible number of bends; a metric value of 1 for minimum angle aesthetic would mean that *all* nodes in the drawing have the optimum angles between its edges (impossible for any cyclic graph). For this reason, a "neutral range" was defined for each aesthetic (based on perception), and for the drawings which varied a particular aesthetic, values of the other four aesthetics were kept within these specified ranges.

- **Location of nodes:** The questions that are asked about the drawings refer to nodes that are highlighted in black on the screen, to distinguish them from the other nodes. The relevant nodes are therefore obvious to the subjects, and the time measured for the subject to answer the question does not include additional time taken for locating the important nodes. The previous study referred to the nodes by labels [11].

2.3 The Online System

Experiments were run online. Each subject interacted with a unique experimental program. These programs were created by a system designed and implemented for the purposes of running experiments relating to graph drawings (called SAGE). The main features of SAGE are:

- Flexibility: so that SAGE can be used for further graph-drawing experimentation, each experiment is specified with an external contents file.

- Randomness: the ordering of graph drawings, their orientation, the ordering of the questions, and the selection of node-pairs for the questions are all able to be randomised.

- Graph and question flexibility: the graph drawings and questions used are defined in separate files, and are easily changed.[1]

- Completeness: all the interface features required for each graph drawing display are provided and specified in the contents file: text, pictures, input fields, pushbuttons.

- Robustness: SAGE can withstand the unexpected input of a novice user, and efficiently and correctly represents the experiment as defined in the contents file.

- Analysable data: the results for each subject are generated automatically as a list of the time between the display of each drawing and question and the entry of an answer, the answer itself, and its correctness.

2.4 The Graphs

The graph for this experiment was carefully designed so that node-pairs could be identified which gave a suitable range of values for the three questions. Thus, a set of node-pairs was defined that would give correct answers to the first question (the shortest path) of either 2, 3, 4 or 5; a set of node-pairs was defined that would give correct answers to the second question (the number of nodes to remove) of either 1 or 2; and a set of node-pairs was defined that would give correct answers to the third question (the number of edges to remove) of either 1, 2 or 3. The graph has 16 nodes and 28 edges.

New metric formulae (all lying within the range 0 to 1) were defined for this experiment, including a more extensive definition of symmetry [12]. Ten experimental graphs were created, two for each of the aesthetics (representing a strong or weak presence of the aesthetic). For convenience, the graph drawings are called after the aesthetic that they consider (**b, c, m, o, s**), and + or - depending on the strength of the aesthetic: + indicates a high aesthetic value (i.e. assumed to be easy to read), and - indicates a low aesthetic value (i.e. assumed to be

---

[1] The graph drawings are in GraphEd format [8], and the questions are in ASCII.

difficult to read). Thus, the s+ drawing has a symmetry metric value closer to 1 than the s- drawing.

Figures 1 and 2 show the ten graph drawings, and their associated metric values. Note that because of the nature of the aesthetics, the metrics cannot be sensibly compared over the aesthetic dimension. Thus, while c- has a cross-less value of 0.87, m- has a value of 0.16; s+ has a symmetry value of 0.96, o+ has an orthogonality value of 0.46. This variation is due to the metric definitions and distributions: it does not affect the results, as the important feature is the variation of the values *within* the aesthetic dimension.[2]

Due to the careful manipulation of aesthetics that was required, some of these drawings may look strangely awkward (e.g. b-, m-). As the aim was to consider the effect of the individual aesthetics (rather than drawings that may feasibly be produced by layout algorithms, or that have been purposefully drawn "neatly"), the artificial nature of some of the drawings was both intentional and necessary.

2.5 Experimental Methodology

The structure of the experiment was similar to the previous paper-based preliminary investigation [11]. The contents file used by SAGE defined experimental programs of the following form:

1. A brief description of graphs, and definitions of the terms *node, edge, path*, and *path length* were presented, followed by an explanation of the three questions that the subjects were required to answer about the experimental graphs. A simple example graph drawing, with the three questions and their correct answers, was shown. At this stage, the subjects were asked if they had any questions about graphs in general, or about the experiment. It was important to ensure that all the subjects knew what was expected of them.

2. The three questions were asked of six "practise" graph drawings, to familiarise the subjects with the nature of graph drawings and the questions, and to ensure that they were comfortable with the task, before tackling the experimental graphs. The subjects were not told that these graph drawings were not experimental.

3. A "filler" task which engaged the subjects' mind on a small problem unrelated to graphs was presented. This ensured that their performance on the subsequent experimental graphs was not affected by any follow-on effect from the practise graphs. A simple logic puzzle, designed to take approximately 1 minute, was used.

4. The ten experimental graph drawings were each displayed three times, once for each question. The order of presentation of the drawings and the questions was random, as was the orientation of the drawings.

---

[2] The metric definitions give more detail on the extremes of the metric values [12].

graph	bend-less	cross-less	minangle	orthog	sym
b+	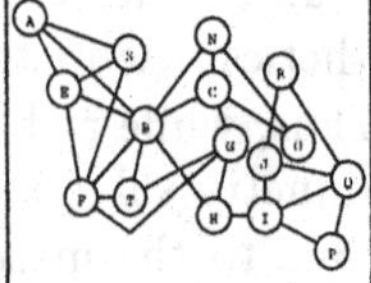 0.96	0.97	0.38	0.27	0.75
b-	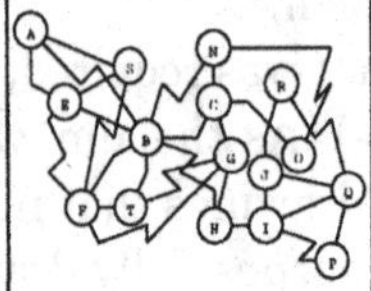 0.47	0.99	0.44	0.28	0.71
c+	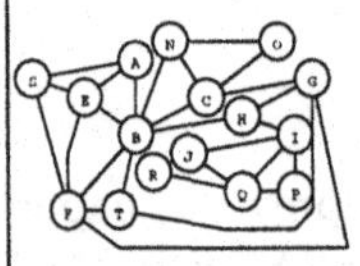 0.82	1	0.46	0.33	0.63
c-	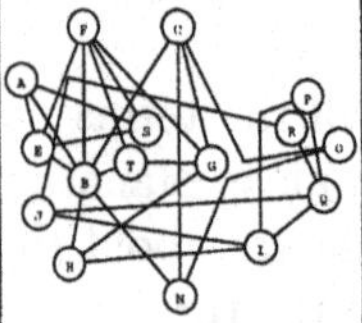 0.87	0.88	0.35	0.29	0.84
m+	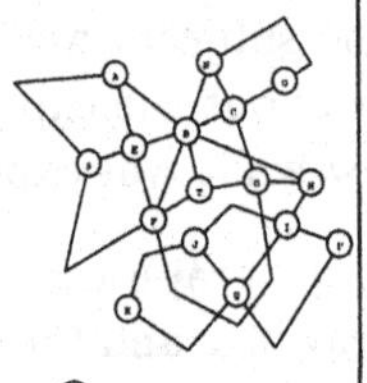 0.71	0.98	0.62	0.22	0.74
m-	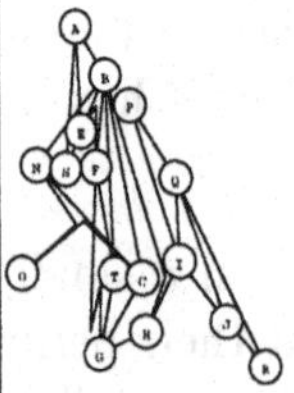 0.82	0.98	0.16	0.26	0.79

Fig. 1. Six of the ten experimental graph drawings, and their aesthetic values.

graph		bend-less	cross-less	minangle	orthog	sym
o+	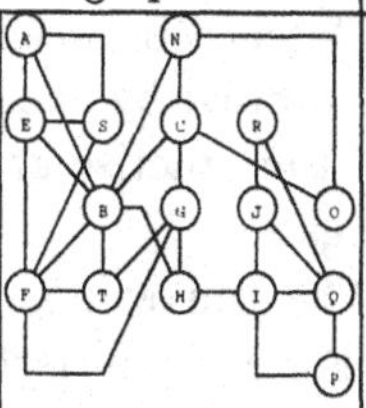	0.82	0.98	0.42	0.46	0.73
o−	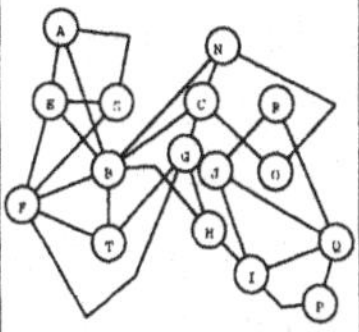	0.82	0.98	0.41	0.21	0.68
s+	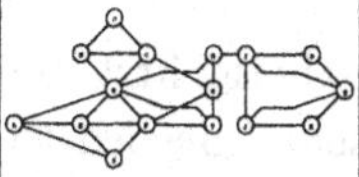	0.77	0.99	0.57	0.29	0.96
s−	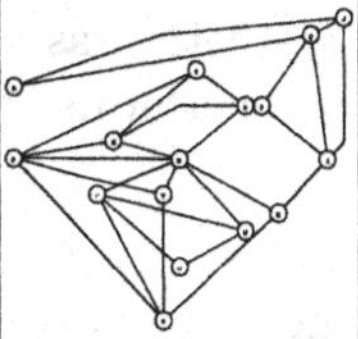	0.87	0.99	0.44	0.25	0.00

Fig. 2. Four of the ten experimental graph drawings, and their aesthetic values.

The questions themselves were randomised too: although the same three questions were asked of each drawing, the pair of nodes chosen for each question was randomly selected from a list of node-pairs (as defined in an external question file). This ensured that any variability in the data could not be explained away by the varying difficulty of the questions. The two relevant nodes for each question were highlighted in black on the screen, ensuring that reaction time did not include time taken to locate the nodes.

The subjects typed their answers to the questions: the time taken for their answer, and the correctness of the answer, was recorded.

The experiment was therefore controlled for the questions and the graphs, the independent variable was the value of the aesthetics in each drawing, and the two dependent variables were the time taken to answer the questions, and the number of errors made for each drawing.

A within-subjects analysis method was used in order to reduce any variability that may have been attributable to the difference between the subjects (e.g. age, experience). Any learning effect was minimised by the large number of graphs used in the experiment, the inclusion of the practise graphs, and the randomisation of the ordering of the graph drawings.

55 second-year computer science students at The University of Queensland took part in the experiment, for a reward of $10. For each subject and for each drawing, the total number of errors was recorded, as well as the total time taken to answer all three questions.

3 Results

The average number of errors and the average reaction time for the ten experimental graph drawings are shown in both tabular and chart form in Fig. 3.

3.1 Testing the Five Individual Hypotheses

To test the five primary hypotheses, one for each aesthetic, first the significance of the effects of the level of difficulty (the +/- dimension) needed to be confirmed. After this confirmation that the +/- dimension had indeed affected the error and reaction time data collected, each individual aesthetic was then tested for its contribution to this overall effect. This analysis was performed for both errors and reaction time.

Results. The 2x5 within-subject analysis of variance showed that:[3]

- The main effect of the level of difficulty (the +/- dimension) was *significant* for both errors ($F_{1,54}=14.89, \alpha=.05$) and reaction time ($F_{1,54}=40.67, \alpha=.05$).

- The simple effect of the **bends** metric was *significant* for errors ($F_{1,54}=14.49, \alpha=.01$) but only *approaches significance* for reaction time ($F_{1,54}=5.84, \alpha=.01$).

- The simple effect of the **crosses** metric was *significant* for both errors ($F_{1,54}=24.25, \alpha=.01$), and reaction time ($F_{1,54}=87.98, \alpha=.01$).

- The simple effect of the **minimum angle** metric was *not significant* for both errors ($F_{1,54}=0.09, \text{NS}$) and reaction time ($F_{1,54}=3.05, \text{NS}$).

- The simple effect of the **orthogonality** metric was *not significant* for both errors ($F_{1,54}=0.00, \text{NS}$) and reaction time ($F_{1,54}=1.44, \text{NS}$).

- The simple effect of the **symmetry** metric was *not significant* for errors ($F_{1,54}=0.09, \text{NS}$), but was *significant* for reaction time ($F_{1,54}=7.57, \alpha=.01$).

---

[3] The statistical analysis used here is a standard ANOVA analysis [9], based on the critical values of the F distribution: α is the level of significance, and results that are not significant are indicated by NS.

	b+	b-	c+	c-	m+	m-	o+	o-	s+	s-
errors	0.24	0.53	0.29	0.80	0.36	0.38	0.36	0.36	0.29	0.31
reaction time	67.18	81.40	66.39	139.78	76.55	68.17	71.37	76.71	55.58	67.74

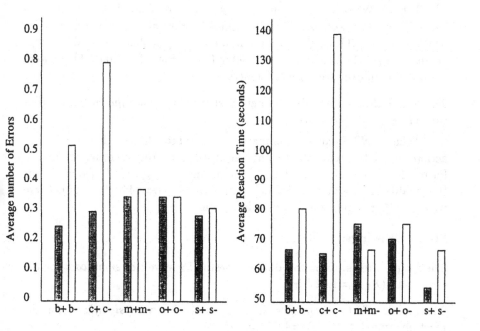

Fig. 3. The average reaction time and average number of errors for each graph drawing.

3.2 Prioritising the Aesthetics

To determine the relative effect of the aesthetics, and attempt to place a priority ordering on their importance, both the set of + drawings and the set of - drawings needed to be tested for the overall effect of the aesthetics. Those sets of drawings for which the effect of the aesthetics were significant were then subject to a Tukey's pairwise comparison [9] to determine which aesthetics differed significantly from one another.

Results. The 2x5 within-subject analysis of variance showed that:

– The main effect of the aesthetics dimension was *significant* for both errors ($F_{4,216}$=4.16,α=.05) and reaction time ($F_{4,216}$=28.49,α=.05).

The - drawings:

– The simple effects of the five different aesthetics were *significant* for the error data ($F_{4,216}$=9.60,α=.025).

The Tukey's WSD pairwise comparisons procedure showed that, for the error data, crosses were significantly different from all other aesthetics: for bends ($F_{5,216}$=9.11,α=.05), minimum angle ($F_{5,216}$=22.05,α=.05), orthogonality ($F_{5,216}$=24.20,α=.05), symmetry ($F_{5,216}$=30.01,α=.05). There were no other significant pairwise differences.

– The simple effects of the five different aesthetics were *significant* for reaction time ($F_{4,216}$=50.89,α=.025).

The Tukey's WSD pairwise comparisons procedure showed that, for the reaction time data, crosses were significantly different from all other aesthetics: for bends ($F_{5,216}$=95.09,α=.05), minimum angle ($F_{5,216}$=143.07,α=.05), orthogonality ($F_{5,216}$=110.98,α=.05), symmetry ($F_{5,216}$=144.79,α=.05). There were no other significant pairwise differences.

The + drawings:

– The simple effects of the five different aesthetics were *not significant* for the error data ($F_{4,216}$=1.02,NS).

– The simple effects of the five different aesthetics were *significant* for the reaction time data ($F_{4,216}$=4.68,α=.025).

The Tukey's WSD pairwise comparisons procedure showed that, for the reaction time data, symmetry was significantly different from the minimum angle ($F_{5,216}$=17.14,α=.05), and orthogonality ($F_{5,216}$=9.72,α=.05).

3.3 Analysis

The error chart in Fig. 3 shows that the average number of errors for the - versions of the drawings (i.e., the "difficult" drawings) was greater than the average number of errors for the + versions, in all cases except orthogonality when the averages were the same. The statistical analysis shows that the level of difficulty of the drawings was only significant for both bends and crosses.

The Tukey's pairwise comparison for the error data showed that the average number of errors for the c- drawing was significantly greater than the errors in the other - versions of the aesthetics, and that there were no significant pairwise orderings for the + drawings.

The reaction time chart in Fig. 3 shows that - versions of the bends, crosses, orthogonality and symmetry drawings all took longer than the + versions. The statistical analysis shows that the level of difficulty of the drawings was only significant for both crosses and symmetry. The unexpected reversal of average reaction time for the two minimum angle drawings is not significant, and can therefore be attributed to chance.

The Tukey's pairwise comparison for the reaction time data showed that the c- drawing took significantly more time than all the other - versions of the aesthetics. In addition, the s+ drawing took significantly less time than the minimum angle m+ and orthogonality o+ drawings.

3.4 Discussion

There is no doubt that the evidence is overwhelmingly in favour of crosses as being the aesthetic that affects human relational graph reading the most, as suggested by the results of the two Tukey pairwise comparison tests performed on the - drawings. The effect of crosses was not noticeable, however, in the + drawings, implying that crosses are only more problematic than the other aesthetics when there are a large number of them.

The results of the other aesthetics are more ambivalent: the bends and symmetry hypotheses were supported either for reaction time or errors, but not both. Orthogonality and minimum angle had no effect on the subjects' relational graph reading at all. The Tukey test for the reaction time data for the + drawings showed that symmetry took significantly less time than the minimum angle and orthogonality, suggesting that symmetry only has a more positive effect than the other aesthetics when it is at a maximum value.

An unusual result was that for the easy drawings, the different aesthetics had no significant effect on the number of errors (even though there was an effect on reaction time). This suggests that the subjects tended to give correct answers on all aesthetics if the drawings were easy, but they used all the time necessary, requiring different amounts of time for the different aesthetics. On the other hand, for the difficult drawings, subjects took the amount of time necessary (which differed for the different aesthetics), but the difficulty of the drawings meant that the number of errors was also differentially affected for different aesthetics.

In interpreting the above result, errors can be interpreted as a measure of the amount of processing required to get the question right, while reaction time can be interpreted as a measure the perceptual processing and comprehension of the drawing.

Limitations. It is common knowledge that all experiments are limited by their parameters, and that the results of any experiment should always be interpreted with respect to the experimental limitations [7]: this is an inevitable consequence of the controlled experimental method. These results can therefore only be interpreted within the context of the graph and tasks specified. There may also be a generalisability restriction on the nature of the subjects, who were all computer scientists, although as a within-subject analysis was performed, any variations in expertise were controlled.

4 Conclusions

These aim of these empirical tests was to indicate to the designers of graph drawing algorithms the most effective aesthetics to use from the point of view of human reading of relational information. The results show that there is strong evidence to support minimising crosses, and weaker evidence for minimising the number of bends and maximising perceptual symmetry. Maximising the orthogonal structure of the drawing, and maximising the minimum angles between edges leaving a node, appear to have little effect.

There is still much work to do in this area: this experiment has only considered the *relational* reading of graph drawings, and different results may be forthcoming from experiments that require an *interpretive* reading of graph drawings in the context of application domains. For example, testing the effect of the different aesthetics when the graph drawings represent object-oriented design diagrams or data-flow diagrams may produce different results.

In addition, another possible study could consider the relational understandability of graph drawings generated by different layout algorithms which aim to maximise the effect of particular aesthetics: it would be interesting to see whether the results obtained from that experiment are compatible with the results of the study reported here.

I am grateful to Robert Cohen (who assisted with the initial experimental definition), to Murray James (who designed and developed SAGE), to David Leonard (who helped define and implement the aesthetic metrics), to Julie McCreddon (who assisted extensively with the statistical analysis), and the Australian Research Council, which funded this work.

References

1. C. Batini, M. Talamo, and R. Tamassia. Computer aided layout of entity-relationship diagrams. *Journal of Systems and Software*, 4:163–173, 1984.
2. S. Bhanji, H.C. Purchase, R.F. Cohen, and M.I. James. Validating graph drawing aesthetics: A pilot study. Technical Report 336, University of Queensland Department of Computer Science, 1995.
3. J. Blyth, C. McGrath, and D. Krackhardt. The effect of graph layout on inference from social network data. In F. Brandenburg, editor, *Proceedings of Graph Drawing Symposium 1995*. Springer-Verlag, Passau, Germany, 1995. Lecture Notes in Computer Science, 1027.
4. G. Booch. *Object-Oriented Design*. Benjamin-Cummings, 1990.
5. P. Eades. A heuristic for graph drawing. *Congressus Numerantium*, 42:149–160, 1984.
6. D. Ferrari and L. Mezzalira. On drawing a graph with the minimum number of crossings. Technical Report 69-11, Istituto di Elettrotecnica ed Elettronica, Politecnico di Milano, 1969.
7. R. Gottsdanker. *Experimenting in Psychology*. Prentice-Hall, 1978.
8. M. Himsolt. GRAPHED user manual. Universität Passau, 1990.
9. P.R. Hinton. *Statistics Explained*. Routledge, 1995.
10. R. Lipton, S. North, and J. Sandberg. A method for drawing graphs. In *Proc. ACM Symp. on Computational Geometry*, pages 153–160, 1985.
11. H.C. Purchase, R.F. Cohen, and M. James. An experimental study of the basis for graph drawing algorithms. *ACM Journal of Experimental Algorithmics*, 2(4), 1997.
12. H.C. Purchase and D. Leonard. Graph drawing aesthetic metrics. Technical Report 361, University of Queensland Department of Computer Science, 1996.
13. R. Tamassia. On embedding a graph in the grid with the minimum number of bends. *SIAM J. Computing*, 16(3):421–444, 1987.
14. R. Tamassia, G. Di Battista, and C. Batini. Automatic graph drawing and readability of diagrams. *IEEE Transactions on Systems, Man and Cybernetics*, SMC-18(1):61–79, 1988.
15. H. Trickey. Drag: A graph drawing system. In *Proc. Int. Conf. on Electronic Publishing*, pages 171–182. Cambridge University Press, 1988.

Implementing a General-Purpose Edge Router

David P. Dobkin,[1]*

Emden R. Gansner,[2] Eleftherios Koutsofios,[2] Stephen C. North[2]

[1] Princeton University, Princeton NJ 08544, USA
[2] AT&T Laboratories, Florham Park NJ 07974, USA

Abstract. Although routing is a well-studied problem in various contexts, there remain unsolved problems in routing edges for graph layouts. In contrast with techniques from other domains such as VLSI CAD and robotics, where physical constraints play a major role, aesthetics play the more important role in graph layout. For graphs, we seek paths that are easy to follow and add meaning to the layout. We describe a collection of aesthetic attributes applicable to drawing edges in graphs, and present a general approach for routing individual edges subject to these principles. We also give implementation details and survey difficulties that arise in an implementation.

1 Introduction

Edge placement is an important problem in graph drawing. Once nodes are positioned, edges must be added in a way that clearly exhibits the relation between nodes, without adding clutter or deceptive artifacts. For example, it is usually desirable that an edge between two nodes does not pass through a third node. When nodes are not drawn as points but are represented by symbols or shapes that have area, the difficulties of edge placement increase. Layouts need to employ edges that bend significantly to avoid touching non-incident nodes. Bent edges can be drawn as polylines, or as curves such as Bezier splines. Though polylines are often easier to compute, their sharp corners create unwanted visual discontinuities that smooth curves do not have.

Curves have been employed previously in drawing edges in layered graphs. In these layouts, nodes are assigned to discrete layers and edges between nodes that are more than one layer apart are replaced by chains of virtual nodes. Virtual node chains provide a useful framework for spline fitting. In VCG [18] and dag [7], splines are computed by straightening virtual node chains, then connecting the virtual nodes with line segments and replacing the sharp corners with bends that fit within individual virtual node boxes. Though the results are usually acceptable, it is not unusual for edges to curve abruptly because turns are constrained to fit inside virtual node boxes.

* Portions of the work of this author done while visiting AT&T Laboratories. This work supported in part by NSF Grant CCR-9643913 and by the US Army Research Office under Grant DAAH04-96-1-0181

The method used in dot [6] incorporates a more general heuristic with two phases. For a given edge to be drawn, the first phase computes the "white space" or constraint polygon where the edge may be drawn so as not to touch any other nodes or create unwanted edge crossings. The second phase fits a smooth spline connecting the edge's endpoints and staying within the constraint polygon. The endpoints can have optional slope constraints. The spline fitting heuristic computes a trial spline between the endpoints. If the spline goes outside the constraint polygon, and it cannot easily be repaired by slight re-aiming, then the spline fitter subdivides the polygon near the broken constraint and draws the top and bottom halves recursively.

We simplified dot's spline fitter by assuming that the input constraint polygon is represented by a list of connected isothetic rectangles, and that the output spline passes through these sequentially. The list of rectangles is easy to compute from virtual node coordinates and spacing between layers. Its structure also makes it straightforward to test if a Bezier curve[3] stays inside the polygon, and to subdivide it for recursion.

We would naturally like good edge placement not only in layered graphs but in other types of layouts as well, e.g., layouts based on virtual physical models, but the rectangular technique is limited to edges in layered graphs. Even in layered graphs, this technique is not always satisfactory. Problems arise in representing the region for an edge that is constrained by another edge sharing the same endpoint if the angle between the 2 edges is small. In such cases, it takes many "steps" in the boundary rectangles to keep them apart. Other complications arise when drawing self-edge loops, or flat edges between nodes in the same layer, particularly if endpoint "ports" or other path constraints are allowed. These problems suggest re-examining the edge drawing problem in a more general setting.

As an initial simplification, we consider drawing edges independently, i.e., we draw an edge without regard to how any other edges may be drawn. We recognize that this ignores any consideration of the global properties of edge layouts, such as reducing edge crossings or emphasizing edges running in parallel.

In this reduced formulation, edge placement is essentially a routing problem. Routing paths around obstacles has been studied extensively in VLSI layout, computational geometry, and robotics path planning. Dubins' classic result [4] shows how to find shortest paths of bounded curvature between two points in the plane with defined tangents at the endpoints. This result has been extended in various ways [1, 10, 17].

Many path planning problems address limitations of robot carts or manipulators, where physical constraints must be accommodated. Schwartz and Sharir [20] studied planning collision-free paths for obstacles. Fortune and Wilfong [5], Kanayama and Hartman [11], Nelson [14], Laumond [13] and others have studied reachability and path planning problems for a point or object subject to a

---

[3] dot uses piecewise cubic Bezier curves because this family of spline curves is implemented in the PostScript graphics language and in public domain graphics code, and so is convenient to work with.

curvature constraint. Latombe [12] (who also presents a thorough survey of the field) studied potential field methods for path planning. This is an interesting approach, though simulating electrostatic fields created by polygonal antennae seems complicated. Suri [21] showed how to find minimum link paths in simple polygons. These paths are interesting because they have a minimal number of corners. Consequently, they can take extreme routes that deviate significantly from shortest paths.

Physical constraints such as robot size, mass, acceleration, or turning radius do not have a clear relationship to natural-looking curves in graph drawings. In general, the bulk of work on route planning does not seem to capture the properties we feel are desirable in drawing edges in graphs. We turn next to a discussion of these properties, and our approach to achieving them.

2 Problem Definition

Though we cannot formally define what it means for an edge connecting two vertices to appear natural, we believe good solutions avoid other vertices in the graph, stay close to a shortest path between the endpoints, do not turn too sharply, and avoid unnecessary inflections.

Taking one approach to satisfying these criteria, we restate our edge routing problem. As a model for the problem, we consider a graph layout to consist of a polygon $\mathcal{P}$ consisting of a simple polygon containing a collection of disjoint simple polygonal holes, corresponding to the node obstacles. Given two points p and q on or in the interior of $\mathcal{P}$, possibly with tangents v_p and v_q, respectively, we want to find a smooth piecewise cubic Bezier curve from p to q, satisfying v_p and v_q, that stays on or within $\mathcal{P}$.

As a first attempt at solving this problem, we considered constructing a simple path L between p and q (typically a shortest path) within the interior of $\mathcal{P}$. We could then compute some simple polygon Q that surrounded L and did not contain any holes. Finally, we could apply some procedure for fitting Bezier curves within the simple polygon Q.

One question was how to define "good" choices of Q. We speculated that good choices might be ones that are as wide as possible at the bisectors of bends of L, so that the output curve has as much room as possible to turn. We experimented with a heuristic that constructs a route constraint polygon on each side of L. On each side, the polygon is calculated by taking each segment of L and building a polygon with the segment as a base. For a segment $s = (p_i, p_{i+1})$ of L, let b_i and b_{i+1} be the bisectors of the corners at p_i and p_{i+1}. Find the set of obstacles between b_i and b_{i+1} on the given side of s. If there are no obstacles, then the polygon formed by s, b_i and b_{i+1} is added. Otherwise, the convex hull of the obstacles is removed from the polygon first.

We found this heuristic complicated to implement because of degeneracies, and because if L is allowed to be any simple path (e.g. one entered interactively), ensuring that Q does not intersect itself involved many cases. Because of these difficulties, we abandoned this approach and evolved the following heuristic.

3 Spline Fitting Heuristic

Let $\mathcal{P}$ be a polygon and S a set of forbidden segments. (Typically, these segments will be the edges of the polygonal holes and the bounding polygon.) Let p and q be points in $\mathcal{P}$. The spline fitter is divided into two algorithms. The first finds a shortest path L within $\mathcal{P}$, connecting p to q, that intersects no edge in S except possibly at an endpoint. There is a choice of algorithms here depending on whether $\mathcal{P}$ can contain holes. The second algorithm fits a Bezier curve along L that intersects no edge in S except at an endpoint.

3.1 Path Finding

If $\mathcal{P}$ does not contain holes (*e.g.*, when emulating the spline router for layered graphs used in dot), we can apply a standard "funnel" algorithm [2, 9] for finding Euclidean shortest paths in a simple polygon. To find a shortest path from point p to point q in a simple polygon, we first triangulate the polygon, using only vertices of the polygon. We then find the triangles that contain points p and q, say t_p and t_q. We then find the sequence of triangles that connect t_p and t_q by doing a depth-first search starting from t_p and searching for t_q. This induces a list of triangle sides (a_i, b_i) interior to $\mathcal{P}$ and crossed by the shortest path from p to q. We then iteratively build a funnel composed, at each step, of the two shortest paths from p to a_i and b_i. Once q is added to the funnel, we have the shortest path from p to q. Given the requisite triangulation, the funnel construction phase runs in linear time.

If $\mathcal{P}$ may contain holes, then we compute the visibility graph of its points plus the two endpoints, and apply Dijkstra's algorithm to find a shortest path between the endpoints. The details of this algorithm are omitted.

3.2 Spline Fitting

Using the shortest path L as a guide, we recursively attempt to fit a curve to it that avoids the obstacles. We use an approach based on the curve-fitting method introduced by Schneider [19]. As input, we start with a collection of points $p_0, p_1, ..., p_n$ that we wish to fit, initially using the vertices of L, plus tangents t_0 and t_n specified at p_0 and p_n. Applying Schneider's method once, we compute a single cubic Bezier segment corresponding to four control points w_0, w_1, w_2, w_3, where $p_0 = w_0$, $p_n = w_3$, and the segments $[w_0, w_1]$ and $[w_2, w_3]$ are parallel to the given tangents at p_0 and p_n, respectively.[4] If this curve does not intersect any of the obstacles, we are done.

If there is an intersection, we attempt a series of local adjustments, in which we move w_1 and w_2 closer to w_0 and w_3, respectively, along the appropriate tangents. If, at any stage, we obtain a viable curve, we are done. Otherwise,

---

[4] If there are only two input points, we put w_1 and w_2 along the appropriate tangents at a distance of $d/3$ away from p_0 and p_n, respectively, where d is the distance between p_0 and p_n.

we pick the point p_i that is furthest from the Bezier curve, compute a tangent t_i that bisects the angle turned at p_i, and recursively apply the algorithm to the points $p_0, p_1, ..., p_i$ with tangents t_0 and t_i, and to the points $p_i, ..., p_n$ with tangents t_i and t_n.

Note that, using this construction, there is no guarantee that the resulting curve is topologically equivalent to L. At some point, a Bezier segment may "leap over" an intervening obstacle. The approximation is good enough, however, that the resulting curve would still be adequate for our purposes and, in practice, this situation does not seem to arise.

4 Implementation

With general edge routing as a goal, we have implemented the path planner as a C library. Its main primitives are:

- *shortestpath(P, p, q)*: finds a shortest path between p and q in a simple polygon P.
- *obspath(obstacles, p, p_{poly}, q, q_{poly})*: finds a shortest path between p and q not intersecting the interior of any polygons in the list *obstacles*. When an endpoint is inside an obstacle (as in Figure 2), the obstacle must be ignored for the route. It is possible that the caller knows when this happens. For example, a graph layout algorithm may know the endpoint nodes, or an interactive diagram editor may perform hit detection on canvas objects. When an endpoint is known to be inside an obstacle, the obstacle may be passed as an argument, viz. p_{poly} or q_{poly}.
- *splinefit(barriers, L, v_p, v_q)*: returns a piecewise cubic Bezier that fits around the input path L and avoids the list of segments in *barriers*. The arguments v_p and v_q provide tangent vectors for the first and last points of L. Generally, L was obtained from one of the two previous primitives. Note that *barriers* can be any collection of segments, not necessarily forming closed polygons.

The current implementations provide room for increased efficiency. The *obspath* routine uses a naive $O(n^3)$ visibility graph algorithm. We intend to replace this with a more efficient algorithm [8, 16]. For interactive or incremental layout, incremental visibility computation would obviously be desirable. The current version of *splinefit* tests each spline against all barrier edges separately for intersections. This is a quadratic algorithm. When this becomes too slow, a bucketing technique can be implemented to get near-linear behavior. Finally, we use an $O(n\log n)$ triangulation algorithm in *shortestpath*.

5 Observations and Conclusions

Figures 1 - 4 are screen dumps from sample runs of the edge router. Although the obstacles in these examples are different from what would be expected in graph drawing, they exhibit the wide applicability of our technique and the quality of the resulting paths. With each figure we give the time in milliseconds needed to compute the route, measured on a Silicon Graphics Indigo2 computer with a 250MHz MIPS R4400 processor and R4010 floating point unit. The measurement is divided into the time to compute the visibility graph (which may be amortized) plus the time to find a shortest path and fit a spline.

Figures 5 and 6 show the router in a graph editor, its intended domain. Figure 5 was made with manual node placement, and figure 6 with the *dynadag* hierarchical layout manager [15].

We are not aware of any formal studies about what kind of edge routes are most effective for information visualization. Without a formal definition of what it means for a spline to be "good," an evaluation of our results must be subjective. We speculate that when drawing edges manually, humans try to interpolate between a shortest path and a path of minimum curvature while avoiding obstacles.

We have ignored the problem of routing multiple edges. Instead, we route individual edges independently. Further work is needed to understand how edge routes interact. One simple approach, foreshadowed by the edge layout in dot, is to route edges consecutively, determining the "territory" available to an edge based on neighboring edges already routed. Clearly, the order of edge drawing becomes significant, as an edge route can affect others drawn afterward. How should the order be planned? Should previously drawn edges ever be moved? Would a more global approach to routing multiple edges offer better results?

Polygons only approximate regions bounded by curved edges and nodes with curved boundaries (such as ellipses or boxes with rounded corners). Re-implementation of the router with splinegons [3] is worth consideration.

6 Acknowledgments

We thank John Ellson from Lucent Corp. for the TCL/tk interface.

References

1. J.-D. Boissonnat, A. Cerezo, and J. Leblond. Shortest paths of bounded curvature in the plane. *J. Intell. and Robotics Systems*, 11:5–20, 1994.

2. B. Chazelle. A theorem on polygon cutting with applications. In *Proc. 23rd IEEE Symp. Foundations of Computer Science*, pages 339–349, 1982.

3. D. P. Dobkin, D. L. Souvaine, and C. J. Van Wyk. Decomposition and intersection of simple splinegons. *Algorithmica*, 3:473–486, 1988.

4. L. E. Dubins. On curves of minimal length with a constraint on average curvature and with prescribed initial and terminal positions and tangents. *Amer. J. Math.*, 79:497–516, 1957.

5. S. Fortune and G. Wilfong. Planning constrained motion. *Annals of Mathematics and Artificial Intelligence*, 3:21–82, 1991.

6. E.R. Gansner, E. Koutsofios, S.C. North, and K.P. Vo. A technique for drawing directed graphs. *IEEE Transactions on Software Engineering*, March 1993.

7. E.R. Gansner, S.C. North, and K.P. Vo. Dag – a program that draws directed graphs. *Software – Practice and Experience*, 18(11):1047–1062, 1988.

8. S. K. Ghosh and D. M. Mount. An output-sensitive algorithm for computing visibility graphs. *SIAM J. Computing*, 20(5):888–910, 1991.

9. J. Hershberger and J. Snoeyink. Computing minimum length paths of a given homotopy class. In *Proc. 2nd Workshop Algorithms Data Struct.*, volume 519 of *Lecture Notes in Computer Science*, pages 331–342. Springer-Verlag, 1991.

10. P. Jacobs. Minimal length curvature constrained paths in the presence of obstacles. Technical Report 90042, Laboratoire d'Automatique et d'Analyse des Systemes, 7 Avenue du Colonel Roche - 31077 Toulouse, France, February 1990.

11. Y. Kanayama and B. I. Hartman. Smooth local path planning for autonomous vehicles. In *Proc. IEEE Intl. Conf. on Robotics and Automation*, volume 3, pages 1265–1270, 1989.

12. J.-C. Latombe. *Robot Motion Planning*. Kluwer Academic Publishers, Boston, 1991.

13. J. P. Laumond. Finding collision-free smooth trajectories for a non-holonomic mobile robot. In *Proc. Intl. Joint Conf. on Artificial Intelligence*, pages 1120–1123, 1987.

14. W. Nelson. Continuous-curvature paths for autonomous vehicles. In *Proc. IEEE Intl. Conf. on Robotics and Automation*, volume 3, pages 1260–1264, 1989.

15. S.C. North. Incremental layout in dynadag. In F.J. Brandenburg, editor, *Symp. on Graph Drawing GD'95*, volume 1027 of *Lecture Notes in Computer Science*, pages 409–418, 1996.

16. M. H. Overmars and E. Welzl. New methods for computing visibility graphs. In *Proc. 4th Annu. ACM Sympos. Comput. Geom.*, pages 164–171, 1988.

17. J. A. Reeds and L. A. Shepp. Optimal paths for a car that goes both forwards and backwards. *Pacific J. of Mathematics*, 145(2), 1990.

18. G. Sander, M. Alt, C. Ferdinand, and R. Wilhelm. Clax, a visualized compiler. In F.J. Brandenburg, editor, *Symp. on Graph Drawing GD'95*, volume 1027 of *Lecture Notes in Computer Science*, pages 459–462, 1996.

19. Philip J. Schneider. An algorithm for automatically fitting digitized curves. In Andrew S. Glassner, editor, *Graphics Gems*, pages 612–626. Academic Press, Boston, Mass., 1990.

20. J. T. Schwartz and M. Sharir. Algorithmic motion planning in robotics. In J. van Leeuwen, editor, *Algorithms and Complexity*, volume A of *Handbook of Theoretical Computer Science*, pages 391–430. Elsevier, Amsterdam, 1990.

21. S. Suri. A linear time algorithm for minimum link paths inside a simple polygon. *Computer Vision and Graphical Image Processing*, 35:99–110, 1986.

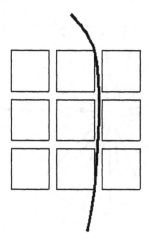

Figure 1 (28.1+7.8 ms.)

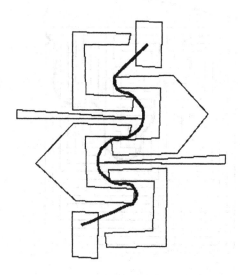

Figure 2 (31+38 ms.)

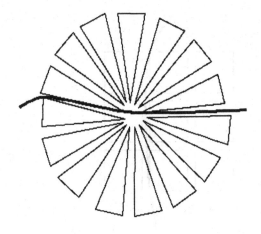

Figure 3 (46.8+15.0 ms.)

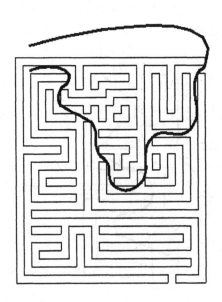

Figure 4 (742.0+123.0 ms.)

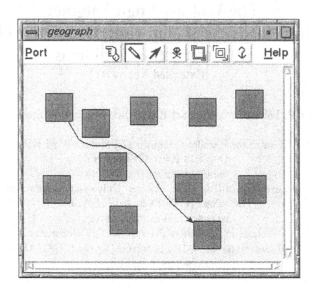

Figure 5

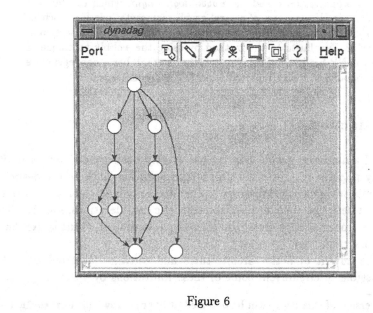

Figure 6

The Wobbly Logic Engine:
Proving Hardness of Non-rigid Geometric Graph Representation Problems
(Extended Abstract) *

Sándor P. Fekete[1] **, Michael E. Houle[2] ***, Sue Whitesides[3] **†

[1] Center for Parallel Computing, Universität zu Köln
D–50923 Köln, GERMANY
sandor@zpr.uni-koeln.de
[2] Department of Computer Science, University of Newcastle
Callaghan NSW 2308, AUSTRALIA
mike@cs.newcastle.edu.au
[3] School of Computer Science, McGill University
3480 University St. #318, Montréal, Québec, CANADA
sue@opus.mcgill.ca

Abstract. The *logic engine* technique has been used in the past to establish the NP-hardness of a number of graph representations. The original technique can only be applied in those situations in which subgraphs exist for which the only possible layouts are rigid. In this paper we introduce an extension called the *wobbly logic engine* which can be used to prove the NP-hardness of several graph representations for which no such rigid layouts exist, representations by visibility and intersection in particular. We illustrate the method by using the wobbly technique to show the NP-hardness of deciding whether a graph has a nondegenerate z-axis parallel visibility representation (ZPR) by unit squares.

1 Introduction

The area of automated graph drawing has largely been shaped by its limitations. Many popular aesthetics for graph representation, such as the desire for planar layouts without edge crossings, are realizable only in the most restricted of application settings. Relaxation of the aesthetic considerations, such as allowing a small number of edge crossings, often leads to difficult (that is, NP-hard) computational problems.

With the advent of better architectures and software for graphics, it has become possible to circumvent some of these limitations by visualizing graphs

---

* A fuller version of this paper can be obtained at http://www.zpr.uni-koeln.de.

** Parts of this work were done while visiting the University of Newcastle, supported by a Visiting Researcher Grant.

*** Parts of this work were done while visiting the Universität zu Köln.

† Supported by NSERC.

in three dimensions instead of just two. For example, whereas edge-crossings are usually unavoidable in two dimensions, in a 3-dimensional setting they may generally be eliminated altogether. Also, by navigating through a 3-dimensional image, the user can often avoid views of the data which are aesthetically unappealing or confusing.

Although the 3-dimensional setting allows for a richer variety of graph representations, it is also true that the increase in dimensionality brings with it new mathematical and computational challenges. In order to open the way for good heuristics, or to justify the relaxation of some of the objectives, it is often a sensible first step to determine whether computation of the representation is an NP-hard problem.

One of the tools for establishing NP-hardness of graph realization problems is the so-called *logic engine*. It was first used by Bhatt and Cosmadakis [6] to show that it is an NP-complete problem to determine whether a tree of maximum degree 4 can be embedded in a planar grid, where tree vertices must be positioned at grid vertices and tree edges must occupy unit length grid edges. The logic engine was later employed to obtain lower bounds for checking whether a graph can be embedded in the plane as a proximity graph [14, 16, 17].

The technique can be summarized as follows (see the above references for more details; an overview will be contained in the upcoming book [13]). The reduction is from the NP-complete problem "Not-All-Equal-3SAT" (NAE-3SAT), whose instances consist of c 3-element clauses on s literals. "Yes" instances are ones with truth assignments such that each clause contains at least one true literal and at least one false literal ([22]).

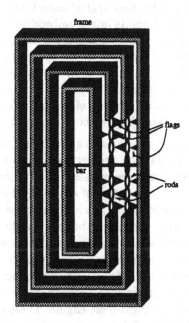

Fig. 1. A logic engine for the NAE-3SAT instance $(x_1 \lor x_2 \lor \overline{x}_3) \land (\overline{x}_1 \lor x_3 \lor \overline{x}_4) \land (x_1 \lor x_3 \lor x_4)$. The shown position corresponds to the truth setting $x_1 = 1$, $x_2 = 1$, $x_3 = 1$, $x_4 = 0$.

For every instance of the NAE-3SAT problem, a machine can be built such that a non-self-intersecting layout of the machine in the plane exists, if and only if the answer to the corresponding NAE-3SAT instance is "yes". In its most general sense, the machine consists of a rigid frame with vertical left and right sides joined at their midpoints by a horizontal bar. Attached to this bar are vertical rods that can rotate independently about the bar. One half of each rod corresponds to some variable X and the other half to its complement X'. Each half-rod has c positions where flags may be placed. The flags may rotate independently about the vertical rods. Clauses are represented by pairs of rows of flag positions. Clause C_i is represented by the row i units above the bar and the row i units below the bar. The contents of each row depends on how the vertical rods are flipped around the bar. The region above the bar is interpreted as the "true" region, the region below is interpreted as the "false" region.

Once this has been established, it is simple to encode a particular instance: if a literal $Y = X$ or X' fails to appear in a given clause C_i, then a flag is positioned on the vertical rod for (X, X'), in the half corresponding to Y and at distance i from the bar. An unfilled position in the ith row above the bar corresponds to the occurrence of a true literal in C_i; similarly, an unfilled position in the ith row below the bar corresponds to the occurrence of a false literal in C_i. Rods and flags are to be turned and flipped so that flags on the left-most and right-most vertical rods point inward, and so that no flags positioned at the same distance i from the bar on two adjacent half-rods face each other. This can be done if and only if the NAE-3SAT instance is a "yes" instance.

To apply the logic engine to a geometric graph representation problem, one first has to find a special graph whose spatial embedding is "rigid"; that is, one in which relative positionings of points in the embedding is fixed. The frame, rods and flags of the logic engine are then constructed using this graph as a building block. The problem with this approach is that for a large number of representations, there are no graphs for which the relative positioning of any points in the embedding is necessarily fixed. For example, no geometric realization of an intersection graph of open squares is rigid, as any of the squares can be perturbed to some degree relative to the others without changing the intersection graph.

The main contribution of this paper is a variant of the logic engine which can be applied to graph representation problems in which no rigid embeddings are possible. In place of rigid structures, we show how "wobbly" equivalents can be built such that the variation in position of any element of the engine, relative to any other element, is bounded. More precisely, we consider particular elements, called *springs*, having upper and lower bounds on their size. By building an appropriate structure from several springs, we can force some of them to be stretched close to their upper limits, while some are compressed to their lower limits. As a consequence, the overall variation can be made arbitrarily small with respect to the size of the engine, in such a way that the movements of the wobbly logic engine mimic those of a rigid logic engine. These springs can be designed for many problems, allowing NP-hardness proofs by constructing wobbly logic engines.

In this paper, we apply the wobbly logic engine technique to intersection graph intersection problems, and visibility representation problems. In particular, in Sections 2 to 4, we give an NP-hardness proof for the problem of deciding whether a graph has a nondegenerate z-axis parallel visibility representation (ZPR) by unit squares. Section 3 describes properties of representations of a class of graphs called k-extensors. These graphs are used as springs in Section 4 for the construction of a wobbly logic engine. The paper concludes in Section 5 with a discussion of other applications of the wobbly logic engine.

2 A 3-dimensional Visibility Representation

The representation for which we establish NP-hardness is the so-called z-parallel visibility representation (ZPR) by unit squares, where vertices are represented by axis-aligned, z-orthogonal closed unit squares, and two vertices are adjacent if and only if the corresponding squares have a z-axis-parallel "cylinder of visibility". Also, we allow only nondegenerate layouts, in that no two edges of two different squares may have identical x- or y-coordinate.

We formally define what is meant by the term "cylinder of visibility". Given a square s, let $\pi(s)$ be the projection of the interior of s onto the xy-plane, and $\zeta(s)$ be its z-coordinate. Let s_1 and s_2 be squares such that $\zeta(s_1) < \zeta(s_2)$. For a cylinder of visibility to exist between s_1 and s_2, $\pi(s_1) \cap \pi(s_2)$ must contain a disk which does not intersect the projection $\pi(s)$ for any square s such that $\zeta(s_1) < \zeta(s) < \zeta(s_2)$.

This representation is a special case of the ZPR for rectangles introduced in [3], variants of which have been considered by a number of authors [1, 4, 5, 8, 19, 21, 23, 25]. Moreover, ZPR for unit squares was considered in [19, 5], where it was shown that K_8 has no layout, whereas K_7 has a nondegenerate layout.

A ZPR for a given graph can be quite useful for an overview of the structure of the graph. In addition, the rectangles representing the nodes can carry extra information by color, shape, or size; furthermore, the orientation of arcs in a directed acyclic graph can be represented by the order in z-direction of the rectangles involved. See [5] for a practical example for a complicated graph representing a software package [2].

From a practical point of view, it would be quite useful to have an efficient (that is, polynomial-time) algorithm that decides whether a given graph can be represented. We will show that the existence of such an algorithm is unlikely. More precisely, we show that the following decision problem is NP-hard:

PROBLEM *ZPR-NDUS*:

Given a graph $G = (V, E)$, does G admit a non-degenerate ZPR layout using unit squares?

In the remainder of the paper, we will often use the same notation for vertices of a graph and the unit squares which represent them. The context in which the notation is used will make it clear whether we are discussing the graph or its layout.

3 Springs for a 3-dimensional Visibility Representation

In this section, we describe a basic structure that is used to build springs for a wobbly logic engine, which will be used in the following section to prove the NP-hardness of Problem ZPR-NDUS.

An *extensor graph of k links* (k-extensor) is a tripartite graph of the form $G_k = (U_k, V_k, W_k, E_k)$, where

$$U_k = \{u_0, u_1, \ldots, u_k\},$$
$$V_k = \{v_0, v_1, \ldots, v_{k-1}\}, \quad \text{and}$$
$$W_k = \{w_0, w_1, \ldots, w_{k-1}\}.$$

Edges of E_k are of the form

$$(v_{i-1}, u_i), (w_{i-1}, u_i) \text{ for } 1 \leq i \leq k, \text{ and}$$
$$(u_i, v_i), (u_i, w_i) \qquad \text{for } 0 \leq i \leq k-1.$$

The vertices u_0 and u_k are called the *tabs* of G_k (see Figure 2.)

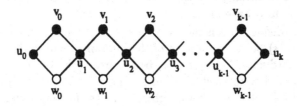

Fig. 2. A k-extensor.

Lemma 3.1 (Helly) *Let R be a collection of n axis-aligned open rectangles in the plane. If every pair of rectangles has a common intersection point, then there is a point common to all rectangles of R.*

Lemma 3.2 *Let G be a graph with the 2-extensor $G_2 = (U_2, V_2, W_2, E_2)$ as an induced subgraph. Furthermore, let the only edges joining vertices of $G \setminus G_2$ with those of G be incident to the tabs u_0 and u_2. Then for any valid non-degenerate layout of G, there exist axis-parallel lines l_0 and l_1 such that*

- *$\pi(u_0)$ is separable from $\pi(u_1)$ by l_0, and $\pi(u_1)$ is separable from $\pi(u_2)$ by l_1,*
- *l_0 and l_1 are parallel,*
- *$\pi(u_1)$ lies between l_0 and l_1, and*
- *all other axis-parallel lines separating $\pi(u_0)$ and $\pi(u_1)$ (or $\pi(u_1)$ and $\pi(u_2)$) are parallel to l_0 and l_1.*

Proof. Contained in the full version of the paper.

Lemma 3.3 *Let G be a graph with the k-extensor $G_k = (U_k, V_k, W_k, E_k)$ as an induced subgraph, for some $k \geq 2$. Furthermore, let the only edges joining vertices of $G \setminus G_k$ with those of G be incident to the tabs u_0 and u_k. Then for any valid non-degenerate layout of G, there must exist a set of axis-parallel lines $L = \{l_0, l_1, \ldots, l_{k-1}\}$ such that for all $0 \leq i < k$,*

- $\pi(u_i)$ is separable from $\pi(u_{i+1})$ by l_i,
- the lines of L are parallel,
- $\pi(u_i)$ lies between l_{i-1} and l_i, and
- all other axis-parallel lines separating $\pi(u_i)$ and $\pi(u_{i+1})$ are parallel to l_i.

Proof. Contained in the full version of the paper.

Given a non-degenerate layout of a graph G, we say that an induced k-extensor is *vertical* if the lines separating the projections of squares of U are orthogonal to the y-axis, and *horizontal* otherwise. Throughout the following lemmas, the notation $(a, b) + \{(c, d) \times (e, f)\}$ denotes an open $(d - c) \times (f - e)$ rectangle, with lower left corner at $(a + c, b + d)$.

Lemma 3.4 *Let G be a graph with the k-extensor $G_k = (U_k, V_k, W_k, E_k)$ as an induced subgraph, for some $k \geq 2$. Furthermore, let the only edges joining vertices of $G \setminus G_k$ with those of G be incident to the tabs u_0 and u_k. Let (x_i, y_i) be the xy-coordinates of the centres of symmetry of u_i, for $0 \leq i \leq k$. For any valid non-degenerate layout of G:*

- *If G_k is vertical, then $(x_k, y_k) \in (x_0, y_0) + \{(-k, k) \times (k, 2k)\}$ or $(x_k, y_k) \in (x_0, y_0) + \{(-k, k) \times (-2k, -k)\}$.*
- *If G_k is horizontal, then $(x_k, y_k) \in (x_0, y_0) + \{(k, 2k) \times (-k, k)\}$ or $(x_k, y_k) \in \{(-2k, -k) \times (-k, k)\}$.*

Proof. Contained in the full version of the paper.

Two cases which come close to the bounds stated in the lemma are shown in Figure 3.

4 A Wobbly Logic Engine for a 3-dimensional Visibility Representation

We can think of a layout of a k-extensor as a flexible *spring* that can be compressed and stretched within the limits described in Lemma 3.4. See Figure 3. In this section, we describe how these springs can be used to construct a wobbly logic engine for the problem ZPR-NDUS.

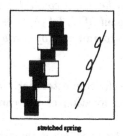 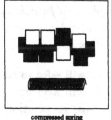

stretched spring   compressed spring

Fig. 3. The layouts "stretched spring" and "compressed spring" for a k-extensor, and their symbols

The idea for the construction is as follows – see Figure 6 for an overview. Using "connector rings", we can combine several springs to "blocks" that have very small relative flexibility; these blocks can be used to build frame, rods, and flags of a logic engine. To allow free rotation of the rods and flags in the logic engine, we use special "pivot connector rings". The overall engine has the same basic properties as a rigid logic engine.

Figure 4 shows the subgraphs C_r (regular connector ring) and C_p (pivot connector ring) and the symbols that are used to represent them in Figure 6. The squares are nodes that may also be tabs of k-extensors of adjacent springs, if necessary in the construction. Figure 5 shows the top view of a feasible layout for both types of connector rings. Note that the top part of a pivot connector ring (containing three tabs) can be transposed with respect to the bottom part (containing five tabs) without violating the layout of the ring. The centre square acts as a pivot for this transposition.

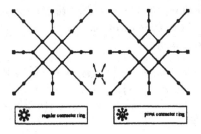

Fig. 4. A standard connector ring (left) and a pivot connector ring (right)

Fig. 5. Feasible representations for the connector rings

The next lemma shows that the tabs of a connector ring cannot be very far from its center; the proof is immediate.

Lemma 4.1 *Let* $G = (V, E)$ *be a graph having a connector ring* C *(* C_r *or* C_p *) as an induced subgraph. In a feasible layout of* G*, let* $(0,0)$ *be the location of the central node of* C*. If* (x_i, y_i) *are the xy-coordinates of one of the tabs, then we have* $(x_i, y_i) \in \{(-4, 4) \times (-4, 4)\}$*.*

In the following, we will refer to specific connector rings in Figure 6 by their reference numbers; we will also use the location of the canonical center of a connector ring to describe its position.

Next we describe the structure of a *block*, which is the arrangement induced by the connector rings 1-9 in Figure 6. A block consists of nine connector rings,

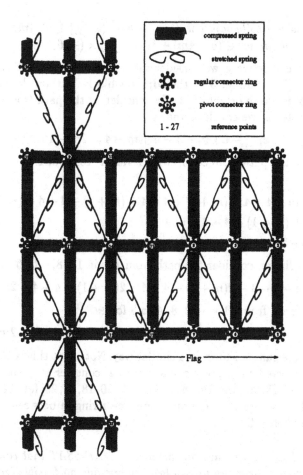

Fig. 6. A flag in a wobbly logic engine

forming a 3×3 lattice, in one direction connected by six $2m$-extensors (shown as the six vertical compressed springs $(1,2)$, $(2,3)$, $(4,5)$, $(5,6)$, $(7,8)$, $(8,9)$), in the other direction connected by six m-extensors (the horizontal compressed springs $(1,4)$, $(2,5)$, $(3,6)$, $(4,7)$, $(5,8)$, $(6,9)$), and diagonally connected by four $(m+1)$-extensors (the stretched springs $(2,4)$, $(2,6)$, $(4,8)$, $(6,8)$). Note that the four diagonal extensors form a diamond shape around the canonical center of the block (location 5 in the figure).

Finally, blocks are used for building the higher-order structures, by joining them as shown in the figure. As the following lemma shows, the relative location of two adjacent block centers is dominated by the size of the parameter m, within error bounds of constant size.

Lemma 4.2 *Let $G = (V, E)$ be a graph consisting of blocks as described above. Let $(0,0)$ be the canonical center of a block and let (x_i, y_i) be the canonical center of an adjacent block.*

- *If (x_i, y_i) is horizontally adjacent, then $(x_i, y_i) \in (2m, 0) + \{(-16, 52) \times (-36, 36)\}$ or $(x_i, y_i) \in (-2m, 0) + \{(-52, 16) \times (-36, 36)\}$.*

– If (x_i, y_i) is vertically adjacent, then we have $(x_i, y_i) \in (0, 4m) + \{(-34, 34) \times (-16, 56)\}$ or $(x_i, y_i) \in (0, -4m) + \{(-34, 34) \times (-56, 16)\}$.

Sketch: Representatively, we consider the horizontal case. In Figure 6, let the position 11 be $(0,0)$ and let 5 be the adjacent block. Without loss of generality assume that 5 lies to the right of 11. By considering the path from connector 11 via 8 to connector 5, we conclude that

$$x_i \geq -4 + m - 4 - 4 + m - 4 = 2m - 16.$$

By considering the path from connector 11 via 14, 10, 8, 6, 2 to connector 5, we conclude that

$$x_i \leq 4 - m + 4 + 4 + (m+1) + 4 + 4 + (m+1) + 4 + 4 + (m+1) + 4$$
$$+ 4 + (m+1) + 4 + 4 - m + 4$$
$$= 2m + 52.$$

For the y-coordinate, consider the path from 11 via 10, 8, 6 to 5, implying

$$x_i \geq -4 + 2m - 4 - 4 - 2(m+1) - 4 - 4 - 2(m+1) - 4 - 4 + 2m - 4 = -36.$$

Similarly, the path from 11 via 12, 8, 4 to 5 forces

$$x_i \leq 4 - 2m + 4 + 4 + 2(m+1) + 4 + 4 + 2(m+1) + 4 + 4 - 2m + 4 = 36.$$

The vertical case can be shown in a similar way. Note that this is also true if the blocks are connected by a pivot connector ring: consider the paths $[17, 18, 19]$, $[17, 16, 14, 18, 27, 29, 26]$, $[17, 14, 18, 27, 26]$, $[17, 20, 18, 25, 26]$ for the distance between 17 and 26, where 18 is the pivot connector ring in between, and 29 is the connector ring below 26. □

Lemma 4.3 *Suppose we have an instance of NAE-3SAT that cannot be satisfied. Then the corresponding wobbly logic engine has no feasible layout.*

Sketch: Suppose we have an NAE-3SAT instance which cannot be satisfied. This means that for any layout of a rigid logic engine, we have a collision between flags. By choosing an appropriate representation of the flags by blocks, there will be an overlap of at least $O(m)$ if we keep all the blocks rigid. As we showed above, allowing for wobble will produce variations of $O(dist(p, q))$ for the relative position of two block centers p and q, where $dist(p, q)$ is the rectilinear distance of p and q in the canonical grid graph of block centers. Since this variation is independent of m and polynomial in the input size, choosing the free parameter m to be of sufficiently large value that is still polynomial in the input size forces an overlap of two flags in any layout of the wobbly logic engine. □

Lemma 4.4 *Suppose we have an instance of NAE-3SAT that can be satisfied. Then there exists a feasible layout of the corresponding wobbly logic engine.*

Proof. Contained in the full version of the paper. See Figure 6 for an example.

Summarizing, we state:

Theorem 4.5 *The problem ZPR-NDUS is NP-hard.*

5 Other Applications

The original logic engine has been used for a variety of "rigid" graph layout problems. One of them is the representation of a graph as the intersection graph of closed objects such as unit squares or unit disks. It is not hard to see how degenerate arrangements can be forced that can be used as rigid building blocks for a logic engine. On the other hand, these degenerate arrangements are necessarily unstable if perturbed even slightly. This makes it more attractive, when considering which graph representations to use, to require objects to intersect in a "fat" manner. This is equivalent to considering open objects only.

The wobbly logic engine can be used to prove NP-hardness of a number of these problems. In this section, we sketch how we can prove NP-hardness of the problem of deciding whether a given graph is an intersection graph of open unit squares or open unit cubes.

The first result is an immediate corollary of Theorem 4.5; note that all constructions of the two preceding sections can be used without any changes:

PROBLEM *Open Unit Square Intersection Graph*:
Given a graph $G = (V, E)$, is G the intersection graph of axis-aligned open unit squares in the plane?

Corollary 5.1 *The problem* Open Unit Square Intersection Graph *is NP-hard.*

With some modifications, the approach may be extended to a 3-dimensional variant of this problem.

PROBLEM *Open Unit Cube Intersection Graph*:
Given a graph $G = (V, E)$, is G the intersection graph of axis-aligned open unit cubes in 3-space?

Theorem 5.2 *The problem* Open Unit Cube Intersection Graph *is NP-hard.*

Consider the modified k-extensor shown in Figure 7. For this type of k-extensor, it is straightforward to prove lemmas analogous to those in Sections 3 and 4. For constructing a wobbly logic engine, we use a 3-dimensional $3 \times 3 \times 3$ lattice as a block, with appropriate diagonal springs forming a diamond shape around the center. Most of the engine can be built from these blocks analogously to the 2-dimensional case. However, since in the 3-dimensional setting flags can otherwise be oriented in four instead of two directions, we exclude two undesired directions by building appropriate protrusions near the pivots.

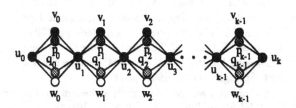

Fig. 7. A k-extensor for the problem *3-dimensional unit box intersection graph*

In the same way, we can generalize even further:

PROBLEM *Open Unit Hypercube Intersection Graph*:
Given a graph $G = (V, E)$, is G the intersection graph of axis-aligned open unit hypercubes in d-space?

Theorem 5.3 *The problem* Open Unit Hyperube Intersection Graph *is NP-hard.*

The wobbly logic engine technique can be applied to classes of objects other than squares or cubes. In particular, the following can be shown:

Theorem 5.4 *The problem* Open Unit Disk Intersection Graph *is NP-hard.*

References

1. H. Alt, M. Godau, and S. Whitesides. Universal 3-dimensional visibility representations for graphs. *Proc. Graph Drawing 95*, Passau 1996. Lecture Notes in Computer Science Vol. 1027, Springer-Verlag, 1996, pp. 8–19.
2. A. Bachem, S. P. Fekete, B. Knab, R. Schrader, I. Vannahme, I. Weber, R. Wegener, K. Weinbrecht, and B. Wichern. Analyse großer Datenmengen und Clusteralgorithmen im Bausparwesen. In *Geld, Finanzwirtschaft, Banken und Versicherungen*. Editors C. Hipp et al., VVW Karlsruhe, 1997.
3. P. Bose, H. Everett, S. P. Fekete, A. Lubiw, H. Meijer, K. Romanik, T. Shermer, and S. Whitesides. On a visibility representation for graphs in three dimensions. *Proc. Graph Drawing '93*, Paris (Sèvres), 1993, pp. 38–39.
4. P. Bose, H. Everett, S. P. Fekete, A. Lubiw, H. Meijer, K. Romanik, T. Shermer, and S. Whitesides. On a visibility representation for graphs in three dimensions. Snapshots of Computational and Discrete Geometry, **3**, *eds.* D. Avis and P. Bose, McGill University School of Computer Science Technical Report SOCS-94.50, July 1994, pp. 2–25.
5. P. Bose, H. Everett, S. P. Fekete, M. E. Houle A. Lubiw, H. Meijer, K. Romanik, G. Rote, T. Shermer, S. Whitesides, and C. Zelle. A visibility representation for graphs in three dimensions. Technical Report ZPR 97-253, Center for Parallel Computing, 1997. Available at ftp://ftp.zpr.uni-koeln.de/pub/papers/zpr97-253.ps.gz.
6. S. Bhatt and S. Cosmadakis. The complexity of minimizing wire lengths in VLSI layouts. *Information Processing Letters*, v. 25, 1987, pp. 263-267.
7. F. J. Brandenburg. Nice drawings of graphs and trees are computationally hard. Technical Report MIP-8820, Fakultät für Mathematik und Informatik, Univ. Passau, 1988.
8. F.J. Cobos, J.C. Dana, F. Hurtado, A. Marquez, and F. Mateos On a visibility representation of graphs. *Proc. Graph Drawing 95*, Passau 1996. Lecture Notes in Computer Science Vol. 1072, Springer-Verlag, 1996, pp. 152–161.
9. R. Cohen, P. Eades, T. Lin, and F. Ruskey. Three-dimensional graph drawing. *Proc. Graph Drawing '94*, Princeton NJ, 1994. Lecture Notes in Computer Science LNCS Vol. 894, Springer-Verlag, 1995, pp. 1–11.
10. L. Danzer, B. Grünbaum, and V. Klee. Helly's theorem and its relatives. *Convexity, Proc. 7th ACM Symp. Pure Math.*, 1963, pp. 101–181.

11. A. Dean and J. Hutchinson. Rectangle visibility representations of bipartite graphs. *Proc. Graph Drawing '94*, Princeton NJ, 1994. Lecture Notes in Computer Science LNCS Vol. 894, Springer-Verlag, 1995, pp. 159–166.

12. G. Di Battista, P. Eades, R. Tamassia, and I. G. Tollis, Algorithms for automatic graph drawing: an annotated bibliography. *Comput. Geometry: Theory and Applications*, 4, 1994, pp. 235–282. Also available from wilma.cs.brown.edu by ftp.

13. G. Di Battista, P. Eades, R. Tamassia, and I. G. Tollis. *Graph Drawing. Algorithms for geometric representation of graphs*. Prentice Hall, to appear.

14. G. Di Battista, G. Liotta and S. Whitesides. The strength of weak proximity. *Proc. Graph Drawing 95*, Passau 1996. Lecture Notes in Computer Science Vol. 1072, Springer-Verlag 1996.

15. G. Di Battista and R. Tamassia. Algorithms for plane representations of acyclic digraphs. *Theoretical Computer Science*, **61**, 1988, pp. 175–198.

16. P. Eades and S. Whitesides. The realization problem for Euclidean minimum spanning trees is NP-hard. *in* Proc. 10th ACM Symp. on Computational Geometry, 1994, pp. 49-56.

17. P. Eades and S. Whitesides. The logic engine and the realization problem for nearest neighbor graphs. *Theoretical Computer Science*, 169 (1), 1996, pp. 23-37.

18. H. ElGindy, G. Liotta, A. Lubiw, H. Meijer and S. Whitesides. Recognizing rectangle of influence drawable graphs. *Proc. Graph Drawing '94*, Princeton NJ, 1994, Lecture Notes in Computer Science LNCS #894, Springer-Verlag, 1995 pp. 352-363.

19. S. P. Fekete, M. E. Houle, and S. Whitesides. New results on a visibility representation of graphs in 3D. *Proc. Graph Drawing 95*, Passau 1996. Lecture Notes in Computer Science Vol. 1072, Springer-Verlag 1996, pp. 234–241.

20. S. P. Fekete, M. E. Houle, and S. Whitesides. The wobbly logic engine: proving hardness of non-rigid graph representation problems. (Full version). Technical Report ZPR 97-273, Center for Parallel Computing, 1997. Available at ftp://ftp.zpr.uni-koeln.de/pub/papers/zpr97-273.ps.gz.

21. S. P. Fekete and H. Meijer. Rectangle and box visibility graphs in 3D. To appear in *International Journal of Computational Geometry and Applications*. Available at ftp://ftp.zpr.uni-koeln.de/pub/papers/ zpr96-224.ps.gz.

22. M. R. Garey and D. S. Johnson. Computers and Intractability: A Guide to the Theory of NP-Completeness. W. H. Freeman, San Francisco, 1979.

23. J. Hutchinson, T. Shermer, and A. Vince. On representations of some thickness-two graphs. *Proc. Graph Drawing 95*, Passau 1996. Lecture Notes in Computer Science Vol. 1072, Springer-Verlag 1996, pp. 324–332.

24. P. J. Idicula. Drawing trees in grids. Master's thesis, Department of Computer Science, University of Auckland, 1990.

25. K. Romanik. Directed VR-representable graphs have unbounded dimension. *Proc. Graph Drawing '94*, Princeton, NJ, 1994, Lecture Notes in Computer Science LNCS Vol. 894, Springer-Verlag, 1995, pp. 177–181.

26. R. Tamassia and I. G. Tollis. A unified approach to visibility representations of planar graphs. *Discrete Comput. Geom.* 1, 1986, pp. 321–341.

3DCube: A Tool for Three Dimensional Graph Drawing*

Maurizio Patrignani[1] and Francesco Vargiu[2]

[1] Dipartimento di Informatica e Automazione,
Università di Roma Tre, Rome, Italy
patrigna@inf.uniroma3.it
[2] Autorità per l'Informatica nella Pubblica Amministrazione, Italy
vargiu@aipa.it

Abstract. In this paper we describe a tool that is a general frame for the three-dimensional representation of graphs, especially devoted to the algorithms evaluation, refinement and development. 3DCube (3D Diagram Drawer) offers innovative features in the user interaction and contains a set of three-dimensional algorithms both taken from the literature and proposed by the authors.

1 Introduction

Three dimensional graph drawing is an emerging field in the graph drawing area. Several tools are already available for the representation of graphs in the plane, and the most sophisticated of them also allow some three-dimensional representation, either as an additional presentation feature of basically $2D$ results (see GMB [13] and PLUM [17]), or as the result of an actual $3D$-drawing algorithm (see 3DSA [4], COMAIDE [6], the ffGraph library [9], GEM-3D [2], GIOTTO3D [11], GOLD [10], GOVE [18]), and PARSA [14]).

The tool we describe in this paper offers to the user a general frame for the representation of graphs entirely devoted to $3D$-drawing algorithms, and especially to their evaluation, development and refinement. It complies to the requirements of an algorithm independent tool, by supporting a set of $3D$-algorithms taken from the literature and proposed by the authors. Furthermore 3DCube proposes an innovative interaction paradigm to the user, both for graph browsing and for inspection and understanding of the layout process.

2 Requirements and Functionalities

The aim of 3DCube is not restricted simply to a representation system. The tool provides also a support i) to the visual analysis of the behavior of the layout algorithms during their execution; ii) to the refinement of algorithms, especially those based on heuristics or including tunable steps; and iii) to the direct comparison among different algorithms.

The following list summarizes the main features of 3DCube:

---

* Research supported in part by the ESPRIT LTR Project no. 20244 - ALCOM-IT

3D-**perception.** We used any known feature to allow the user to recognize distances and dimensions of the *3D* objects mapped on the screen, so enhancing the comprehension of object distribution in space.

The user of 3DCube is able to browse through the representation, by using the three primitives rotation, shifting, and zooming, which allow to simulate any real movement. Depth cuing, perspective viewing and surface characteristics are used to obtain a realistic *3D* effect.

The tool implements also a *Snapshot* utility that allows to store a collection of meaningful views of a diagram. By selecting a view from a film-like sequence, the user asks the tool to recreate the same conditions in which the snapshot was captured, improving the user-friendliness of the system.

Parametric Representation. Different shapes and properties for nodes and edges can be specified according to the user requirements.

We adopt a separation between the abstract description of the graphs and the geometrical properties of nodes and edges; we use a synthetic and adaptable model to describe them, compatible with the file formats of Diagram Server [5]. A list of graphic primitives (such as spheres, cubes, polygons) and associated attributes (colors, dimensions, etc.) defines each type of node. Analogously, for each type of edge it is possible to define the shape of bends and of segments.

Algorithms Visual Inspection and Understanding. By understanding the operations involved in a layout algorithm and how such operations reflect in the final results, the user is supported to possibly improve the algorithm itself or to conceive new algorithms.

3DCube offers an *algorithm animation* facility that allows to inspect the intermediate results of a diagram construction by means of its graphical representation. This facility permits to animate the passage from consecutive representations (both forward and backwards), so that the user may easely follow the evolution of the drawing through a continuous movement of the involved objects.

Extensibility. The capability of adding new algorithms and new features without a main implementation effort is achieved by using a modular approach in the architecture design.

3 The Architecture of 3DCube

The architecture of 3DCube was conceived according to an object oriented methodology. It consists of a set of independent modules that exchange informations and services to each other, as shown in Figure 1.

The *User Interface* manages the interaction with the user by means of the windows environment, menus, buttons and controls. A particular attention was placed to support the navigation (a control panel makes available different features) and to support the depth perception, using perspective, shades and colors.

The *Graphic Driver* filters each graphic system call from the User Interface to the computing platform thus improving the portability of the tool.

3DCube is not limited to the showing of the results of a specific layout algorithm. The *Algorithm Manager* is the repository of algorithms; in the current version it mainly contains *3D*-orthogonal drawing algorithms. We cite the algo-

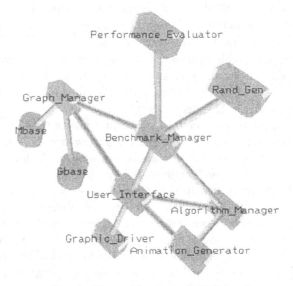

Fig. 1. The architecture of 3DCube.

rithm proposed by Therese Biedl [1], the Komolgorov and Bardzin algorithm [7], the Compact and the Three Bends [8], the algorithm proposed by Papakostas and Tollis [15], and Reduce Forks, proposed by the authors [16]; moreover it contains the 3*D*-straight-line drawing algorithm, called Momentum Curve, of Cohen, Eades, Lin and Ruskey [3].

The *Animation Generator* module, by using the services of the Algorithm Manager, is able to simulate the continuos movement of nodes and edges during the algorithm execution.

The *Graph Manager* implements the storing and loading functions for the graphs, their representations, the snapshots, and the associated models in the *GBase* and *MBase*.

The *Benchmark Manager* allows the automatic execution of a list of algorithms on a set of graphs, and is able to generate reports about the behavior of the algorithms with respect to a given set of aesthetic criteria.

For this purpose the Benchmark Manager interacts with the *RandGen* module which is able to generate a base of random graphs according to some user-specified features. It also communicates with the *Performances Evaluator* to obtain the computation of the results with respect to the selected criteria and to generate reports and graphics.

4 An example of graph browsing

In this section we describe, with some pictures, the main features of the user interface of 3DCube, by simulating a typical work session. The user may load and visualize one or more different graphs at the same time. For every loaded graph 3DCube shows on a window the list of the drawing methods applyable to the given graph (see Figure 2). The selection of a method triggers its execution on the graph and the obtained representation is displayed on a new window, called the *diagram window*. In the above mentioned picture it is shown a diagram window with the result of Reduce Forks applied to a graph with 24 nodes.

Figure 3 shows a K_6 graph drawn by using the algorithms of Biedl, Compact, Komolgorov and Bardzin, Papakostas and Tollis, Reduce Forks, and Three Bends, respectively. The control panel allows the user to navigate through the representation to inspect the objects displacement. During the navigation, the user can save meaningful views of the representation by clicking on the snapshot button. The collection of all previously saved snapshots for the given representation is shown in the snapshot window (see Figure 2).

5   An example of drawing algorithm refinement

In this section we describe how 3DCube was effectively used in the development of a new $3D$-orthogonal drawing algorithm.

The starting idea was to obtain a $3D$-orthogonal drawing through the following iterative procedure: firstly the same coordinates are assigned to all the nodes of the graph; then at every step a plane (orthogonal to one of the axes and hosting at least two nodes with the same coordinates) is split in two, and its nodes are redistributed between the two resulting planes. The procedure terminates when all nodes have different coordinates. The procedure sketched above is not sufficient to produce an orthogonal drawing: when a split occurs, an edge may become slant. So slant edges are split in two, as they appear, and a dummy node (that eventually will become a bend) is inserted and positioned in such a way that its two adjacent edges are orthogonal. Furthermore, we defined four sufficient conditions, that, once they are satisfied one after the other, guarantee the termination of the algorithm, namely:

– the separation between original nodes,
– for each original node, the coherence of the directions of its incident edges,
– the separation between original and dummy nodes, and
– the separation between dummy nodes.

In the first version of the algorithm, called Four Targets, we defined and implemented four different split criteria, based on heuristics, each devoted to satisfy one of the four corresponding conditions of the previous list.

By extensively using the tool in order to verify the effectiveness of Four Targets we noticed that, while the algorithm satisfied the convergence conditions, it made very costly choices with respect to the final number of bends and volume occupation. So we reviewed the last three split criteria by introducing a ranking among alternative split operations, aimed to minimize the number of bends introduced. The new algorithm significantly outperformed the previous one.

Finally we noted that a specific configuration, called *fork*, consisting of a pair of adjacent edges both with an end in each of the plane generated by a split, was responsible for the insertion of an extra bend either at the end of the split operation, or in a successive split. Thus we conceived a new heuristic for the first split criterium, aimed to reduce the number of forks.

Experiments showed that the performance of this method, called Reduce Forks, are interesting with respect to the number of bends and space occupation. By using 2000 random test graphs approximately, with a number of nodes between 10 and 100, we obtained an average of less than 2.5 bends per edge and a volume occupation of $0.6n^3$.

While the drawings produced by the Reduce Forks algorithm are, in our opinion, aesthetically pleasant, we are aware that the algorithm may be further improved both by tuning the existent steps, and by possibly adding further steps (such as a bend stretching post processing step).

6 Conclusions and Future Improvements

The current version of the tool is developed in C++ language, on a UNIX RISC 6000 workstation, using the Motif libraries for the management of the windows environment and the graPHIGS libraries for the three-dimensional visualisation. We plan to rewrite the tool by using the VRML language to make available the implementation of the graph drawing algorithms on the Internet. We aim to increase the number of implemented graph drawing algorithms. Finally, in order to increase the interoperability with other tools, we aim to add support for new formats such as GML [12].

Acknowledgements

We are grateful to Giuseppe Di Battista for initiating our interest in the $3D$ graph drawing area, for his comments, suggestions and encouragement. We are also grateful to Walter Didimo, Antonio Leonforte, Sandra Follaro and Loredana Bottaro for their support.

References

1. T.C. Biedl. Heuristics for 3D-Orthogonal Graph Drawing. Presented at the 4th Twente Workshop, Enschede, June 1995.
   http://rutcor.rutgers.edu:80/ therese/ps/3D.twente.ps
2. I. Bruß and A. Frick. Fast Interactive 3-D Graph Visualization. *Proc. GD '95*, LNCS 1027, pp. 99-110, Springer-Verlag, 1995.
   ftp://i44ftp.info.uni-karlsruhe.de/pub/papers/frick/gd95p.ps.gz
3. R. F. Cohen, P. Eades, T. Lin and F. Ruskey. Three-dimensional graph drawing. *Proc. GD '94*, LNCS 894, pp. 1-11, 1994.
4. I.F. Cruz and J.P. Twarog. 3D Graph Drawing with Simulated Annealing. *Proc. GD '95*, LNCS 1027, pp. 162-165, Springer-Verlag, 1995.
5. G. Di Battista, G. Liotta, and F. Vargiu. Diagram Server. *JVLC* (special issue on Graph Visualization, I. F. Cruz and P. Eades, editors), 6(3), 1995.
6. D. Dodson. COMAIDE: Information Visualization using Cooperative 3D Diagram Layout. *Proc. GD '95*, LNCS 1027, pp. 190-201, Springer-Verlag,1995.
7. P. Eades, C. Stirk and S. Whitesides. The Techniques of Komolgorov and Bardzin for Three Dimensional Orthogonal Graph Drawing. *TR 95-07, Dept. of Computer Science, University of Newcastle NSW, Australia*, October 1995.
   ftp://ftp.cs.newcastle.edu.au/pub/techreports/tr95-07.ps.Z

8. P. Eades, A. Symvonis and S. Whitesides. Two Algorithms for Three Dimensional Orthogonal Graph Drawing. *Proc. GD '96*, LNCS 1190, pp. 139-154, Springer-Verlag,1996.

9. C. Friedrich. The ffGraph Library. *Lehrstuhl für Informatik*, Universität Passau, December 1995. http://kaolin.unice.fr/Documentation/Doc.html

10. A. Frick, C. Keskin and V. Vogelmann. Integration of Declarative Approaches (System Demonstration). *Proc. GD '96*, LNCS 1190, pp. 184-192, Springer-Verlag,1996.

11. A. Garg and R. Tamassia. GIOTTO3D: A System for Visualizing Hierarchical Structures in 3D. *Proc. GD '96*, LNCS 1190, pp. 193-200, Springer-Verlag, 1996.

12. M. Himsolt. The Graphlet System (System Demonstration). *Proc. GD '96*, LNCS 1190, pp. 233-240, Springer-Verlag, 1996.

13. D. Jablonowsky and V.A. Guarna. GMB: A Tool for Manipulating and Animating Graph Data Structures. *Softw. - Pract. Exp.*, 19(3), pp. 283-301, 1989.

14. B. Monien, F. Ramme and H. Salmen. A Parallel Simulated Annealing Algorithm for Generating 3D Layouts of Undirected Graphs. *Proc. GD '95*, LNCS 1027, pp. 396-408, Springer-Verlag, 1995.

15. A. Papakostas. Information Visualization: Orthogonal Drawings of Graphs. *PhD thesis*, Department of Computer Science, University of Texas at Dallas, Dec. 1996.

16. M. Patrignani. Visualizzazione di Diagrammi in Tre Dimensioni. *Ms. Sc. Degree thesis*, Dipartimento di Informatica e Sistemistica, Università degli Studi di Roma "La Sapienza", Oct. 1996.

17. S.P. Reiss. E-D Visualization of Program Information. *Proc. GD '94*, LNCS 894, pp. 12-24, Springer-Verlag, 1995.

18. R. Webber and A. Scott. GOVE Grammar-Oriented Visualisation Environment. *Proc. GD '95*, LNCS 1027, pp. 516-519, Springer-Verlag, 1995.

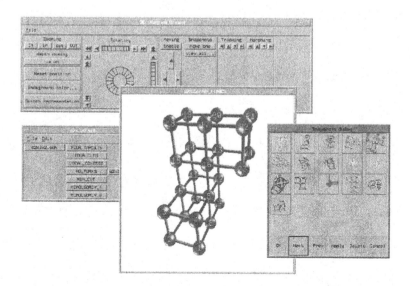

Fig. 2. A session of 3DCube displaying a graph of 24 nodes drawn by the Reduce Forks algorithm and the snapshot window.

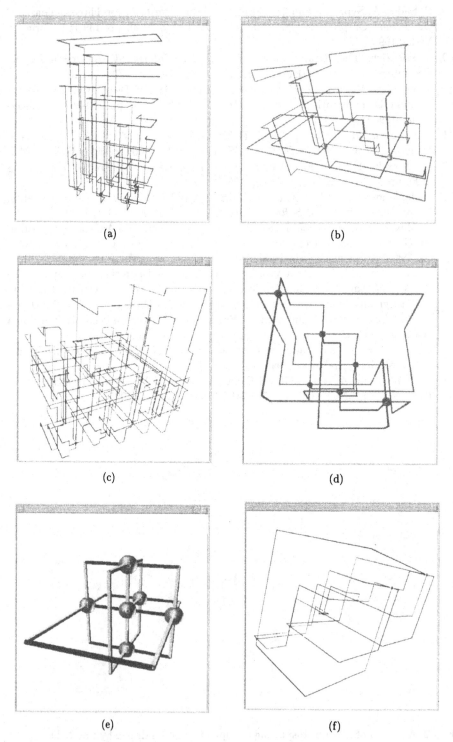

Fig. 3. The Biedl (a), Compact (b), Komolgorov (c), Papakostas-Tollis (d), Reduce Forks (e), and the Three Bends (f) algorithms applied to a K_6 graph.

Graph Clustering Using Multiway Ratio Cut (Software Demonstration)

Tom Roxborough and Arunabha Sen*
Department of Computer Science and Engineering
Arizona State University
Tempe, AZ 85287
USA

Abstract. Identifying the *natural clusters* of nodes in a graph and treating them as *supernodes* or *metanodes* for a higher level graph (or an abstract graph) is a technique used for the reduction of visual complexity of graphs with a large number of nodes. In this paper we report on the implementation of a clustering algorithm based on the idea of *ratio cut*, a well known technique used for circuit partitioning in the VLSI domain. The algorithm is implemented in WINDOWS95/NT environment. The performance of the clustering algorithm on some large graphs obtained from the archives of Bell Laboratories is presented.

1 Introduction

Graphs are frequently used to model problems from various diverse domains such as telecommunication networks, VLSI circuit design, databases and computational chemistry. In these domains the nodes are used to represent certain entities of that domain and the edges represent the relationships between them. The relationship between the entities can be very effectively conveyed visually by a *nice* layout of the graph. However, in most of the realistic problem instances, the number of nodes and edges is far too many for a nice layout and also for comprehension of the information the graph was supposed to convey. In such situations, an abstract graph is constructed where each node represents a set of nodes of the original graph and the edges represent the relationship between these sets of nodes. Thus an abstract graph construction problem reduces to the problem of partitioning the node set of the original graph $G = (V, E)$, into a subset of nodes $V_1, V_2, ..., V_k$, such that $\cup_{i=1}^{k} V_i = V$ and $V_i \cap V_j = \emptyset$ for $i \neq j$.

The subsets V_is $(1 \leq i \leq k)$ are known as the *clusters* of the graph $G = (V, E)$. One problem with this approach is that there is no consensus among the researchers as to what constitutes a cluster. There is some intuitive understanding of what constitutes a cluster but there is no universally accepted formal definition of a cluster. In case the nodes and edges of the graph have some semantic information associated with them (in the form of labels), such information can

---

* Corresponding author, Telephone: 602-965-6153, Fax: 602-965-2751, e-mail: arunabha.sen@asu.edu. This work was partially supported by a grant from Tom Sawyer Software under NIST Advanced Technology Program.

be used for the purpose of clustering (or *grouping*) the nodes. An example of such information could be the IP addresses associated with the nodes in a telecommunication network. In case the graph has no such information, then the structural properties of the graph have to be utilized for the purpose of generating the clusters. We will refer to such graphs as the *flat graphs*. Several candidates for the structures have been proposed in the literature. These include *biconnected components* [2], *paths and triangles* [1], *circles of cliques* [7]. The reader is referred to [5] for discussions of some other possible structures for clustering. In spite of the differences of opinion as to what constitutes a cluster, one idea is universally accepted: the nodes belonging to a cluster must have a *strong relationship* between them in comparison with the nodes outside the cluster. In case of a flat graph this translates to finding a partition that minimizes the number of inter-cluster edges (or maximizes the intra-cluster edges).

2 Ratio Cut Technique

The Ratio Cut technique was proposed in [4] for the purpose of identifying the *natural clusters* of a graph. The technique was proposed in the VLSI domain for the circuit partitioning problem. Both in the case of circuit partitioning as well as graph clustering, minimization of the *cut edges* is a very important objective. In case the node set needs to be partitioned into only two subsets V_1 and V_2, the minimum cut partition can easily be computed using the *max-flow* techniques. However, the technique does not have any control on the size of subsets V_1 and V_2. In the VLSI domain each of the subsets has to fit into an integrated circuit chip and as such the size of each subset has to conform to some pre-specified maximum limit. Therefore, the max-flow technique is not very useful in the circuit partitioning problem. The Kernighan-Lin technique, a well known heuristic for circuit partitioning, requires that the size of the two partitions V_1 and V_2 of the node set V be equal. This technique heuristically tries to find a partition with a small cut value, all the while keeping the size of the two subsets V_1 and V_2 equal.

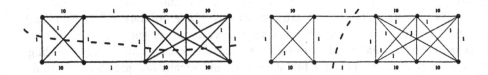

Fig. 1. Partitions produced by Kernighan-Lin Algorithm and Ratio Cut Algorithm respecively

As shown in [4], such a strict size requirement often forces a partition with a high cut size. The K-L technique produces the partition shown in figure 1, whereas a partition into natural clusters with a much better cut size is shown in figure 2. To attain the twin objectives of (i) minimizing the cut value and (ii) minimizing the difference in the size of the subsets, the authors of [4] proposed a new metric called the *ratio cut* to measure the quality of a partition. The ratio cut is defined as follows:

Consider a graph $G = (V, E)$. Suppose $c_{i,j}$ is the capacity (or the weight) of the edge connecting the nodes i and j. Suppose V_1, V_2 is a partition of the node set V. $(V_1 : V_2)$ denotes a cut that separates the nodes of V_1 from the nodes of $V_2 = V - V_1$. The capacity of this cut is equal to

$$C(V_1, V_2) = \sum_{i \in V_1} \sum_{j \in V_2} c_{ij}$$

The corresponding *ratio* value is given by

$$R(V_1, V_2) = C(V_1, V_2) / \mid V_1 \mid \times \mid V_2 \mid$$

The *ratio cut* is defined to be the cut that has the minimum ratio among all possible cuts of the graph, i.e., a cut (V_1, V_2) will be known as a ratio cut if

$$R(V_1, V_2) = \min_{X \subset V; Y = V \backslash X; X, Y \neq \emptyset} R(X, Y)$$

3 Implementation

As seen in the discussion in the previous section, the ratio cut technique proposed in [4] is applicable for a two-way partition of the graph. The authors dealt with the multi-way partition problem by repeated application of the two-way partition. For the graph clustering problem, we adapted the two-way ratio cut principle to a multi-way partition problem. In case of a k-way partition we compute the ratio as follows:

$$R(V_1, V_2, ..., V_k) = C(V_1, V_2, ..., V_k) / \mid V_1 \mid \times \mid V_2 \mid \times ... \times \mid V_k \mid$$

where

$$C(V_1, V_2, ..., V_k) = 1/2 \sum_{p=1}^{k} \sum_{i \in V_p} \sum_{j \notin V_p} c_{ij}$$

The *ratio cut* is defined exactly the same way as before, that is the cut that has the minimum ratio among all possible cuts of the graph.

In our implementation, we did not put a limit on the maximum number of nodes in a cluster, as we felt that such a restriction is artificial. However, our algorithm requires the user to specify the number of clusters in which the node set should be partitioned. The rational for this requirement is the following: The reason for clustering the nodes of the graph is to reduce the visual complexity and as such it should be left to the user to determine what level of complexity is acceptable for his application.

4 Experimental Results and Discussions

We tested our implementation of the multiway ratio cut based clustering algorithm with a wide range of graphs - small, medium and large. We created some example graphs for testing purposes and extensively used the Bell Laboratories graph library. This library has a large collection of graphs of wide range of variation in terms of number of nodes, edges and node degrees. Some representative examples of the output of our clustering algorithm is attached. The clustering algorithm runs on both the UNIX and the PC environment. We used Tom Sawyer Software's Graph Layout Toolkit for the layout of the graphs. The clustering algorithm computed the clusters in less than a few seconds in almost all of the Bell Laboratories graphs. Only in a small number of cases it required a few minutes for clustering.

5 Demo Environment

The clustering algorithm was implemented on a WINDOWS95/NT environment using Microsoft Visual C++ version 4.0. All graph layouts were produced by the Graph Layout Toolkit version 2.3 of Tom Sawyer Software Corporation. A PC with a 486 or Pentium processor along with 16MB of RAM is the only hardware requirement for the demo. The demo currently runs under either WINDOWS95 or NT.

6 Acknowledgements

The authors wish to acknowledge Jay Noh for his work on the initial development of the clustering technique.

References

1. R. Sablowski and A. Frick, "Automatic Graph Clustering," *Proceedings of Graph Drawing'96*, Berkeley, California, September,1996.
2. U. Dogrusoz, B. Madden and P. Madden, "Circular Layout in the Graph Layout Toolkit," *Proceedings of Graph Drawing'96*, Berkeley, California, September,1996.
3. P. Eades, "Multilevel Visualization of Clustered Graphs," *Proceedings of Graph Drawing'96*, Berkeley, California, September,1996.
4. Y. C. Wei and C. K. Cheng, "Ratio Cut Partitioning for Hierarchical Designs," *IEEE Transactions on Computer Aided Design*, vol. 10, no. 7, pp.911-921, July 1991.
5. C. J. Alpert and A. B. Kahng, "Recent Directions in Netlist Partitioning: A Survey," *INTEGRATION: the VLSI journal*, vol. 19, pp.1-81,1995.
6. D. Kimmelman, B. Leban, T. Roth, D. Zernik "Dynamic Graph Abstraction for Effective Software Visualization," *The Australian Computer Journal*, vol. 27, no. 4, pp.129-137, Nov 1995.
7. F. J. Brandenburg, M. Himsolt and K. Skodinis, "Graph Clustering: Circles of Cliques," submitted to *Graph Drawing'97*.

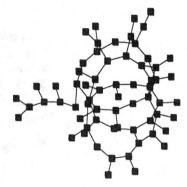

Fig. 2. Bell Lab graph 1572 before clustering

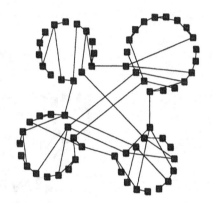

Fig. 3. Bell Lab graph 1572 with 4 clusters

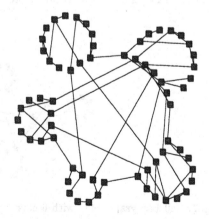

Fig. 4. Bell Lab graph 1572 with 5 clusters

296

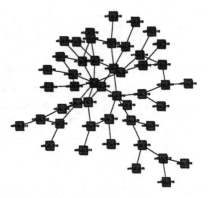

Fig. 5. Bell Lab graph 1487 before clustering

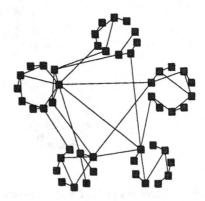

Fig. 6. Bell Lab graph 1487 with 5 clusters

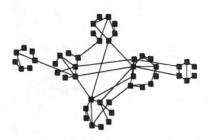

Fig. 7. Bell Lab graph 1487 with 6 clusters

ArchE: A Graph Drawing System for Archaeology

Christoph Hundack *
Petra Mutzel **
Igor Pouchkarev
Stefan Thome

Max-Planck-Institut für Informatik, Saarbrücken (Germany)***

Abstract. We present *ArchE* (Archaeological Editor), a system for processing and displaying archaeological data. *ArchE* checks these data for consistency, simplifies and displays them; for each of these steps *ArchE* offers a number of different algorithms. The interactive features (eg input, data editing and modification of the layout) are easy to handle. Furthermore, *ArchE* contains algorithms for focusing on user-defined aspects of the data.

Apart from archaeological applications, *ArchE* can be used as a general graph drawing system.

1 Introduction

In archaeology the dating of objects is not only obtained by absolute dating evidence (as radiocarbon dating) but also by relative dating evidence. This relative dating within archaeological contexts (eg, excavations) is known as stratigraphy. Practical experience has shown that in almost all cases a spatial relationship ("is lying deeper than") directly translates into a chronological relationship ("is earlier than").

Stratigraphy was put on a systematic basis in the early 70s by E. Harris when he was confronted with an extremely complex and dense excavation in Winchester (compare [4]). Harris developed the so-called "Harris–Matrix" (or Harris-Diagram) to describe chronological relationships between the archaeological objects.

Since then a number of programs have been developed to facilitate the display and the analysis of the enormous amount of data. One of these programs is part of WinBASP (Bonn Archaeological Software Package). WinBASP was designed

* Research supported by a Graduiertenkolleg graduate fellowship of the Deutsche Forschungsgemeinschaft

** partially supported by DFG-Grant Ju204/7-1, Forschungsschwerpunkt "Effiziente Algorithmen für diskrete Probleme und ihre Anwendungen"

*** {hundack,mutzel,pouchkar,thome}@mpi-sb.mpg.de

in largly by the Rheinisches Amt für Bodendenkmalpflege, Bonn (Germany) and is a system for the support of various archaeological tasks (eg, mapping, seriation, and correspondence analysis of archaeological data). Unfortunately the present program of WinBASP for generating a Harris Matrix (see [5]) still runs under DOS, is only capable of displaying planar structures (or a slight extension of them) and has only limited layout features. *ArchE* will replace this program in the near future.

A *Harris Matrix* is defined as a set with the following relationships between its elements:

(*i*)   *a* is *later* than *b*;
(*ii*) *a* is *earlier* than b;
(*iii*) *a* and *b* are *contemporary*;
(*iv*) *a* and *b* are *equal*;
(*v*)   *a* and *b* have *no direct relationship*.

Considering only relationships (i),(ii), and (v) a Harris Matrix is a partially ordered set. The relationship (iii) is used to combine objects found in the same layer, whereas (iv) states that parts of one object are found in different locations. From the properties (i) to (v), consistency checks can be derived, eg, testing for directed cycles.

In addition to these checks, the task is to compute and display a simple diagram of a Harris Matrix revealing the chronological order of its elements. In the course of this abstract we adopt standard graph terminology, refering to a Harris Matrix as a directed graph (for a survey on properties and layout of ordered sets compare [9]).

2   The *ArchE* program

At present *ArchE* achieves three goals:

1. it allows easy collection of data which are automatically checked for consistency;
2. it simplifies the input data;
3. it produces a layout which can be further edited.

All these goals have to be achieved sufficiently fast as the input which is given online usually consists of large graphs with up to several thousand vertices. Figure 1 shows the flow diagram of *ArchE*.

As most of the data is collected during the course of an excavation, input errors are very frequent. By searching for (and possibly displaying) directed cycles *ArchE* performs a consistency check of the input graph. Due to properties of the contemporary relationship we had to modify standard techniques and have implemented algorithms for testing the whole input graph and online insertion of edges.

In general the given graphs are not planar. In order to facilitate reading we minimize the number of edges to display all existing relationships between vertices; this is done by computing the transitive reduction of the input graph. We have implemented different algorithms using depth-first search, the Floyd-Warshall algorithm, and boolean matrix multiplication.

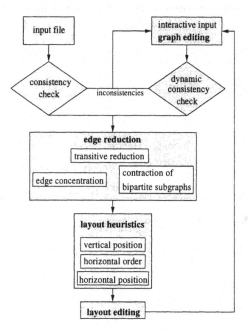

Fig. 1. Flow diagram of *ArchE*

The first one is intended for sparse graphs while the second is based on the well-known Floyd-Warshall algorithm for all-pairs shortest paths on arbitrary graphs. For dense graphs the third algorithm is the fastest; it relies on the fact that boolean matrix multiplication, transitive closure, and transitive reduction are equivalent problems with the same asymptotic running time [1].

Further reduction of edges is obtained by either *edge concentration* or search for complete bipartite subgraphs. We choose two arbitrary levels A and B and consider the induced bipartite subgraph $G, V(G) = A \cup B$. In both cases we search for a covering of G by complete bipartite subgraphs $(A_1, B_1), \ldots, (A_k, B_k)$, such that each edge belongs to at least one (A_i, B_i). For a complete bipartite subgraph (A_i, B_i) we insert an additional dummy vertex v' between A_i and B_i, and replace the edges between vertices of A_i and B_i by edges between A_i and v', B_i and v', respectively. We thereby reduce the number of edges from $|A_i||B_i|$ to $|A_i| + |B_i|$ and increase the readability of the layout.

In the first case we try to minimize $\sum_{i \le k} |A_i| + |B_i|$, ie, the number of replacement edges. In the second case we try to minimize k, ie, the number of covering bipartite subgraphs. As these problems are NP-hard, NP-complete respectively, we have implemented heuristics (for details on edge concentration compare [8]).

Additionally *ArchE* includes an algorithm for finding small complete bipartite subgraphs (see [3]) as the existing complete bipartite subgraphs often possess only few vertices.

The main layout procedure uses extensions of the Sugiyama algorithm [11], which consists of three phases. First using topological sort, we determine the vertical order of the vertices (contemporary vertices are placed on the same level). Then their horizontal order is computed by a heuristic (eg, barycenter, median). Finally the coordinates of the vertices are determined following [11].

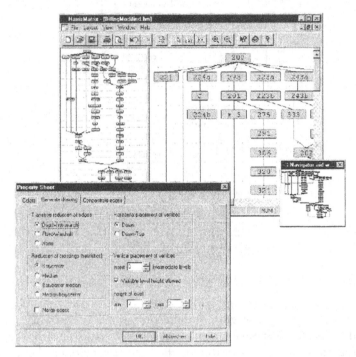

Fig. 2. A screenshot of *ArchE*

Graph editing features include the contraction of subgraphs; they are displayed as single vertices. Further *ArchE* has most of the advanced windows program properties including easy interactive editing, customizing of all user interface objects, zooming function (for different windows), undo/redo functions, printing and print preview functions, export/import facilities (for the archaeological format as well as for several graph editor formats), and a detailed help function (Figure 2).

We have chosen Windows95 and WindowsNT as platforms because these are the most frequently used among archaeologists. The need for an event based environment, the extendable design, and the required speed were reasons for designing *ArchE* in Visual C++ (4.2) using MFC application frame.

3 Practical experience and comparison with other programs

ArchE is one of the few graph drawing systems running under Windows and thereby usable for archaeologists. It has been successfully tested by several archaeological groups in Europe and is currently adapted to the WinBASP package. Due to the fact that *ArchE* computes edge concentrations and reduces bipartite subgraphs it decreases the number of edges of the graph and therefore receives drawings with a considerably lower number of crossings.

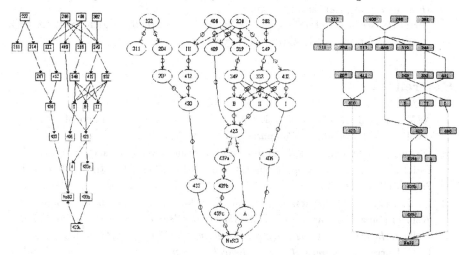

Fig. 3. Layouts of the data set *billing* by *VCG*, *dotty*, and *ArchE*

An example comparing the layouts of *VCG* [10], *dotty* [6], [7], and *ArchE* is shown in Figure 3.

We have tested *ArchE* on real data with up to 1056 vertices, 2624 edges and on even larger randomly generated graphs (up to 3000 vertices, 7000 edges).

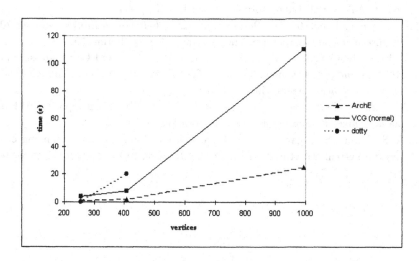

Fig. 4. run time comparisions with *VCG*, *dotty*, and *ArchE*

Comparisons with other graph drawing tools are listed below. They show that *ArchE* is one of the fastest existing systems (Figure 4).We received the data courtesy of I. Herzog (Rheinisches Amt für Bodendenkmalpflege, Bonn), S. Lütgert, and A. Fuller (Museum of London Archaeology Service). The tests were performed on a Pentium 133 with 40 MB RAM under Windows 95. For *VCG* we used layout speed "normal"; for layout speed "fast and ugly" *VCG* and *ArchE* achieved similar running times.

Currently we are extending *ArchE* beyond the archaeological context to a user interface for the AGD–Graph–Library [2].

References

1. A. V. Aho, M. R. Garey, and J. Ullman. The transitive reduction of a directed graph. *SIAM Journal on Computing*, 1(2):131–137, 1972.
2. D. Alberts, C. Gutwenger, P. Mutzel, and S. Näher. AGD–library: A library of algorithms for graph drawing. Technical report, Max-Planck-Institut für Informatik, Saarbrücken, 1997.
3. M. de Berg, M. Overmars, and M. van Kreveld. Finding complete bipartite subgraphs in bipartite graphs. *ALCOM: Algorithms Review, Newsletter of the ESPRIT II Basic Research Actions Program Project no. 3075 (ALCOM)*, 1, 1990.
4. E. C. Harris. *Principles of Archaeological Stratigraphy*. Academic Press, London, San Diego, 2nd edition, 1989.
5. I. Herzog. Computer-aided harris matrix generation. In E. Harris, editor, *Practices of archaeological stratigraphy*, pages 201–217, London, 1993.
6. E. Koutsofios and S. C. North. Drawing graphs with dot. Technical report, AT&T Bell Laboratories, Murray Hill NJ, 1992.
7. E. Koutsofios and S. C. North. Applications of graph visualisation. Technical report, AT&T Bell Laboratories, Murray Hill NJ, 1994.
8. F. Newbery Paulisch. *The Design of an Extendible Graph Editor*. Number 704 in Lecture Notes in Computer Science. Springer-Verlag, Berlin, 1993.
9. I. Rival. Reading, drawing, and order. In I. Rosenberg and G. Sabidussi, editors, *Algebras and orders*, volume 389 of *NATO ASI Series [Series C: Mathematical and physical Sciences]*, pages 359–404, Dordrecht, 1993.
10. G. Sander. Graph layout through the VCG tool. Technical report, FB14 Informatik, Universität des Saarlandes, Saarbrücken, 1994.
11. K. Sugiyama, S. Tagawa, and M. Toda. Methods for visual understanding of hierarchical system structures. *IEEE Transactions on Systems, Man, & Cybernetics*, 11(2):109–125, Feb. 1981.

InteractiveGiotto: An Algorithm for Interactive Orthogonal Graph Drawing[*]

Stina S. Bridgeman[1], Jody Fanto[1], Ashim Garg[2], Roberto Tamassia[1], Luca Vismara[1]

[1] Center for Geometric Computing, Department of Computer Science
Brown University, USA
{ssb,jrf,rt,lv}@cs.brown.edu
[2] Department of Computer Science
State University of New York at Buffalo, USA
agarg@cs.buffalo.edu

Abstract. We present INTERACTIVEGIOTTO, an interactive algorithm for orthogonal graph drawing based on the network flow approach to bend minimization.

1 Introduction

The last fifteen years have seen an impressive growth of the number of existing graph drawing algorithms. Most of them have been devised with the implicit assumption of being used in "batch mode" and are not well suited for being a component of an interactive system. In fact, the typical use of an interactive system consists of the repetition of the following phases: the user is presented with a drawing of a graph (the *current* drawing); he/she performs a slight modification (an *update*) of the drawing, by inserting/deleting some vertices and/or edges; a graph drawing algorithm is executed to obtain another drawing (the *new* drawing). It is important that the new drawing be as similar as possible to the current one, in order to preserve the *mental map* the user has of the drawing [3, 7]. Motivated by this type of application, various results on interactive graph drawing algorithms have been recently presented. See, for example, [1, 2, 4, 6, 8, 9, 13].

We focus our attention on interactive *orthogonal* graph drawing algorithms. We recall that in an orthogonal drawing, each vertex is represented as a rectangle and each edge is represented as a chain of alternating horizontal and vertical segments. Papakostas and Tollis [12] describe four scenarios for interactive orthogonal graph drawing, and present algorithms for two of them. The results of an extensive experimental study comparing the performances of two of those scenarios are reported in [10]. Biedl and Kaufmann [1] present a linear time incremental algorithm for orthogonal drawings, where the vertices are given one at a time and their position cannot be changed after the initial placement.

In this paper, we present INTERACTIVEGIOTTO, an interactive algorithm for orthogonal graph drawing based on the network flow approach to bend mini-

---

[*] Research supported in part by the U.S. Army Research Office under grant DAAH04-96-1-0013, by the National Science Foundation under grant CCR–9423847, and by a National Science Foundation Graduate Fellowship.

mization. Our algorithm does not appear to fit into any of the scenarios described in [12]; in particular, the general shape of the drawing is preserved, as in the Relative-Coordinates scenario, but the coordinates of some vertices and/or edges may change by more than a small constant after each update.

The rest of the paper is organized as follows: Section 2 describes INTERACTIVEGIOTTO, Section 3 illustrates the features of INTERACTIVEGIOTTO with two examples, and Section 4 outlines plans for future work.

2   A Brief Description of InteractiveGiotto

In this section we first describe GIOTTO, then we present the innovative features of INTERACTIVEGIOTTO, and finally we discuss how these features are realized.

GIOTTO [16] is a successful general-purpose algorithm for producing orthogonal drawings of graphs, and has been widely used in software visualization systems. It is composed of three steps:

Planarization The input graph is, in general, *non-planar*, i.e., it cannot be drawn in the plane without edge crossings. In this step the graph is converted into a planar graph by replacing each edge crossing with a fictitious vertex. A heuristic is used to reduce the number of edge crossings.

Orthogonalization In this step, GIOTTO constructs an orthogonal representation of the planarized graph. This representation consists of a description of the bends along each edge and of the angles between consecutive edges around each vertex. In particular, GIOTTO uses the algorithm described in [15], which produces an orthogonal representation with the minimum number of edge bends by means of a reduction to a minimum cost flow problem.

Compaction In this step an orthogonal drawing is constructed out of the orthogonal representation. The total length of the edges is minimized, again using a reduction to a minimum cost flow problem.

Compared to traditional orthogonal graph drawing algorithms, INTERACTIVEGIOTTO can preserve the following properties of the current drawing during an update:

- the embedding of the graph, i.e., the circular ordering of the edges around each vertex;
- the edge crossings;
- the edge bends and, for 90° edge bends, also their type (left or right);
- for each vertex, if R is the rectangle representing the vertex, the number of corners of R between any two consecutive edges.

Examples illustrating these properties are presented in Section 3. In addition, our implementation of INTERACTIVEGIOTTO has morphing capabilities that allow the transition from the current drawing to the new one to gradually take place on the screen. This gives the user a better idea of the modifications that are taking place in the drawing.

We now discuss how the previous properties are preserved. INTERACTIVEGIOTTO is composed of three steps, similarly to GIOTTO:

Planarization The input to the planarization step is no longer a combinatorial description of a graph, as in GIOTTO, but rather a drawing of a graph. Accordingly, GIOTTO's planarizer has been replaced by an algorithm that replaces each edge bend and edge crossing with a fictitious vertex. Edge crossings are detected using a segment intersection algorithm on the set of segments representing the edges of the graph. The embedding and the edge crossings of the current drawing are preserved in this step. In the implementation we make use of LEDA, the Library of Efficient Data Structures and Algorithms [5].

Orthogonalization This step takes the embedded planar graph produced during the planarization and constructs an orthogonal representation preserving the edge bends, the type of the 90° edge bends, and the number of corners between consecutive edges around each non-fictitious vertex. This goal is achieved through a new algorithm inspired by that described in [14]. The orthogonal representation of the current graph is given by the optimal solution of a minimum cost flow problem in a particular network associated with the current graph. The edge bends, the type of the 90° edge bends, and the number of corners are preserved by setting a target value for the flow in some arcs of the network and solving the constrained minimum cost flow problem. Note that new bends may be introduced by the algorithm, if needed.

Compaction Same as in GIOTTO.

3   Using InteractiveGiotto

In this section we illustrate the main features of INTERACTIVEGIOTTO through two examples, shown in Figs. 1 and 2.

In the first example, the user incrementally constructs the graph in Fig. 3 of [11]. The initial graph drawn by the user is shown in Fig. 1a using grey vertices and non-orthogonal polylines. INTERACTIVEGIOTTO is executed and the new drawing is shown in Fig. 1b using white vertices. Note how all the bends are preserved and how all the vertices, except one, are drawn vertically aligned to preserve the number of corners. The user draws other four vertices and seven edges, and executes INTERACTIVEGIOTTO again. Note how the crossing introduced by the user is preserved (see Fig. 1c). After the next update, a new row and a new column are created in the middle and at the bottom of the drawing, respectively, in order to accommodate two new edges (see Fig. 1d). Similar updates are performed in the remaining two steps (see Fig. 1e–1f).

In the second example, we describe some other features of INTERACTIVEGIOTTO. This time the initial drawing, shown in Fig. 2a using white vertices, is not empty. In the first step, the user adds one vertex and two edges, each containing one bend. Note how one of the edges drawn by the user crosses three existing edges. In the new drawing, shown in Fig. 2b, one more bend is introduced by INTERACTIVEGIOTTO along that edge, in order to preserve the crossings and the number of corners. In the next step, the vertices shown in grey in Fig. 2b are deleted by the user. In Fig. 2c, the user, without having redrawn the graph,

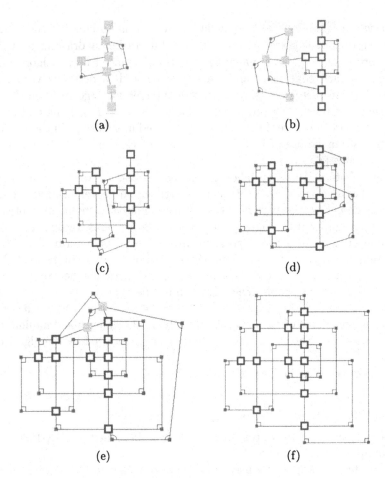

Fig. 1. Example of use of INTERACTIVEGIOTTO: an incremental construction of the graph in Fig. 3 of [11].

moves the subgraph induced by the five vertices show in grey on the outside of the current drawing, reroutes the two existing edges connecting the subgraph to the rest of the graph, and adds one new edge. In the new drawing, shown in Fig. 2d, the shapes of the subgraph and, more importantly, of the rest of the graph are preserved, even though the metric has changed. Note how one of the (non-90°) edge bends introduced by the user is drawn by INTERACTIVEGIOTTO as a 180° edge bend, i.e., the bend is preserved but its type is not. Finally, the user adds three vertices and nine edges. The drawing produced by INTERACTIVE-GIOTTO is shown in Fig. 2e. It should be contrasted with the drawing produced by GIOTTO, shown in Fig. 2f. The latter has fewer crossings and bends, a smaller area, and a smaller total edge length, but its shape does not resemble that of the drawing in Fig. 2d at all, thus not preserving the mental map the user has of the drawing.

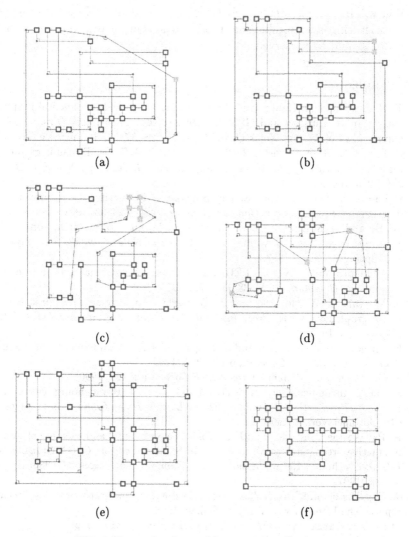

Fig. 2. Example of use of INTERACTIVEGIOTTO.

4 Future Work

The system presented in this paper is a first prototype. We plan to improve it
and expand it in the following directions:

- As in previous works on interactive orthogonal graph drawing, we consider
  graphs where the vertices have degree at most four. In a future implementation, we will allow arbitrary degree by using a vertex expansion mechanism.
- Currently the edge bends, the type of the 90° edge bends, and the number
  of corners are always preserved; this may be too restrictive. In a future implementation, it will be possible to choose whether to favor the preservation
  of the drawing or the reduction of the number of bends. In particular, the

user will be able to interactively specify those portions of the drawing whose restructuring is acceptable, and those portions of the drawing whose current shape should be preserved.

References

1. T. C. Biedl and M. Kaufmann. Area-efficient static and incremental graph darwings. In R. Burkard and G. Woeginger, editors, *Algorithms (Proc. ESA '97)*, volume 1284 of *Lecture Notes Comput. Sci.*, pages 37–52. Springer-Verlag, 1997.
2. R. F. Cohen, G. Di Battista, R. Tamassia, and I. G. Tollis. Dynamic graph drawings: Trees, series-parallel digraphs, and planar *ST*-digraphs. *SIAM J. Comput.*, 24(5):970–1001, 1995.
3. P. Eades, W. Lai, K. Misue, and K. Sugiyama. Preserving the mental map of a diagram. In *Proceedings of Compugraphics 91*, pages 24–33, 1991.
4. W. He and K. Marriott. Constrained graph layout. In S. North, editor, *Graph Drawing (Proc. GD '96)*, volume 1190 of *Lecture Notes Comput. Sci.*, pages 217–232. Springer-Verlag, 1997.
5. K. Mehlhorn and S. Näher. LEDA: a platform for combinatorial and geometric computing. *Commun. ACM*, 38:96–102, 1995.
6. K. Miriyala, S. W. Hornick, and R. Tamassia. An incremental approach to aesthetic graph layout. In *Proc. Internat. Workshop on Computer-Aided Software Engineering*, 1993.
7. K. Misue, P. Eades, W. Lai, and K. Sugiyama. Layout adjustment and the mental map. *J. Visual Lang. Comput.*, 6:183–210, 1995.
8. S. Moen. Drawing dynamic trees. *IEEE Softw.*, 7:21–28, 1990.
9. S. North. Incremental layout in DynaDAG. In F. J. Brandenburg, editor, *Graph Drawing (Proc. GD '95)*, volume 1027 of *Lecture Notes Comput. Sci.*, pages 409–418. Springer-Verlag, 1996.
10. A. Papakostas, J. M. Six, and I. G. Tollis. Experimental and theoretical results in interactive orthogonal graph drawing. In S. North, editor, *Graph Drawing (Proc. GD '96)*, volume 1190 of *Lecture Notes Comput. Sci.*, pages 371–386. Springer-Verlag, 1997.
11. A. Papakostas and I. G. Tollis. Interactive orthogonal graph drawing. Technical report, The University of Texas at Dallas, 1996. http://www.utdallas.edu/~tollis/papers/interactive_J.ps.
12. A. Papakostas and I. G. Tollis. Issues in interactive orthogonal graph drawing. In F. J. Brandenburg, editor, *Graph Drawing (Proc. GD '95)*, volume 1027 of *Lecture Notes Comput. Sci.*, pages 419–430. Springer-Verlag, 1996.
13. K. Ryall, J. Marks, and S. Shieber. An interactive system for drawing graphs. In S. North, editor, *Graph Drawing (Proc. GD '96)*, volume 1190 of *Lecture Notes Comput. Sci.*, pages 387–394. Springer-Verlag, 1997.
14. R. Tamassia. New layout techniques for entity-relationship diagrams. In *Proc. 4th Internat. Conf. on Entity-Relationship Approach*, pages 304–311, 1985.
15. R. Tamassia. On embedding a graph in the grid with the minimum number of bends. *SIAM J. Comput.*, 16(3):421–444, 1987.
16. R. Tamassia, G. Di Battista, and C. Batini. Automatic graph drawing and readability of diagrams. *IEEE Trans. Syst. Man Cybern.*, SMC-18(1):61–79, 1988.

each node and each bend on a different point of the grid, (i.e. defines integer coordinates for each of them) and maps each edge into a chain of horizontal and vertical segments.

Orthogonal grid drawings are widely used for graph visualization in many applications, including database design (entity-relationship diagrams) software engineering (data flow-diagrams) and circuit design (circuit schematics).

A very elegant algorithm [14] allows to generate, in polynomial time, orthogonal grid drawings with the minimum number of bends of a planar graph, within the given embedding. Unfortunately, the bends-minimization problem over all the possible planar embeddings is NP-complete [6].

We propose an interactive tool, GRID, implementing a new algorithm for computing orthogonal drawings with the minimum number of bends. We call this algorithm *Slow Orthogonal*, due to exponential time needed by the computation. It is based on a branch-and-bound technique with an efficient enumeration schema of the planar embeddings and several new methods for comuputing lower bounds of the number of bends. Slow selects the *best* embeddings taking advantage of an SPQR-tree [4], a data-strucure representing the decomposition of the given graph into its triconnected components. As far as we know, GDToolkit (available at www.inf.uniroma3.it/people/gdb/wp12/) is the only library currently providing an SPQR-tree class. A complete description of the Slow Orthogonal algorithm, with experimental results, can be found in [1].

GRID also provides two classical algorithms [14, 15] which compute an orthogonal drawing preserving the given embedding.

The user can select the algorithm that is better suitable for his needs in terms of a compromise between effectiveness and efficiency or/and to compare the results produced by the Slow Orthogonal against those produced by classical algorithms. A complete description of the GRID tool is given in the next section.

2 GRID: Graph Interactive Drawer

In this section we describe the main features of GRID, along with some extensions we plan to implement in the near future.

The tool provides a user-friendly graph editor (Fig. 1). You can create/edit a multi-graph drawing, apply one of the available layout algorithms on the underlying graph, and edit again the resulting drawing. In particular, a new layout algorithm, *Slow Orthogonal*, is available in the tool, which allows to compute an orthogonal drawing with the minimum number of bends; all the tool's features with respect to this algorithm are described in the next subsection. In addition, GRID provides two classical layout algorithms: *Quick Orthogonal* computes an orthogonal layout of the graph, using a heuristic method to reduce the number of bends within the given embedding [15]. *Optimal Orthogonal* computes an orthogonal layout of the graph in polynomial time $O(n^2 \log n)$, minimizing the number of bends within the given embedding. We implemented a minimum cost flow algorithm searching minimum cost paths with the Dijkstra algorithm, as described in the original paper of Tamassia [14].

GRID: An Interactive Tool for Computing Orthogonal Drawings with the Minimum Number of Bends *

Walter Didimo [†]   *Antonio Leonforte* [‡]

didimo@inf.uniroma3.it   leonfort@inf.uniroma3.it

Abstract

In this paper we present a new interactive tool for computing orthogonal grid drawings of planar graphs. The tool is based on GDToolkit, an object-oriented library of classes for handling graphs and computing their layout. GDToolkit is built on LEDA (an efficient library of data types and algorithms) and currently implements three orthogonal layout methods. Especially, we provide a new branch-and-bound algorithm choosing a planar embedding in order to minimize the number of bends. The enumeration schema of the branch-and-bound algorithm is based on the GDToolkit SPQR-tree class (as far as we know, the only existing SPQR-tree implementation). The tool offers an interactive graphical interface to the branch-and-bound algorithm, which allows to edit the embedding, to execute the algorithm step by step and to view partial results. It also gives quality measures on the drawing, and quantitative measures on the algorithm's perfomance.

1 Introduction

The graph drawing research has received increasing attention recently. Variuos graphic standards have been proposed to draw graphs, in order to fit specific application fields. An extensive literature on the subject can be found in [2]. Several libraries of data structures and algorithms supporting graph drawing have been developed, e.g. LEDA [11], Tom Sawyer Graph Layout Toolkit [16] and GraphBase [10]. Many graph visualization and editing systems are also available, e.g. Diagram Server [3], daVinci [5], Graphlet [8], GIOTTO3D [7].

This paper is about *orthogonal grid drawing* of 4-planar graphs (i.e. planar graphs with degree of the vertices at most 4). An orthogonal grid drawing places

---

*Research supported in part by the ESPRIT LTR Project no. 20244 - ALCOM-IT.

[†]Dipartimento di Informatica e Automazione, Università di Roma Tre, via della Vasca Navale 84, 00146 Roma, Italy

[‡]Dipartimento di Informatica e Automazione, Università di Roma Tre, via della Vasca Navale 84, 00146 Roma, Italy

2.1 Features of the Slow Orthogonal algorithm

The Slow Orthogonal algorithm computes an orthogonal layout with the minimum number of bends by applying the Optimal Orthogonal algorithm while exploring all the relevant embeddings with a branch-and-bound approach. This algorithm takes advantage of new lower-bounds recently found [1] for the number of bends of an orthogonal drawing.

The tool offers several features you can use freely and interactively, according to the graph you want to draw/inspect, and to the goals you want to achieve. Nevertheless, on a general basis, you are likely to cycle on a sequence of operations we describe hereunder.

Create a 4-planar biconnected graph (or load and edit one).

Decompose the graph into its triconnected components. Clicking the DECOMPOSE button, the tools prompts for a reference edge e, then performs a decomposition of the current graph into triconnecetd components. As a result of such a decomposition, an SPQR-tree is built and displayed on a dedicated window. Series, parallel, rigid and Q nodes are respectively drawn as triangles, ellipses, rectangles and circles. Q-nodes display can be disabled when their number makes the SPQR-tree drawing too wide.

Evaluate the graph structure by playing with the corresponding SPQR-tree. You can change the root of the SPQR-tree in, simply clicking at the button EVERT. You can display an orthogonal drawing of any S/P/Q/R component by clicking on the corresponding SPQR-tree node: a dedicated window pops up displaying the skeleton subgraph and highlighting with different colors the separation-pair (orange nodes), virtual (green) and non-virtual (black) edges (Fig. 3. a). You can cycle across all the possible embeddings of a given skeleton, with respect to its reference edge, and thus, changing the currently planar embedding of the graph. Please note that the skeleton of a rigid component has only two possible embeddings (flipping around the separation-pair), while the skeleton of a parallel component has a number of embeddings equal to the number of the possible permutations of its non-reference edges.

Optionally adjust the parameters affecting the Branch-and-Bound embedding search. You can adjust several parameters according to the graph structure, in order to reduce the search time. You can decide to compute the upper bounds in a non-repetitive way or in a random way, and you can set how to calculate the lower bounds: efficient lower bounds need more computation time, but they very often reduce the search domain more quickly. Please note that all these options can be changed even once the embedding search is started, by pausing it temporarily.

Start the Branch-and-Bound embedding search. Clicking the START button, the Branch-and-Bound embedding search starts. During the embedding search, the SPQR-tree drawing is updated to reflect each search step (animation and printouts, however, can be disabled to minimize the running time). A complete redraw is performed each time a new reference-edge is choosen. Node colors change to reflect the algorithm's embedding choices: each node is yellow, red or green when the embedding of the corresponding skeleton is respectively unkown, fixed or under construction (Fig. 3. b). Given an edge e and its expansion graph G_e, the label of e reflects the preprocessing lower-bound available for the number of bends of any orthogonal drawing of G_e.

Optionally pause or stop the Branch-and-Bound embedding search. You can optionally pause the search in order to adjust some algorithm parameters. You can also execute the search algorithm step by step, or stop it definitively when you think the current orthogonal drawing is suitable.

Qualitative and quantitative measures. Once the Branch-and-Bound embedding search is completed, you can display the final orthogonal drawing, and evaluate the quality and performance measures provided by the tool. The quality measures available are: *bends*: total number of bends; *maxedgebends*: max number of bends on an edge; *unifbends*: standard deviation of the number of edge bends; *area*: area of the smallest rectangle with horizontal and vertical sides covering the drawing; *totaledgelen*: total edge length; *maxedgelen*: maximum length of an edge; *uniflen*: standard deviation of the edge length; *screenratio*: deviation of the ascpect ratio of the drawing (width/height or height/width) with respect to typical screen ratio (4/3). The performance measures are: cpu time, total number of search tree nodes, number of visited search tree nodes.

Experimenting GRID on a large number of reference graphs, we observed that the layouts computed by Slow Orthogonal tipically have a smaller area (and a smaller number of bends) with respect to the ones computed by Quick and Optimal. (Fig. 2 and Fig. 4).

2.2 Extensions

GRID layout algorithms currently run with 4-planar graphs only. We plan to make them run with nodes of any degree by using, for instance, the expansion techniques cited above. Moreover, the current implementation of the SPQR-trees only supports biconnected graphs; we plan to enhance the SPQR-tree class to make it handle any connected planar graph [4]. Slow Orthogonal, however, can be already applied on all the biconnected components of a graph; this yields a powerful heuristic for reducing the number of bends in connected graphs.

Akcnowledgements

We'd like to thank Paola Bertolazzi and Giuseppe Di Battista for their original ideas. We are also grateful to Sandra Follaro, Armando Parise and Maurizio Patrignani for their technical support.

References

[1] P. Bertolazzi, G. Di Battista and W. Didimo. Computing Orthogonal Drawing with the Minimum Number of Bends. In *Proc. Workshop Algorithms Data Struct.*, 1997 (to appear).

[2] G. Di Battista, P. Eades, R. Tamassia and I. G. Tollis. Algorithms for drawing graphs: an annotated bibliography. *Comput. Geom. Theory Appl.*, 4:235–282, 1994.

[3] G. Di Battista, G. Liotta, M. Strani and F. Vargiu. Diagram Server. *Proceedings of Advances Visual Interfaces*, 36:415–417, 1992.

[4] G. Di Battista and R. Tamassia. On-line planarity testing. *SIAM J. Comput.*, 25(5):956–997, 1996.

[5] M. Fröhlich and M.Werner. Demonstration of the Interactive Graph-Visualization System da Vinci. *Lecture Notes in Computer Science*, 894:266–269, 1994.

[6] A. Garg and R. Tamassia. On the computational complexity of upward and rectilinear planarity testing. Submitted to *SIAM Journal on Computing*, 1995.

[7] A. Garg and R. Tamassia. GIOTTO3D: A System for Visualizing Hierarchical Structures in 3D. In *Symposium Graph Drawing, GD'96, LNCS*, 1190, 1996

[8] M. Himsolt. The Graphlet System. In *Symposium Graph Drawing, GD'96, LNCS*, 1190, 1996

[9] J. Hopcroft and R. E. Tarjan. Dividing a graph into triconnected components. *SIAM J. Comput.*, 2:135–158, 1973.

[10] D.E. Knut. The Stanford GraphBase: a platform for combinatorial algorithms. Stanford University, 1993.

[11] K. Mehlhorn and S. Näher. LEDA: a platform for combinatorial and geometric computing. *Commun. ACM*, 38:96–102, 1995.

[12] T. Nishizeki and N. Chiba. Planar graphs: Theory and algorithms. *Ann. Discrete Math.*, 32, 1988.

[13] A. Scott. A Survey of Graph Drawing Systems. *Technical Report 95-06.*, Department of Computer Science University of Newcastle, Australia, 1995.

[14] R. Tamassia. On embedding a graph in the grid with the minimum number of bends. *SIAM J. Comput.*, 16(3):421–444, 1987.

[15] R. Tamassia and I. G. Tollis. Planar grid embedding in linear time. *IEEE Trans. on Circuits and Systems* CAS-36:1230–1234, 1987.

[16] Tom Sawyer Software. Tom Sawyer Graph Layout Toolkit. *Tom Sawyer Software Corporation*, 1824B Fourth STreet, Berkley, CA94710, USA.

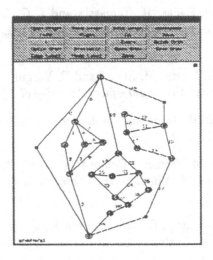

Figure 1: A 20-nodes graph created by the GRID editor

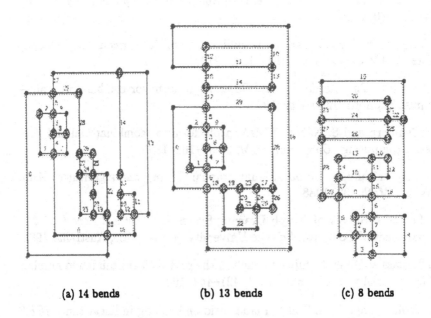

(a) 14 bends   (b) 13 bends   (c) 8 bends

Figure 2: Graph drawn by: (a) Quick Orthogonal, (b) Optimal Orthogonal, (c) Slow Orthogonal

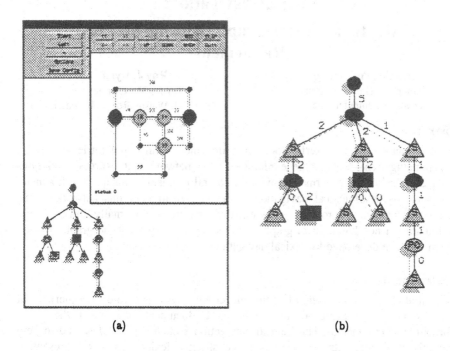

(a)   (b)

Figure 3: (a) An SPQR-tree with the skeleton of an R-node. (b) The same SPQR-tree during the computation

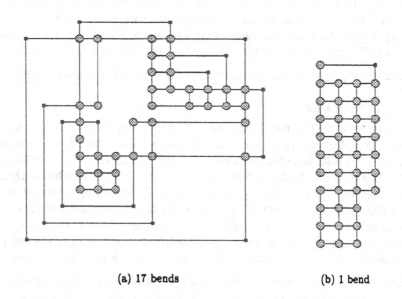

(a) 17 bends   (b) 1 bend

Figure 4: Graph drawn by: (a) Optimal Orthogonal, (b) Slow Orthogonal

Lexical Navigation:
Using Incremental Graph Drawing for Query Refinement

Daniel Tunkelang
Carnegie Mellon University,
School of Computer Science

Roy J. Byrd
James W. Cooper
IBM T. J. Watson Research Center

Abstract

Query refinement is a powerful tool for a document search and retrieval system. Lexical navigation—that is, the exploration of a network that expresses relations among all possible query terms—provides a natural mechanism for query refinement. An essential part of lexical navigation is the visualization of this network. This dynamic visualization problem is essentially one of incrementally drawing and manipulating a non-hierarchical graph. In this paper, we present the graph-drawing system we have developed for lexical navigation.

1. Introduction

The central challenge for users of document retrieval systems is to create queries that are effective at retrieving all and only those documents that meet the users' information requirements. The Lexical Navigation system [2] was designed to help meet this challenge. Its approach is to construct a lexical network representing concepts and relationships found in the entire document collection, and to use this network to prompt users with candidate terms for query refinement.

The subject of this paper is a graph drawing system for displaying and navigating large lexical networks. The next section describes lexical networks and presents our motivation for using them for query refinement. Section 3 describes our graph drawing system. Section 4 illustrates the system in action and compares it to DEC's Visual LiveTopics. Section 5 outlines the status of our work and plans to extend it.

We note that [6] describes a related but limited approach. Our work is independent.

2. Lexical Networks

When translating an information requirement into a query for a document retrieval system, a user must construct a query using terms that match those found in database documents. The user faces the "vocabulary problem" described in [7]. One aspect of this problem is specificity; for example, the query term "fruit" will not be as helpful as "pear," "nectarine," or "raisin," assuming that the latter concepts are part of the information requirement. Another aspect is ambiguity. An otherwise unconstrained query containing the term "nut" could return documents related to hardware, seeds, and eccentric people. Finally, there is the problem that (according to [1] and [17]) large full-text databases contain many different expressions for the same concept.

Manual and automatic techniques for enhancing queries are described in [8], [11], [17], and [18]. The techniques follow two themes: query expansion using terms obtained from thesauruses or co-occurrence lists, and relevance feedback—that is, adding query terms obtained from documents known to be relevant. Despite some successes (see [18]), automatic techniques have serious limitations. Thesauruses rarely contain the most useful types of information for query expansion. Published references miss current names and technical terms. Also, thesauruses often contain only synonym, hypernym, and hyponym relations. Actual relations are richer and often domain-specific. At the same time, there is no reliable means for automatically disambiguating query terms. In general, these techniques ignore the best sources of good query terms: the document collection itself and the user.

We address the above problems using a technique called "lexical navigation," which we developed for the LexNav document query and retrieval system. In LexNav, we address the vocabulary problem by analyzing all of the collection's documents and by leveraging the user's intuition about concept relations. Before queries begin, we use a suite of text analysis tools called "Talent" to develop a vocabulary of the ordinary words, proper names, and largely multi-word technical terms from the entire collection. Additional Talent tools organize the vocabularies into an extensive set of relationships consisting of triples of <vocabulary-item; relation; vocabulary-item>.

Once extracted, the entire set of relationships for a collection is organized into a "lexical network," in which the vocabulary items are the nodes and the relations are the links. The nodes, which represent concepts in the subject domain of the document collection, contain a number of attributes, including: the canonical form and other variant forms of the vocabulary item, the category (term, person, place, organization, etc.), and occurrence statistics. The links in the lexical network represent the domain relationships among concepts. They contain the name of the relationship, if known, the strength of the relationship, represented as a number between 0 and 100, and the direction of the relationship. These relationships and the networks they produce are overwhelmingly non-hierarchical. Lexical networks are stored in a database, from which individual nodes and links can be efficiently accessed. Please see [2] for more details about the properties of lexical networks.

Lexical navigation allows the user to explore the contents of lexical networks derived from the collection being queried. By showing the user domain-specific names and terms related to his/her original query, lexical navigation reminds the user of alternative concepts and labels for original query concepts that might be useful for improving the query. We model this exploration of the domain's concept space as a navigation through the lexical network from some starting point related to the query.

3. Visualizing the Lexical Network

LexNav's first user interface displayed lexical networks in tabular form. Users could click on a displayed node in order obtain a new list box containing relations which linked it to other nodes. Although in principle powerful enough to explore a lexical network, the list box interface was not adequate for meeting our goal of providing an intuitive navigation mechanism for lexical networks.

The lexical network described in the previous section is a large, non-hierarchical network. Its size—which might range from thousands to millions of terms—precludes the possibility of visualizing the entire network simultaneously. The lack of hierarchy rules out approaches that assume a taxonomy of terms. We need a way to navigate relevant portions of the network and portray its general class of relations among terms.

The literature (see [4]) categorizes graph drawing algorithms according to whether they presume a hierarchy (or near hierarchy) among the nodes. Algorithms for drawing hierarchies are mostly based on [12]. For non-hierarchical networks, most approaches use simulated physical models based on [5]. They are generally slow and unreliable, since the lack of a hierarchy makes the problem space unwieldy. There is some consensus regarding the desirable aesthetics for drawings of non-hierarchical networks: nodes connected by short paths in the network should be drawn near each other, and edge crossings should be avoided. The former tells the user that physical proximity in the drawing corresponds to logical proximity in the network. The latter helps ensure the clarity of the drawing. In addition, there is the common-sense requirement that node and edge labels should be legible, and that nodes not overlap.

Recall that we are dynamically presenting the user with a small part of the network. The interactivity of the application raises the need for a dynamic algorithm that can draw the network incrementally with enough constancy for the user to maintain a sense of the state of the network and his/her current position in it. Constancy is the vague notion that small modifications to the network should not substantially change the drawing. Only recently have researchers been paying serious attention to the problem of incremental graph drawing (see [10] and [13]). Our present application illustrates its importance.

We have developed two drawing algorithms based on [14] and [15]. Both algorithms use a cost function based on [5] and performs incremental placement in such a way as to achieve a reasonable response time and preserve constancy of the drawing.

The cost function has three components: a node-node "electrical" repulsion, an edge "spring" energy, and, for the first algorithm, edge-crossing avoidance. The repulsion models Coulomb's Law: the energy for each pair of nodes is proportional to the inverse square of the distance between them. The spring energy models Hooke's Law: the energy for each edge is proportional to the square of the difference between its length and its rest length, which corresponds to the strength of the relation. For the first algorithm, there is a constant energy for each edge crossing in the drawing.

The heart of both algorithms is inserting the neighbors of already placed nodes. The first algorithm splits node insertion into two cases, depending whether the new node has at least two neighbors already placed in the current drawing. If the new node has only one placed neighbor, then the algorithm inserts the new node by splitting the largest angle between two edges incident to that neighbor; otherwise, it uses sampling techniques from [14]. In both cases, it then "makes room" for the new placement by creating a rectangle of space around it to avoid overlap with the other nodes while minimally perturbing their positions. The second algorithm is discussed in [15].

Having established an initial position in the lexical network, the user navigates it on a scrollable canvas. Operations include "growing" nodes to see their neighbors (color indicates whether a node has unseen neighbors), dragging nodes to modify the drawing, combining nodes into clusters, and accessing a node's attributes.

The graph drawing system has been implemented as a Java 1.1 client/server application: the client can be an applet while the server is an application using Java RMI and JDBC to connect to the data base containing the lexical network.

4. Example Scenario and Comparison with Visual LiveTopics

We illustrate the operation of our system with the scenario in Figures 1 through 3, in which the user is seeking corporations whose products involve "artificial intelligence."

To our knowledge, DEC's Visual LiveTopics (VLT) system (see [3], [16]) is the only other implemented system which uses graphical layout of lexical networks for query reformulation. Significant differences between VLT and LexNav, include:

- VLT establishes position in the network based on the hit-list obtained by processing the original query, whereas LexNav obtains its initial position directly from the query.

- VLT drawings are static, while LexNav permits incremental movement within the network without requiring the processing of intermediate queries.

- VLT network nodes are single words, while LexNav's nodes are often multi-word proper names and domain terms. Furthermore, by construction, a lexical network node represents the canonical form and all variants of the underlying term.

- VLT's inter-word relationships are precomputed, unnamed, and (apparently) static, whereas LexNav uses Talent to compute new lexical networks for each collection, containing domain-specific names, terms, and relationships.

5. Status and Future Plans

Our system has been incorporated in the user interface client of the LexNav document query and retrieval system. We have demonstrated it to researchers and developers at IBM, as well as to IBM customers.

As explained earlier, lexical networks are too large to be viewed in their entirety. In the LexNav system, our largest network was extracted from a 140 M collection of financial and technical documents and comprises 51,000 nodes and over 85,000 links. Since performance is independent of the total size of the network, we expect to scale up to significantly larger collection and network sizes.

Scale and performance may become issues for local complexity. If a node has too many neighbors, both the aesthetics and performance of layout deteriorate. We plan to introduce summarizing conventions to address this issue. The user could expand summary nodes using display techniques more appropriate than graph layout, e.g. a list box. We are also developing filtering strategies. The node categories, relation types, relation strengths, and other attribute data stored in lexical networks provide a rich basis for controlling the size and complexity of local network neighborhoods.

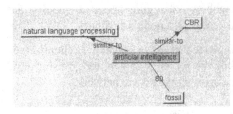

Figure 1 - We start with "artificial intelligence" and expand it to show the surrounding named and unnamed relations.

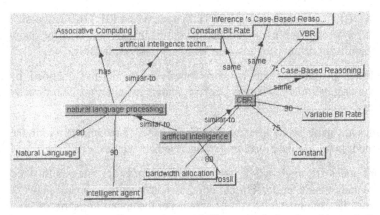

Figure 2 - Second-generation neighbors include the "Associative Computing" company. Note that "Case-Based Reasoning" disambiguates the ambiguous "CBR."

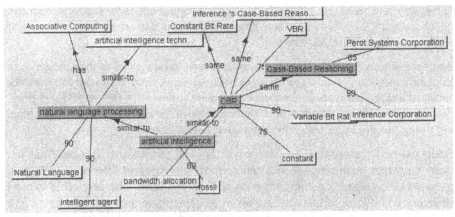

Figure 3 - Finally, we reach "Inference Corporation" and "Perot Systems Corporation," who use case-based reasoning in their products.

Bibliography

[1] Bates, Marcia J. "Human, Database, and Domain Factors in Content Indexing and Access to Digital Libraries and the Internet," Allerton, 1996

[2] Cooper, James W. and Byrd, Roy J., "Lexical Navigation: Visually Prompted Query Expansion and Refinement" in *Proceedings of the 2nd ACM International Conference on Digital Libraries*, July 1997.

[3] DeJesus, Edmund X., "The Searchable Kingdom" in *Byte*, June 1997.

[4] Di Battista, Giuseppe et al., "Annotated Bibliography on Graph Drawing Algorithms" in *Computational Geometry: Theory and Applications 4*, 1994.

[5] Eades, Peter, "A Heuristic for Graph Drawing," in *Congressus Numerantium 42*, 1984.

[6] Fowler, Richard H., Wilson, Bradley A., and Fowler, Wendy A. L. "Information Navigator: An information system using associative networks for display and retrieval", Report NAG9-551, No, 92-1, Dept of Computer Science, University of Texas - Pan American, Edinburg, TX.

[7] Furnas, G. W, Landauer, T. K., Gomes, L. M., and Dumais, S. T. "The Vocabulary Problem in Human-System Communication," in *Communications of the ACM*, vol. 30, no. 11, November 1987, pp. 964-971.

[8] Harman, D. "Relevance Feedback and Other Query Modification Techniques," in W. B. Frakes and R. Baeza-Yates, eds., *Information Retrieval: Data Structures and Algorithms*, Prentice-Hall, 1992.

[9] NIST *TIPSTER Information-Retrieval Test Research Collection*, on CD-ROM, published by the National Institute of Standards and Technology, Gaithersburg, MD, 1993.

[10] North, Stephen C., "Incremental Layout in DynaDAG," in *Proceedings of Symposium on Graph Drawing*, 1995.

[11] Schatz, Bruce R., Johnson, Eric H., Cochrane, Pauline A., and Chen, Hsinchun," Interactive Term Suggestion for Users of Digital Libraries," in *Proceedings of ACM Digital Libraries Conference*, 1996.

[12] Sugiyama, Kozo et al., "Methods for Visual Understanding of Hierarchical Systems" in *IEEE Transactions on Systems, Man, and Cybernetics 11*, No. 2, 1981.

[13] Tom Sawyer, on the World-Wide Web at http://www.tomsawyer.com

[14] Tunkelang, Daniel, "A Practical Approach to Drawing Undirected Graphs", Technical Report CMU-CS-94-161, Carnegie Mellon University, June 1994.

[15] Tunkelang, Daniel and Wegman, Mark, "Applying Numerical Approximation to Graph Drawing," unpublished manuscript. To request, please send email to quixote@cmu.edu.

[16] Visual LiveTopics, on the World-Wide Web at http://www.altavista.digital.com

[17] Voorhees, E. M. "Query Expansion using Lexical-Semantic Relations," in *Proceedings of the 17th Annual ACM-SIGIR Conference*, 1994, pp. 61-69.

[18] Xu, Jinxi and Croft, W. Bruce "Query Expansion Using Local and Global Document Analysis," *Proceedings of the 19th Annual ACM-SIGIR Conference*, 1996, pp. 4-11.

Design Gallery Browsers Based on 2D and 3D Graph Drawing (Demo)

Brad Andalman[1], Kathy Ryall[2], Wheeler Ruml[3],
Joe Marks[1], and Stuart Shieber[3]

[1] MERL — A Mitsubishi Electric Research Laboratory
201 Broadway
Cambridge, MA 02139, U.S.A.
E-mail: {andalman,marks}@merl.com
[2] Department of Computer Science
University of Virginia
Charlottesville, VA 22903, U.S.A.
E-mail: ryall@cs.virginia.edu
[3] Division of Engineering and Applied Sciences
Harvard University
Cambridge, MA 02138, U.S.A.
E-mail: {ruml,shieber}@eecs.harvard.edu

Abstract. Many problems in computer-aided design and graphics involve the process of setting and adjusting input parameters to obtain desirable output values. Exploring different parameter settings can be a difficult and tedious task in most such systems. In the Design Gallery™ (DG) approach, parameter setting is made easier by dividing the task more equitably between user and computer. DG interfaces present the user with the broadest selection, automatically generated and organized, of perceptually different designs that can be produced by varying a given set of input parameters. The DG approach has been applied to several difficult parameter-setting tasks from the field of computer graphics: light selection and placement for image rendering; opacity and color transfer-function specification for volume rendering; and motion control for articulated-figure and particle-system animation. The principal technical challenges posed by the DG approach are *dispersion* (finding a set of input-parameter vectors that optimally disperses the resulting output values) and *arrangement* (arranging the resulting designs for easy browsing by the user). We show how effective arrangement can be achieved with 2D and 3D graph drawing. While navigation is easier in the 2D interface, the 3D interface has proven to be surprisingly usable, and the 3D drawings sometimes provide insights that are not so obvious in the 2D drawings.

1 Introduction

Many problems in computer-aided design and graphics involve the process of setting and adjusting input parameters to obtain desirable output values. Exploring different parameter settings can be a difficult and tedious task in most such systems. Managing and organizing the exploration of input-parameter space is usually the responsibility of the user; the computer is used as a passive instrument, not unlike a brush or pencil. Design Galleries were introduced [4] as a general paradigm for designing computer graphics and animations in which the parameter-setting task is divided more equitably and more appropriately between user and computer. In the Design-Gallery (DG) approach, the computer generates a representative set of graphics or animations for perusal by the user; the representative set is dispersed uniformly in the space of possible graphics. Achieving uniform dispersion is a novel and challenging technical problem, which is discussed in detail in the original paper. Once satisfactory dispersion has been achieved, the other technical problem in the DG paradigm is arrangement: the automatically generated graphics or animations — typically there will be several hundred of them — must be presented in a logical and accessible way to facilitate browsing by the user.

One solution to the arrangement problem uses graph drawing.[1] If we map each generated image or animation to a graph node, and use a perceptual distance metric (see [4] for the details of several such metrics) to generate edge weights for each edge in a complete graph, a graph drawing that correlates the perceptual distances between nodes with distances in a 2D or 3D embedding should provide an intuitive visualization of the set of graphics or animations.

This idea is illustrated in Figure 1. The figure contains a 2D Design Gallery of 584 ray-traced images of the same geometric scene model, each illuminated by a single light source. The differences in the images result from different light parameters, e.g., ones relating to light type and position. These input parameters are dispersed automatically by the system to yield a diverse selection of light sources. A user might browse this Design Gallery to assemble a set of lights that combine to communicate a particular mood or to focus attention on certain scene elements. 2D graph drawing is used to arrange low-resolution thumbnails of the images in the center display area. A full-size image appears when a thumbnail image is selected; full-size images can be moved to the surrounding gallery for convenience. In the figure, the gallery has been populated with a representative set of images, and the association between thumbnail and gallery images is indicated by overlaid lines.[2] The user can pan and zoom the center display area to examine subsets of the images in more detail. In Figure 2, the user has concentrated attention on the leftmost images of the display in Figure 1.

In Figure 3, 3D graph drawing is used to arrange the same set of images. The images are placed on self-orienting *billboard* polygons that float in space, and among which the user can navigate. The polygons are represented in the Virtual Reality Modeling Language (VRML), so navigation can be done using

---

[1] We have also investigated another solution based on uniform-depth hierarchical arrangement, which requires solving a graph-partitioning problem to form the hierarchy [4].

[2] In the interactive system, the association between thumbnail and full-size images is indicated dynamically: placing the mouse over an image in the gallery highlights its associated thumbnail, and vice versa.

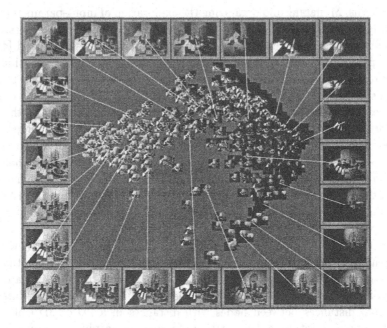

Fig. 1. A 2D Design Gallery for light selection and placement.

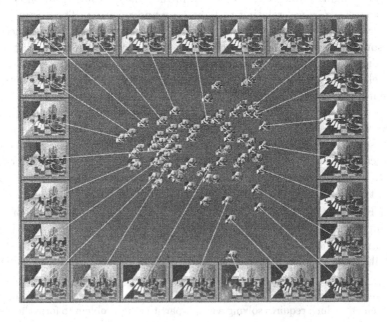

Fig. 2. The same Design Gallery after some panning and zooming.

any VRML browser. These browsers offer a variety of navigation modes, including *walk* (free movement subject to gravity), *fly* (free movement without gravity), and *examine* (free movement with additional capabilities for examining and manipulating individual objects) [1]. We have augmented the standard VRML-browsing functionality by providing a semi-transparent plane orthogonal to the user's viewing direction that can be used to focus attention on foreground images. Figure 4 concentrates on approximately the same images featured in Figure 2; this view was achieved by rotating the 3D view, moving forward, and using our semi-transparent "curtain" to hide background images. We also allow the user to tag certain images as "lighthouses": a flashing beacon is attached to images so designated, making them easier to find again later.

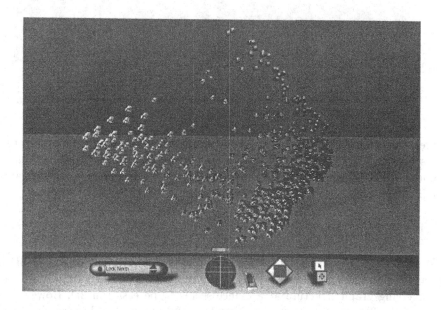

Fig. 3. A 3D Design Gallery for light selection and placement.

The 2D and 3D locations of the image nodes in the respective graph layouts were computed using Torgerson's classical multidimensional scaling (MDS) method [6].[3] From a matrix of perceptual distances between each pair of images, MDS computes the appropriate 2D or 3D layout coordinates for each image such that the distances in the drawing closely match those in the input matrix. Classical scaling is one of the simplest and fastest algorithms for MDS, but it is less general than iterative methods. When the input distances come from measure-

---

[3] Rubner et al. have independently investigated the use of MDS techniques for arranging a collection of images [5]. The use of MDS for graph drawing was pioneered by Kruskal and Seery [3].

Fig. 4. The same Design Gallery after some 3D navigation and manipulation.

ments in a high-dimensional Euclidean space (which is true for the applications discussed here, although it need not be the case in other DG applications), the algorithm can be viewed as an efficient technique for principal-component analysis [2].

2   Comparisons

We expected the 2D Design Gallery interface to be far superior to the 3D interface: panning and zooming in 2D is much easier than navigating in 3D, and perceiving graph structure at a glance seems to be easier in 2D. However, we found the 3D interface to be surprisingly useful and usable. In some instances, the 3D interface supplied insights that were not apparent in 2D. For example, Figures 5 and 6 show Design Galleries for 256 volume-rendered images of a computed-tomography (CT) data set representing a human pelvis. The differences in the images result from the use of different opacity transfer functions for the various issue types. The graph drawings in both figures appear essentially the same, modulo a reflection. However, a slight rotation of the 3D graph drawing (see Figure 7) reveals some additional structure: images that depict predominantly bone and muscle lie approximately in a plane (left), while images that depict varying amounts of fatty tissue lie off this plane (right). The additional structure evident in 3D can make it easier to find and locate images with specific characteristics.

Several factors make a 3D interface more successful in our application than one might expect:

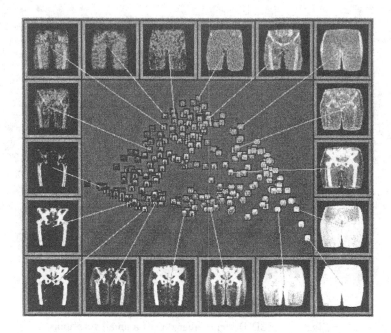

Fig. 5. A 2D Design Gallery for volume-rendered images.

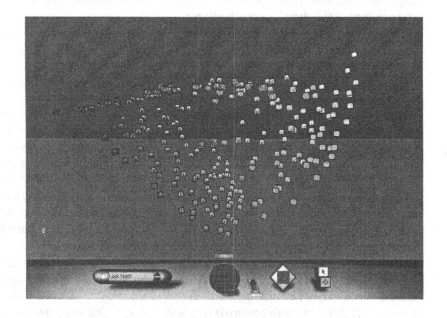

Fig. 6. A 3D Design Gallery for the same volume-rendered images.

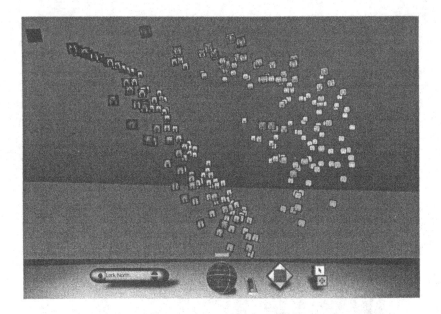

Fig. 7. The 3D Design Gallery after a small rotation.

- Edges and text labels are not needed to convey any useful information; their absence makes the drawings much clearer.
- Because each node is an easily perceived image, and because similar images cluster together, it is easier for the user to discern and identify regions of the graph, which aids navigation.
- Because we are trying to visualize distances in a high-dimensional space (the perceptual similarities for the volume-rendered images were expressed in 24 dimensions, for example), having three dimensions instead of two for our graph drawings means that less information is lost when projecting from the higher-dimensional perceptual-similarity space onto the lower-dimensional drawing space. Even though the first two embedding dimensions captured around 40% of the variance in the CT images, compared with about 15% for the third dimension, the extra information is more useful in combination with the other dimensions than one would expect from considering the variance alone. We speculate that this advantage of the third dimension may grow in importance as the number and variety of graph nodes increase.
- Because each graph node is a single texture-mapped rectangle, we can take advantage of graphics workstations that have been optimized specifically to display texture-mapped polygons efficiently.[4] This permits smooth motion through 3D drawings with hundreds of graph nodes, which is crucial for the

---

[4] Most of our experimentation was done on a high-end Silicon Graphics workstation with multiple processors and raster managers, and lots of memory.

usability of the interface.
– The VRML interaction metaphors suffice for most of our needs (this might
  not be true for other graph-drawing applications, e.g., ones in which graph
  manipulation is required), so we can take advantage of mature, standard
  methods for navigating 3D virtual environments.

3 Conclusion

Browsing automatically generated Design Galleries is a new application for graph
drawing. We have developed browsing interfaces based on both 2D and 3D graph
drawing. The 2D interface has proven to be usable and useful; counter to our
initial expectations, so has the 3D interface. However, we have identified several
factors peculiar to our application that makes 3D graph drawing easier and more
appropriate. While the future of 3D graph drawing as a general visualization
paradigm may be problematic, we speculate that it may prove useful for niche
applications such as ours.

Acknowledgements

This material is based upon work supported by the National Science Foundation
under Grant Numbers IRI-9350192 and IRI-9618848 and by a grant from MERL
— A Mitsubishi Electric Research Laboratory to SMS. This work was done while
KR was at the Division of Engineering and Applied Sciences, Harvard University.

References

1. R. Carey and G. Bell. *The Annotated VRML 2.0 Reference Manual*. Addison Wesley Developers Press, Reading, Massachusetts, 1997.
2. J. C. Gower. Some distance properties of latent root and vector methods used in multivariate analysis. *Biometrika*, 53:325–338, 1966.
3. J. B. Kruskal and J. B. Seery. Designing network diagrams. In *Proceedings of the First General Conference on Social Graphics*, pages 22–50, Leesburg, Virginia, October 1978.
4. J. Marks, B. Andalman, P. Beardsley, W. Freeman, S. Gibson, J. Hodgins, T. Kang, B. Mirtich, H. Pfister, W. Ruml, K. Ryall, J. Seims, and S. Shieber. Design galleries: A general approach to setting parameters for computer graphics and animation. In *SIGGRAPH 97 Conf. Proc.*, pages 389–400, Los Angeles, California, August 1997.
5. Y. Rubner, L. J. Guibas, and C. Tomasi. The earth mover's distance, multi-dimensional scaling, and color-based image retrieval. In *Proc. of the DARPA Image Understanding Workshop*, New Orleans, Louisiana, May 1997.
6. W. S. Torgerson. *Theory and Methods of Scaling*. Wiley, New York, 1958. See especially pages 254–259.

Online Animated Graph Drawing for Web Navigation

Peter Eades, Robert F. Cohen and Mao Lin Huang

Department of Computer Science and Software Engineering
The University of Newcastle, NSW 2308, Australia
{eades,rfc,mhuang}@cs.newcastle.edu.au

Abstract. This paper describes an online animated graph drawing method. The method deals with huge graphs which are partially unknown. It is instantiated in OFDAV, a web diagram visualizer.

1 Introduction

Traditional graph drawing techniques assume that the complete graph can reasonably be represented in a readable and understandable manner on the display medium. However, there are important situations where this assumption does not hold. For example, it is unlikely that there is a readable and understandable technique to visualize the complete WWW on any display medium. Other examples include large file systems, and graphs arising in information retrieval.

This paper presents techniques to visualize very large amounts of relational information. We assume that the amount of data that can be effectively displayed at one time is only a tiny subset of the graph. Our aim is to provide tools to help humans understand large graphs while only being able to view a small portion of the graph at a time.

Most graph drawing systems to date approach this problem in the following manner:

1. Layout the graph on a virtual (very large) page.
2. Provide a small window and scroll bars to allow the user to navigate the visualization.

However, this technique has a number of major problems that impinge upon human understandability:

- The whole graph may not be known. In some cases, the local node in a distributed system may know only a part of the graph; in other cases, at the time of viewing only a small subgraph is known.
- The layout is predefined and views are extracted of this layout. This means that changing views is a geometrical operation and not a logical operation. The user naturally thinks in terms of the logical relations in the application domain (for example, hypertext links), not in terms of the synthesized geometry of the layout; thus navigation by scroll-bars is difficult.

- The approach is *structure orientated* in the sense that the information structure drives most aspects of the display. Note that in a Geographical Information System, the topology on the screen is closely related to the physical topology in the application domain. However, in graph drawing applications, the topology is synthetic and created by an algorithm with fixed aesthetic criteria. This problem is severe for large graphs because the user cannot see the whole graph. A user orientated approach is more suitable. The user should be able to control logical content of the display.

Further, a number of minor technical problems arise:

- long edges on the large virtual screen are hard to follow,
- it is difficult to navigate out of large empty areas, and
- it costs huge memory to store and display the large virtual screen.

Some alternative techniques have been proposed (see [12, 3, 8, 7, 10, 11, 5]). For example, *fish-eye* [12] views can keep a detailed picture of a part of a graph as well as the global context of the graph. Three dimensional methods, such as *cone trees* [10], increase the density of information on the screen. While these techniques effectively deal with graphs of moderately large size, they do not help where the graph (such as a WWW graph) is not completely known. Further, these techniques still predefine the geometry.

Our *Online Force-Directed Animated Visualization (OFDAV)* technique provides a major departure from traditional methods. Our aim is to seek a new visualization technique that can provide effective navigational views. Our technique does not need the whole graph to be known, it does not predefine the geometry (the user can navigate logically), and it is user-oriented.

In OFDAV, the view of the user is focused on a small subgraph of a large graph G at any point in time. This subgraph is defined by a *focus node v*. We use variations of traditional graph drawing algorithms to draw the subgraph of G and a logical "neighborhood" around this subgraph. We then allow the user to change focus node by selecting another node of G. We use multiple animations to guide the user between views and preserve the mental map [3]. We also adopt a linear "history" that traces the subgraphs that the user has visited. This assists in backtracking through the graph.

The technique is implemented as a system for navigating the WWW. OFDAV is a Java program. It uses a force-directed layout method [2].

A number of researchers have noted that "overview diagrams" [6, 9, 8] provide a reasonable solution to the famous "Lost in hyperspace" problem, where users become disoriented with respect to a complex system of hypertext links. Our system provides such overview diagrams. Some existing overview diagram systems have developed [9]; however, these systems all predefine the layout or geometry, and they only visualize the *history* within very limited context levels. In contrast of this, OFDAV provides an on-line browsing environment in which we can navigate through unlimited context levels.

2   The online graph model

In this section we describe the graph model and the transitions on which our system is based.

Our system aims to explore a huge graph $G = (V, E)^1$.

The exploration of the graph G uses a sequence of subgraphs $F_1, F_2, \ldots$; each subgraph in this sequence is a *logical key frame*. Intuitively, the logical key frame is the subgraph which is currently being viewed. For each i, there is a FIFO queue Q_i of the *focus* nodes of F_i. Intuitively, the queue Q_i consists of the nodes that have been visited during the exploration of G and the logical key frame F_i is the graph induced by nodes near Q_i (in graph theoretic distance).

To change from one logical key frame F_i to the next logical key frame F_{i+1}, the user selects a node $v_{i+1} \in F_i$ with a mouse click. The node v_{i+1} is appended to the queue, and (in a FIFO manner) a node is deleted from the queue. We have experimented with queue lengths between 7 and 10.

3   The graph drawings

For each logical key frame F there is an *animated graph drawing*. This is a continuous sequence of drawings called the *screens* of F; it plays the same role as "in-betweening" in traditional computer animation.

The change in logical key frames from F_i to F_{i+1} is triggered by a mouse click on the new focus node. The first screen of the first logical key frame is a graph drawing of F_1 computed by a modified force directed algorithm. The first screen for F_i is defined as follows. The nodes common to F_i and F_{i-1} stay on the screen as they were in the final screen of F_{i-1}. The new nodes appear radially around and very close to their neighbors in F_{i-1}.

The following screens of the animated graph drawing of F_i are computed using an animated force directed approach. In our implementation, the force directed algorithm is based on Hooke's law springs, and the strength of the springs varies. It tends to arrange the focus nodes in a line in the order that they appear in the queue. Details of the force model and the methods used to compute each screen are given in [4]. The algorithm is tuned to web graphs, which have a tree-like structure. However, in other applications, an application specific force directed method may be suitable.

4   Remarks

The OFDAV system is an effective tool for navigating through web documents. The layout of each logical key frame is a "spring" drawing and has the advantages and disadvantages of spring drawings (see [1]). However, it has other features which are particularly meaningful for visualizing large graphs which are partially unknown. Some of these features are listed below.

---

[1] In our implementation, the graph is directed, connected, and rather tree-like; but our methods apply in general to undirected graphs.

The mental map: A major problem in moving from one logical key frame to the next is preserving the user's mental map of the graph (see [3] for a full study of this issue). Our goal is to preserve the users mental map, while taking best advantage of the view screen. In OFDAV, we use animation to assist the user in understanding the change in view. We use three types of animation:

- – *Fade Animation:* We use shrinking/growing to help the user identify nodes that are disappearing/appearing.
- – *Camera Animation:* this moves the drawing so that the new focus node moves toward the center of the screen.
- – *Layout Animation:* We use a complex system of forces based on a Hooke's law springs.

Visualizing the history: The queue of focus nodes shows the recent history of the exploration of the graph. The force model that we use tends to keep this history in a line; the new nodes are added to one end of the line and the nodes which disappear are at the other end. This aids the user in keeping track of where they are and where they have come from. As well as the queue of focus nodes, OFDAV keeps a track of *past history*: a list of all previous focus nodes. An option in OFDAV is to show all the history.

Text with graphics: OFDAV optionally works in *Show-Document mode:* by clicking on *ShowDocument* button, the animation freezes and a mouse click on a particular node brings up the associated HTML document to the screen.

We believe that the principles of OFDAV can be used for any large relational system, and it is particularly useful for systems which are partially unknown.

5 Examples

The following three figures are screen dumps from OFDAV. All are from a visualization of the web context of the Department of Computer Science and Software Engineering at University of Newcastle. Figure 1 shows the initial key frame F_1, consisting of the focus node $v_1 = $ 'Dept' and a set of nodes surrounding it with maximum distant $d_{init} = 2$ from node v_1. Figure 2 shows the following key frame, formed after a user mouse click on the 'PostStudents' node. Three frames further on we have Figure 3, formed by clicking on the nodes 'Mao', 'links' and then on the 'News' node.

References

1. G. Di Battista, P. Eades, R. Tamassia, and I. Tollis. Algorithms for drawing graphs: An annotated bibliography. *Computational Geometry Theory and Applications*, 4(5):235–282, 1994.
2. P. Eades. A heuristic for graph drawing. *Congr. Numer.*, 42:149–160, 1984.
3. P. Eades, W. Lai, K. Misue, and K. Sugiyama. Preserving the mental map of a diagram. In *Proceedings of Compugraphics 91*, pages 24–33, 1991.

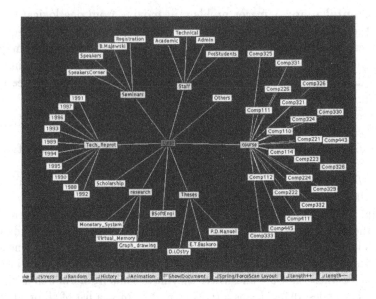

Fig. 1. A visualization of the web context of the Department of Computer Science and Software Engineering at University of Newcastle. This is the initial key frame F_1, consisting of the focus node $v_1 =$ 'Dept' and a set of nodes surrounding it with maximum distant $d_{init} = 2$ from node v_1.

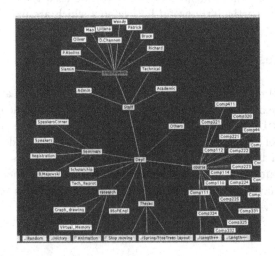

Fig. 2. Navigating the web context of the Department of Computer Science and Software Engineering at University of Newcastle. This is a frame F_2 following the graph in Figure 1. It has been formed after a user mouse click on the 'PostStudents' node, Here, $v_2 =$ 'PostStudents' and the *recent history* consists of the nodes 'Dept' and 'PostStudents'.

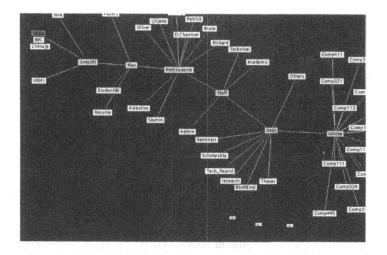

Fig. 3. Three frames further from Figure 2, formed by clicking on the nodes 'Mao', 'links' and then on the 'News'. Note that 3 old nodes are smoothly disappearing and 3 new nodes are smoothly appearing. Here, v_5 = 'News' and the *recent history* consists of the nodes 'Dept', 'PostStudents', 'Mao', 'links' and 'News'.

4. Peter Eades, Mao Lin Huang, and Junhu Wang. Online animated graph drawing using a Modified Spring algorithm. Technical Report 97-05, Dept. of Computer Science and Software Engineering, Uni. of Newcastle, 1997.

5. P. Kahn. Visual clues for local and global coherence in the WWW. *Communications of the ACM*, 38(8):67–69, 1995.

6. Sougata Mukherjea and J. Foley. Visualizing the World-Wide Web with the navigational view builder. *Computer Networks and ISDN Systems. Special Issue on the Third International World Wide Web Conference*, 27(6):1075–1087, 1995.

7. Sougata Mukherjea, J. Foley, and S. Hudson. Interactive clustering for navigating in hypermedia systems. In *Procceedings of the ACM European Conference on Hypermedia Technology*, 1994.

8. J. Nielsen. The art of navigating through hypertext. *Communications of the ACM*, 33(3):296–310, 1990.

9. Chris Pilgrim and Ying Leung. Applying bifocal displays to enhence WWW navigation. In *Procceedings of the Second Austrlian World Wide Web Conference*, 1996.

10. G. G. Robertson, J. D. Mackinlay, and S. K. Card. Cone trees: Animated 3D visualizations of hierarchical information. In *Proc. ACM Conf. on Human Factors in Computing Systems*, pages 189–193, 1991.

11. George G. Robertson, Stuart K. Card, and Jock D. Mackinlay. Information visualization using 3D interactive animation. *Communications of the ACM*, 36(4):57–71, 1993.

12. M. Sarkar and M. H. Brown. Graphical fisheye views. *Commun. ACM*, 37(12):73–84, 1994.

Grappa: A GRAPh PAckage in Java

Naser S. Barghouti[*]
Bear, Stearns & Co., Inc.
New York, NY, USA
naser@bear.com

John M. Mocenigo
AT&T Laboratories
Florham Park, NJ, USA
john@research.att.com

Wenke Lee[†]
Columbia University
New York, NY, USA
wenke@cs.columbia.edu

Abstract

Grappa is an extensible graph drawing package written in Java. The package comprises classes that implement graph representation, presentation and layout services. It provides an application programming interface (API) on top of which Web-based applications that need to visualize information in terms of graphs, such as process flows, business workflows or program dependencies, can be built. Through subclassing, the classes that implement an application inherit graph drawing and layout services provided by *Grappa*; these services can be enhanced and customized for the specific application. To illustrate its utility, a new version of Improvise, a multimedia process modeling environment, was written on top of *Grappa*.

---

[*]This author's contributions were made while employed at AT&T Laboratories
[†]This author's contributions were made during a summer internship at AT&T Laboratories

1 Introduction

In many domains, information can be organized and visualized in terms of graphs. Examples include process flow diagrams, software architectures, business workflows, data dependencies in data mining, and routing information in network management. Many applications supporting these domains include graph drawing modules. Not surprisingly then, there has been a significant amount of research on how to store, layout and display graphs most effectively.

With the increasing importance of Web-based, and, more generally, distributed computing, there has been an interest in providing packages for remote graph drawing services. By remote graph drawing we mean the ability for an application running on one machine to request graph drawing and layout services from a server running on another machine anywhere on the network. North's email-based graph service is an example of a remote graph layout service: it receives a graph specification as input via email and returns, also via email, a graph drawing whose layout is computed by *dot* [3]. The graph server at Brown University [2] is more extensive in that it provides interactive graph drawing and translation (among many graph drawing algorithms) via the WWW.

This paper presents a graph drawing package, *Grappa*, on top of which Web-based applications requiring graph drawing services can be built. The Grappa approach differs from the Brown graph server approach in that it provides an Application Programming Interface (API) so that applications can be developed to not only use its standard services, but also extend and customize them to accomplish application-specific tasks. This alleviates the burden of writing all the code that manages the graph drawing aspects of an application. The application writer need only concentrate on the semantics of the application and how to map those semantics onto graphs.

The advent of Java [1] motivated us to exploit the features it provides, such as portability, graphical user interface generation facilities, and distributed computing features. Therefore, we built *Grappa* in Java, which allows *Grappa*-based applications to be built as applets that can be executed over the WWW via any Java-enabled Web browser; this greatly facilitates the distribution of the application software and enhances its availability.

Other client-server applications (not necessarily over the WWW) can also be built on top of *Grappa*. In this case, a *Grappa* server provides a set of standard graph drawing, editing and viewing functions; the front-end client can be launched from any node in the network and be connected to the server. This architecture has the advantage of saving computing resources since the computational-intensive graph layout process need not be on every client machine; in addition, graph representations can be stored by the server and shared by multiple clients.

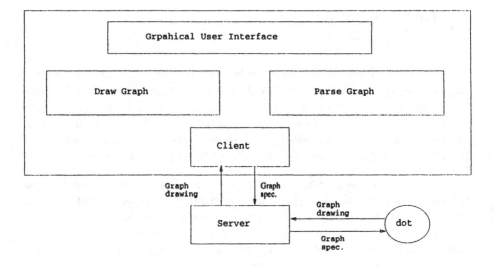

Figure 1: The Architecture of *Grappa*

2   The Design of *Grappa*

We had three main requirements in building *Grappa*:

1. Extensibility, which allows for a natural evolution path, where new services can be incorporated into the package.

2. Portability, which is essential in the multi-platform world in which we live.

3. Customizability, which allows for different kinds of application to be built on top of *Grappa*, re-using its facilities and avoiding re-writing a lot of code.

We discuss how these requirements were met in the design and implementation of *Grappa*.

2.1   The Components of *Grappa*

Grappa is a class package written in Java. The classes implement a client-server architecture that is inherited by applications. Figure 1 shows the main architectural components of *Grappa*. The server processes client requests (received in the form of messages) for graph layout. The server creates a Java thread to process each client message. The header of each message specifies the request type, and the body contains a graph specification (in *dot* format). The server invokes *dot* to compute a graph layout (drawing). The server then sends *dot*'s output, a graph drawing (also in *dot* format), back to the client.

Client classes handle most of the interactions with the end user. Its graphical user interface is constructed using the standard Java Abstract Windowing Toolkit (AWT), with pull-down menus, and canvasses for graph objects. The contents of the menus and the functions to perform menu operations can be changed for specific applications.

The *Grappa* client classes are organized in three hierarchies, as shown in Figure 2. The first hierarchy, whose root is the class *DotGraph*, contains classes for graph *definition* (i.e., for defining graphs, subgraphs, nodes, edges, and associating attributes with nodes and edges). *DotGraph* defines a graph as a set of *DotElements*, each of which is a node, an edge, or a subgraph. Each instance of *DotElement* has set of attributes associated with it, such as *shape*, *style*, *color*, and so on. *DotGraph* also refers to the class *DrawPane*, whose instances contain information about displayed instances of *DotGraph*.

The second hierarchy, whose root is the abstract class *DrawObject*, includes classes for graph *drawing*. Class definitions for drawing 36 node shapes (Box, Circle, Ellipse, etc.) and several edge types are included in this hierarchy. Each instance of *DrawObject* refers to a specific *GraphicContext*, which provides information for drawing an object on specific canvas.

2.2 The *Grappa* API

The third hierarchy, rooted at *AppObject*, is a place holder for application-specific classes, and it constitutes the application program interface (API) for *Grappa*. An instance of *AppObject* refers to an instance of *DotElement* and has a reference to an instance of *DrawObject*; depending on the value of the type attribute of the object (e.g., the node shape), the methods of the appropriate subclass of *DrawObject* are called to draw the object on the client canvas. Application programmers can extend and customize *Grappa* by following the same approach of extending standard Java APIs. In particular, application-specific classes should be added as subclasses of *AppObject*. Application-specific behavior can then be implemented by adding data members and methods to these subclasses, overriding existing behavior.

Grappa application objects may have different drawing styles and operations for different types of application-specific information entities. For example, a program structure application may define a class *Module* as a subclass of *AppObject* and assign the value "Ellipse" to its *shape* attribute (which is automatically defined because every *AppObject* refers to a *DotElement*). The class *Module* may then define new attributes, such as *name*, *description* and *owner*, that are specific to software modules. Finally, it may specialize the *draw* method

340

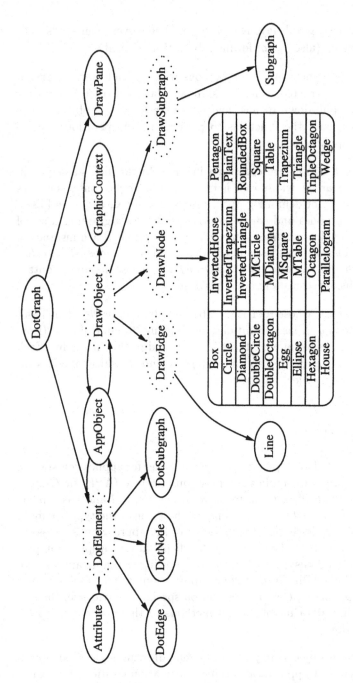

Figure 2: Class Hierarchy of *Grappa* Application Objects

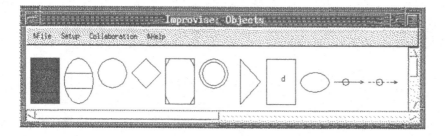

Figure 3: *Improvise* Template Objects

of class *Ellipse* (which is a subclass of *DrawNode*) to draw two horizontal lines to divide up the ellipse into three sections, and display the module's *name*, *description*, and *owner* in the three sections respectively.

3 Example: Improvise

Improvise is a multimedia process (workflow) modeling and analysis tool. It represents a process (a partially-ordered set of activities) is terms of a graph. It was first implemented on top of the graph editor *dotty* [4]. Improvise provides a graphical modeling notation that defines a set of object (node) types that represent various entities in a workflow, such as *ManualTask*, *Data*, and *AbstractProcess*; two edge types are also defined. These object types are displayed in an object palette. The user selects the type of object from the palette and clicks on an un-occupied area of a drawing canvas to create an instance of the type. An edge is drawn by clicking on an existing node and dragging the mouse to another node. Repeated selection and clicking (without worrying about placement of nodes and edges) results in a process flow diagram; on demand, the diagram is sent to *dot* to compute a graph drawing that is then displayed on the canvas, replacing the previous diagram.

Last year, we made a decision to re-implement Improvise in Java and make it executable (in a client-server manner) over the WWW. One alternative was to re-code Improvise in Java from scratch. The second alternative was to build it on top of *Grappa*, which was being designed at that time. We opted for the second option because it would relieve us from writing all the graph drawing and layout code.

Compared with using *dotty*, our experience shows that the *Grappa* API is more elegant because of four reasons: 1) the graph object hierarchy is intuitive and easy to extend; 2) *Grappa* hides the implementation details of graph drawing and layout, which are not the main concern of Improvise; 3) the dynamic features of Java enable *Grappa* to evolve in parallel and independently with application development (i.e., the layout algorithm can be changed without affecting Impro-

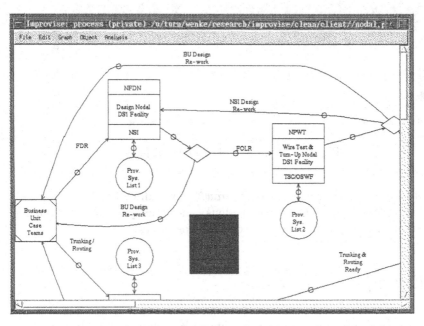

Figure 4: *Improvise* Menu and Canvas Showing a Workflow

vise); and 4) being web-based, the new version of Improvise does not require any software distribution, and new versions are immediately available.

4 Conclusion

We presented *Grappa*, an extensible graph drawing package written in Java. *Grappa* comprises a set of classes that implement graph representation and presentation services. It also provides an API for applications that require graph drawing, editing and browsing services. Application classes are defined as subclasses of *Grappa* classes; the subclasses can specialize the base classes to support application-specific behavior.

One of the motivations of our work was to explore and experiment with Java. Although still very young and evolving, Java has a lot of advanced features that will make it a success. As discussed throughout this paper, Java's object-oriented, multi-threading, and dynamic features are most important to the design of *Grappa*. Its exception handling feature can also be used as a very useful programming aid: an error stack trace which includes file names and line numbers can be printed out when an exception is raised, so that bugs can be easily located and fixed. A worth noting phenomenon that have fueled the rapid adoption of Java is that a lot of Java enthusiasts have made their programs freely available on the Internet. A lot of these programs are not just "cool", but they actually provide needed features that are not yet supported by the standard Java packages. For example, our *Grappa* parser was constructed using *JavaCup*, a *LALR*(1) parser generator developed by Scott Hudson in Georgia Institute of Technology.

Our overall experiences with Java is quite positive. We were able to implement *Grappa* in about 10k lines of Java code in two and a half months. *dot* was not ported to Java; rather, we use it as a *native* program (in the native host environment) that is invoked from the *Grappa* server.

5 Acknowledgment

The authors would like to thank Eleftherios Koutsofios, Stephen North and Robin Chen of AT&T Labs - Research, and Jeff Korn of Princeton University for their helpful discussions.

References

[1] Ken Arnold and James Gosling. *The Java Programming Language*. Addison Wesley, Reading, MA, 1996.

[2] Stina Bridgeman, Ashim Garg, and Roberto Tamassia. A Graph Drawing and Translation Service on the WWW. In *Proc. of Symposium on Graph Drawing GD'96*, Berkeley, CA, USA, September 1999. The URL is http://loki.cs.brown.edu:8081/graphserver/home.html.

[3] Emden Gansner, Eleftherios Koutsofios, Stephen C. North, and Kiem-Phong Vo. A Technique for Drawing Directed Graphs. *IEEE Transactions on Software Engineering*, 19(3), March 1993.

[4] Stephen C. North and Eleftherios Koutsofios. Applications of Graph Visualization. In *Proc. of Graphics Interface '94*, pages 235–245, Banff, Alberta, 1994.

GRAVIS — System Demonstration

Harald Lauer, Matthias Ettrich and Klaus Soukup

Universität Tübingen, Sand 13, 72076 Tübingen, Germany
Email: {lauer, ettrich, soukup}@informatik.uni-tuebingen.de

Abstract. GRAVIS is a powerful, interactive graph visualization system, designed to be generally usable in research and practical applications. The implementation of GRAVIS is based on a flexible object-oriented system architecture, portable to many platforms. The intuitive and efficient user interface is completed by the ability of the base system to meet the requirements of future applications.

1 Features of GRAVIS

Graph drawing research and practical applications working with structured data sets represented as graphs demand a powerful visualization tool. Common for both areas is, that the graph drawing tool must be interactive, extensible and flexible to cover the wide range of application domain specific requirements found in research and praxis. Together with a robust and maintainable realization, this would result in *the* ideal tool for graph drawing. GRAVIS aims to fulfill as many of the above requirements as possible.

Asking practitioners using graph visualization within their applications about their needs reveals a very broad spectrum of requirements covering performance, ease of use, graphical attributes, layout tools and many more. The requirements analysis and thus the design of GRAVIS is based on such a "wishlist" [9], therefore the design philosophy of GRAVIS is to be a generally usable tool for information visualization, only restricted to data representable as graphs. The following list shows a collection of the main features of GRAVIS realizing this design goal:

- Highly modularized architecture.
- Complete set of graphical attributes for all graph elements.
- Easy to use program interfaces to implement extension modules.
- Dynamic loading of extension modules at runtime.
- Multiple views on the same graph structure.
- Highly optimized visualization component.
- 2- and 3-dimensional graph visualization.
- Support for hierarchical graphs.
- Intuitive graphical user interface.
- Active nodes, which perform user defined actions when pressed.
- Unlimited undo and redo.
- Zoom to arbitrary levels.
- Various input/output formats (including GML [8] and PostScript).

There is a number of other grapheditors available, created mostly in the context of research projects. An example of a commercial product is the *Graph Layout Toolkit* [10] by Tom Sawyer Software. Other systems include, but are not limited to, Graphlet [8], daVinci [6], GD-Workbench [2] and the VCG tool [15].

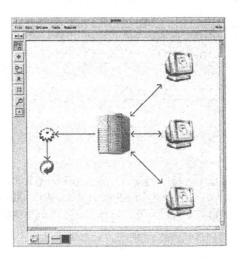

Fig. 1: The graphical frontend of GRAVIS displaying a graph using various bitmaps instead of the standard node types (circle, rectangle, rhomb, etc.).

2 Graphical User Interface

Much work has gone into the design of the user interface of GRAVIS, to produce an intuitively and efficiently usable tool for the interactive manipulation of arbitrary graphs. The philosophy of the user interface of GRAVIS is different from the traditional approach taken by tools like Graphlet [8] and the Graph Layout Toolkit [10] in one important aspect: although GRAVIS supports various input modes, the main mode for editing is *not* further split into modes for node/edge creation, attribute editing and more.

Instead, each frequently executed input or manipulation function is in the edit mode of GRAVIS directly accessible by a context sensitive mouse operation. The benefit of this approach is that there is no need for time consuming and concentration breaking mode changes, forcing the user to search for a "mode change" button if he for example wants to create nodes and edges in an arbitrary order. The *context* of the mouse event is sufficient to unambiguously decide which operation is requested by the user and is defined by

1. the mouse location (over a node, edge, bend, selection or free space) and
2. the mouse movement (a simple click or dragging the mouse after clicking).

Less frequently needed operations are accessible by the standard menu structure or by buttons arranged around the drawing area. Users of GRAVIS however, are seldomly forced to search for menu entries, since even functionality often found in additional panels are available directly at the object in the drawing canvas. Figure 2 for example shows how graph elements can be resized.

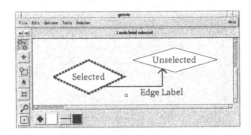

Fig. 2: Resize button in the lower right corner of the selected node, which allows resizing the node to an arbitrary height and width. The edge offsets can be moved from their default position (node center) using a similar mechanism.

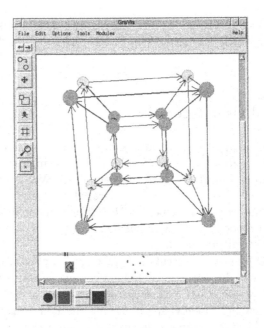

Fig. 3: The 3D graph viewer of GRAVIS. Below the drawing area is a mini-view of the graph, displaying the relative user position (indicated by the eye).

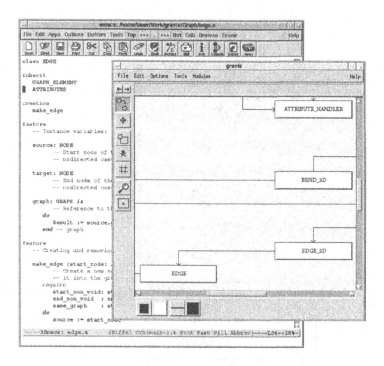

Fig. 4: Active nodes functioning as buttons with associated commands. GRAVIS can be used to layout object-oriented class diagrams and automatically load classes into an editor when the corresponding node is activated. In this case, the class **EDGE** (part of the *Graph* cluster of GRAVIS; the screenshot shows a fraction of the cluster) is now loaded in the editor after the node **EDGE** has been clicked.

3 Layout Modules

GRAVIS is dynamically extensible with new modules using well defined interfaces. Such modules can provide services of any kind, ranging from simple graph to sophisticated layout algorithms and even interfaces to external applications. A number of modules are included in the current version of GRAVIS and more are in development by the GRAVIS project group. Among these modules are:

- The layout algorithm for orthogonal, high-degree graphs *Kandinsky* [4].
- Interactive orthogonal graph layout based on [13] using extensions from [3].
- Symmetric layout (simplified version) based on [11] (see figure 5).
- Layout of series-parallel digraphs according to [1].
- Tree-based layout of general graphs (eg. for object-oriented class diagrams).
- Several graph generators.
- The Gem Springembedder [5] and a tree layout algorithm from GraphEd [7] demonstrating the interface to external applications and libraries.

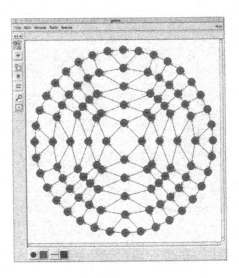

Fig. 5: A 10 × 10 grid drawn using the symmetric layout algorithm from [11].

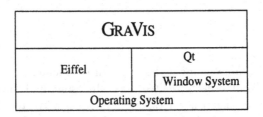

Fig. 6: Dependencies of GRAVIS on the programming language (Eiffel) and user interface toolkit (Qt). Details of the operating and the window system (eg. X Windows) are hidden from the implementation of GRAVIS.

4   Platforms and Portability

Realizing the object-oriented paradigm used in the design phase of GRAVIS, the system itself is implemented in the object-oriented programming language *Eiffel* [12]. Since one of the features of Eiffel is the abstraction of hardware or operating system specific issues and a well defined standard for kernel libraries, GRAVIS is portable to any platform supported by a compiler vendor, which covers most Unix systems and Microsoft Windows 95/NT.

Limiting the portability of graphical applications is often the access to window system functions for display purposes. GRAVIS however, uses the *Qt* library [14] for all user interface related parts of the system, which itself is available for many systems including the platforms mentioned above. The combination of the programming language and the user interface toolkit hides hardware and OS-specific details from the implementation of GRAVIS (see figure 6).

5 Conclusion

GRAVIS is freely available for research purposes and individual, non-commercial use. The most recent and all subsequent versions are available at our ftp site

`ftp.informatik.uni-tuebingen.de` in `/pub/PR/gravis`

Updated and more general information about the ongoing GRAVIS project can be found at the GRAVIS homepage located at the URL:

`http://www-pr.informatik.uni-tuebingen.de/Research/GraVis/`

Helpful comments and suggestions have been provided by Michael Kaufmann and Ulrich Fößmeier. This research was partially supported by DFG-Grant Ka812/4-1, "Graphenzeichnen und Animation".

References

1. P. Bertolazzi, R. F. Cohen, G. Di Battista, R. Tamassia, and I. G. Tollis. How to draw a series-parallel digraph. In *Scandinavian Workshop on Algorithm Theory*, volume 621 of *LNCS*, pages 272–283, 1992.
2. L. Buti, G. Di Battista, G. Liotta, E. Tassinari, F. Vargiu, and L. Vismara. GD-workbench: A system for prototyping and testing graph drawing algorithms. In *Graph Drawing*, volume 1027 of *LNCS*, pages 111–122, 1995.
3. U. Fößmeier. Interactive orthogonal graph drawing: Algorithms and bounds. To appear in Graph Drawing '97.
4. U. Fößmeier and M. Kaufmann. Drawing high degree graphs with low bend numbers. In *Graph Drawing*, volume 1027 of *LNCS*, pages 254–266, 1995.
5. A. Frick, A. Ludwig, and H. Mehldau. A fast adaptive layout algorithm for undirected graphs. In *Graph Drawing*, volume 894 of *LNCS*, pages 388–403, 1994.
6. M. Fröhlich and M. Werner. Demonstration of the interactive graph visualization system daVinci. In *Graph Drawing*, volume 894 of *LNCS*, pages 266–269, 1994.
7. M. Himsolt. Graphed: A graphical platform for the implementation of graph algorithms (extended abstract and demo). In *Graph Drawing*, volume 894 of *LNCS*, pages 182–193, 1994.
8. M. Himsolt. The graphlet system (system demonstration). In *Graph Drawing*, volume 1190 of *LNCS*, pages 233–240, 1996.
9. H. Lauer. Grapheditoren — eine Wunschliste. Technical Report WSI-95-5, Universität Tübingen, 1995.
10. B. Madden, P. Madden, S. Powers, and M. Himsolt. Portable graph layout and editing. In *Graph Drawing*, volume 1027 of *LNCS*, pages 385–395, 1995.
11. J. B. Manning. *Geometric Symmetry in Graphs*. PhD thesis, Purdue University, 1990.
12. B. Meyer. *Eiffel: The Language*. 1992.
13. A. Papakostas and I. G. Tollis. Issues in interactive orthogonal graph drawing. In *Graph Drawing*, volume 1027 of *LNCS*, pages 419–430, 1995.
14. Qt. Information about Qt is available at http://www.troll.no.
15. G. Sander. Graph layout through the VCG tool. In *Graph Drawing*, volume 894 of *LNCS*, pages 194–205, 1994.

Touching Graphs of Unit Balls*

Petr Hliněný

Dept. of Applied Mathematics, Charles University,
Malostranské nám. 25, 118 00 Praha 1,
hlineny@kam.ms.mff.cuni.cz

Abstract. The touching graph of balls is a graph that admits a representation by non-intersecting balls in the space (of prescribed dimension), so that its edges correspond to touching pairs of balls. By a classical result of Koebe [5], the disc touching graphs are exactly the planar graphs. This paper deals with a recognition of unit-ball touching graphs. The 2–dimensional case was proved to be *NP*-hard by Breu and Kirkpatrick [1]. We show in this paper that also unit-ball touching graphs in dimensions 3 and 4 are *NP*-hard to recognize. By a recent result of Kirkpatrick and Rote, these results may be transferred in ball-touching graphs in one dimension higher.

1   Introduction

The intersection graphs of geometrical objects have been extensively studied for their many applications. Formally the *intersection graph* of a set family $\mathcal{M}$ is defined as a graph G with the vertex set $V(G) = \mathcal{M}$ and the edge set $E(G) = \{\{A, B\} \subseteq \mathcal{M} \mid A \neq B,\ A \cap B \neq \emptyset\}$. Then *geometrical intersection graphs* are those in which the set family $\mathcal{M}$ is determined by some geometrical meaning; in that case we can also think about *touching* (or *contact*) *graphs* if we allow the geometrical objects only to touch each other.

A graph H is called a touching graph of a certain geometrical type, if it is isomorphic to the intersection graph G of some touching set family $\mathcal{M}$ of that type. The *recognition problem* of intersection (spec. touching) graphs is the question whether given graph is isomorphic to an intersection (touching) graph of the specified type.

Probably the first interest in touching graphs is represented by a very nice result of Koebe [5], who proved that touching graphs of arbitrary discs in the plane are exactly the planar graphs (and this result was also rediscovered later). Recently, practical applications led to the introduction of more complex classes of intersection and touching graphs, most of which are *NP*-hard to recognize. Among the touching (or contact) graphs, that have attracted attention recently, we may notice the triangle contact graphs [2] or the contact graphs of straight line segments and of simple curves [4] in the plane.

* The author acknowledges the support of the grant GAUK 193/1996, Czech–US science and technology research grant no. 94051, and the DONET grant.

The interest of our paper is in unit-ball touching graphs in Euclidean space. The d–ball touching graphs are those that admit a representation of vertices by balls in E^d, so that their interiors are disjoint and two vertices are adjacent iff the two corresponding balls touch each other. Specially, if all the balls are of the same diameter, say 1, the *unit-ball touching graphs* are obtained. The touching set of balls is called the (unit) ball-touching representation of the related graph.

If we consider the dimension 1, the touching case is trivial, and the intersection case leads to well known interval graphs [7] (or unit-interval graphs). In the dimension 2, the classical result of Koebe is mentioned above. In opposite, the unit-disc touching graphs (and more generally disc touching graphs with bounded ratio of diameters) were considered in [1], and turned out to be *NP*-hard to recognize. Let us mention that also intersection graphs of arbitrary discs are *NP*-hard [6].

A natural question arises what is the complexity of recognition of ball touching graphs in higher dimensions. It is likely that the unit-ball touching graphs are *NP*-hard in any dimension greater than 1. This paper proves the case of unit balls in dimensions 3 and 4.

The next section also mentions that from this result it follows that recognizing touching graphs of arbitrary balls in dimensions 4 and 5 is *NP*-hard, whereas the case of dimension 3 was already shown to be *NP*-hard by Kirkpatrick.

2   Preliminary results

To get a better feeling about touching graphs of balls or of unit balls in higher dimensions, let us first mention two easy (and probably not original) facts about representability of graphs by balls.

Lemma 2.1 *For every $d \geq 3$, there exists a constant $k(d)$ such that the d–ball touching graphs are $k(d)$–degenerated.*

Proof. Let ϱ be a ball of the smallest radius r in the touching representation of a graph G. Then each ball touching ϱ can be reduced to radius r so that it still touches ϱ at the same point (but possibly not other balls). Since we have a touching representation, the interiors of all such balls around ϱ are pairwise disjoint, and they are, after reduction, all contained in a ball $\bar{\varrho}$ of radius $3r$ concentric with ϱ. Now it suffices to consider the volume of $\bar{\varrho}$, thus $k(d) < 3^d$. □

Realize that the exact value of $k(d)$ is closely related to the so called kissing number problem (how many non-intersecting unit balls can touch one central unit ball) which is not completely solved yet even in dimension 4 (!), so it is probably very difficult to determine.

On the other hand, if the dimension is not fixed, any graph can be represented by unit balls:

Lemma 2.2 *Every graph G on v vertices has a unit-ball touching representation in dimension $v - 1$.*

Proof. Any graph G is an induced subgraph of a d–regular graph G' $(d = \Delta(G))$. Let $\bar{G}$ be obtained from G' by subdividing each edge by a new vertex, and $\bar{M}$ be the adjacency matrix of $\bar{G}$. Then the rows of $\bar{M}$ corresponding to the vertices of G can be taken as centres of balls of the touching representation—two centres x_i, x_j have distance $\sqrt{2d}$ if $\{i,j\} \notin E(G)$, and $\sqrt{2d-2}$ if $\{i,j\} \in E(G)$ (e.g. the "unit distance" equals $\sqrt{2d-2}$). $\square$

Recently, Rote and Kirkpatrick [personal communication] discovered an elegant transformation that allows us to prove the NP-hardness of recognizing $(d + 1)$-ball touching graphs from the NP-hardness of recognizing d-unit-ball touching graphs:

Lemma 2.3 (Kirkpatrick and Rote) *The question whether a given graph G has a unit-ball touching representation in dimension d, can be reduced to the question whether the graph $G \oplus K_2$ has a ball touching representation in dimension $d+1$.*

3   Unit-ball touching graphs in E^3

In this section, a touching graph means a unit-ball touching graph in the 3–dimensional space.

Theorem 1 *The unit-ball touching graphs in 3–dimensional space are NP-hard to recognize.*

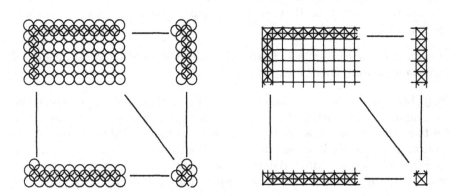

Fig. 1. A touching representation of the $FRAME_3$ graph

The proof of this theorem goes as follows: We form a "firm" global frame (shaped as a large square) that has a unique touching representation, and represent an instance of the $3 - SAT$ problem within this frame. These ideas are contained in the following sequence of lemmas.

Let us denote by $FRAME_3$ the graph that is shown in Figures 1 and 2, sufficiently large to represent a given SAT formula (see later). The same picture

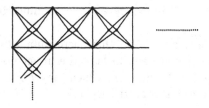

Fig. 2. A detail of a corner of $FRAME_3$

also shows a touching representation of this graph, consisting of a large square layer of unit balls, with additional balls on the boundary in subsequent layers below and above.

Lemma 3.1 *The graph $FRAME_3$ has a unique touching representation in the 3–dimensional space (up to an isometry).*

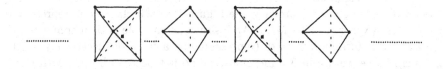

Fig. 3. How the vertices of F "stick together"

Proof. To prove the lemma, see that the boundary can be decomposed into a sequence of tetra- and octahedrons, each successive pair of them sharing a whole triangle, see Figure 3. Since an edge in the graph F means a unit distance between the centres of the corresponding balls, and both the regular tetra- and octahedrons are unique, there is only one possible shape of the touching representation of the boundary in E^3. Then the internal balls are arranged in "chains" tightly stretched between the boundary balls, so have unique positions by the triangle inequality. $\square$

Lemma 3.2 *The maximum number of disjoint unit balls that can touch one of the interior balls of $FRAME_3$ from one side of the central layer, is equal to 3.*

Proof. Imagine the interior balls of $FRAME_3$ as unit balls with centers in $[x, y, 0]$ where $x, y \in Z$. It is shown how to arrange three disjoint unit balls with a positive third coordinate, all touching the ball B_0 with center $[0, 0, 0]$ and disjoint with other balls of the frame: The centres of these balls are in coordinates $[-\frac{1}{2} + \gamma, \frac{1}{2} - \alpha, z_1]$, $[\frac{1}{2} - \beta, \frac{1}{2} - \gamma, z_2]$, $[\frac{1}{2} - \alpha, -\frac{1}{2} + \gamma, z_3]$, where α, β, γ are very small satisfying $0 < \gamma \ll \beta^2 \ll \alpha^4$ and z_1, z_2, z_3 are determined by unit distance from the origin (close to $\frac{1}{2}\sqrt{2}$).

The proof of the upper bound (i.e. that there cannot be 4 such balls) is too technical and is not presented here. In fact, the upper bound of 3 balls is not critical in the presented NP reduction, which may be adapted to any bound greater than 3 (see also the next section). $\square$

For the next classical result see [3].

Lemma 3.3 *The $3 - SAT$ problem is NP-complete.*

Now we are ready to describe the NP reduction to the unit-ball touching problem.

Lemma 3.4 *For every $3 - SAT$ formula φ, there exists a graph (of polynomial size with respect to φ) that has a unit-ball touching representation in E^3, if and only if φ is satisfiable.*

Proof. Once having built the above firm frame, it is easy to represent a given formula φ within it. Each variable v is replaced by two vertices $v, \neg v$, both adjacent to four chosen neighbouring vertices of the internal layer of $FRAME_3$ (forming an octahedron as on the boundary, see Figure 4). Each clause c is represented by one chosen vertex of $FRAME_3$. An instance of variable is then connected to its clause by a suitably long path (represented by a chain of successively touching balls). Since we are in 3 dimensions, we do not need to bother with chain crossings. A scheme of this construction is shown in Figure 4.

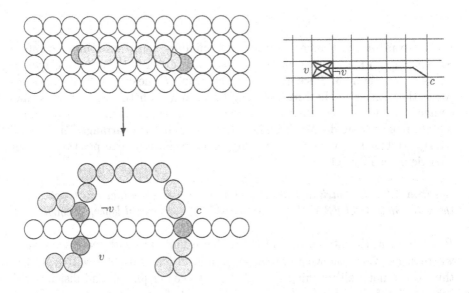

Fig. 4. Representing variables and clauses, and their connection

Of course this is not all, we need to distinguish the false and true sides of each clause vertex (the variable vertices are naturally symmetric). This is done globally for all clauses by introducing a special new vertex o adjacent to some four neighbouring vertices of the layer, and connected by sufficiently long paths to all vertices representing clauses, see Figure 5. This trick also solves the problem of clauses containing only 2 variables, such clauses are connected to o by two paths.

All these paths are then concentrated to the vertex o using a binary tree. The side of a touching representation of $FRAME_3$, which the ball representing o lies in, then becomes the false side for clauses (since every clause has one or two of the 3 touching positions on that side occupied by a path to o).

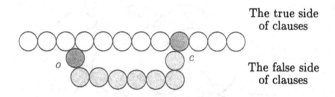

The true side
of clauses

The false side
of clauses

Fig. 5. The common "obstacle" for clauses on the false side

In general, we first position the variable and clause vertices within the internal layer of $FRAME_3$. Then we determine the length of the paths connecting clauses and variables (or the special vertex o) so that they can be realized without crossing in a possible touching representation. It is clearly enough to use a size of the occupied area and of paths quadratic in the input formula size. Then the whole graph $FRAME_3$ is made so large that the paths cannot go behind it. Of course, the paths also cannot go through the internal layer. Thus they can correctly encode logical values as needed in the reduction.

Suppose now that there is a touching representation of the above constructed graph. Then, by Lemma 3.2, each clause vertex must have at least one of the paths connecting it to its variables coming from the true side of the frame layer (i.e. from the side opposite to that containing a ball representing vertex o). The position of the v or $\neg v$ ball representing a variable v on the true or false side then determines the logical value of v, and in that evaluation each clause is true, so φ is satisfied.

Conversely, having a satisfying evaluation for φ, we can construct a touching representation of our graph, using the above-presented ideas in the opposite direction. □

4   Unit-ball touching graphs in E^4

In this section, we briefly show how the ideas of the proof of Theorem 1 can be adapted to 4–dimensional space. However, this is still only a special construction—we cannot proceed into higher dimensions with the same arguments, since the 4–dimensional cube already has a too long diagonal (and we may "insert" another unit ball into it).

Theorem 2 *The unit-ball touching graphs in 4–dimensional space are NP-hard to recognize.*

Sketch of proof. The proof proceeds by the same steps as the previous one, and we only sketch it.

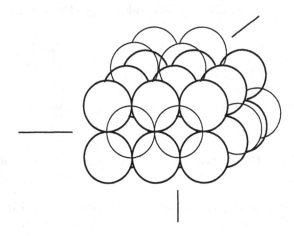

Fig. 6. A scheme of the internal layer of $FRAME_4$

First we construct a (sufficiently large) graph $FRAME_4$, shown in Figures 7,6, with a unique touching representation. This is the most important step of the reduction. The $FRAME_4$ graph is formed from a large "cube" grid (in the sense of edge structure of 3-cube) of vertices. An example of one such cube is shown in Figure 7, vertices denoted by $c_1, c_2, \ldots, c_8$. Each cube on the boundary is also adjacent to two neighbouring vertices, in the example they are denoted by f_1, f_2. We add more some vertices to the cube faces on the boundary, such as the vertices d_1, d_2, d_3 in the example (but all three of them are used only in corners...).

It is easy to represent this structure using the following coordinates for ball centers: $c_1 = [0, 0, 0, 0]$, $c_2 = [1, 0, 0, 0]$, $c_3 = [1, 1, 0, 0]$, $c_4 = [0, 1, 0, 0]$, $c_5 = [0, 0, 1, 0]$, $c_6 = [1, 0, 1, 0]$, $c_7 = [1, 1, 1, 0]$, $c_8 = [0, 1, 1, 0]$, $f_1 = [\frac{1}{2}, \frac{1}{2}, \frac{1}{2}, \frac{1}{2}]$, $f_2 = [\frac{1}{2}, \frac{1}{2}, \frac{1}{2}, -\frac{1}{2}]$, $d_1 = [\frac{1}{2}, -\frac{1}{2}\sqrt{2}, \frac{1}{2}, 0]$, $d_2 = [1 + \frac{1}{2}\sqrt{2}, \frac{1}{2}, \frac{1}{2}, 0]$, $d_3 = [\frac{1}{2}, \frac{1}{2}, 1 + \frac{1}{2}\sqrt{2}, 0]$, and other balls by translating this scheme.

To see that $FRAME_4$ has unique touching representation, realize that all centres of $c_1, \ldots, c_8$ must lie in an intersection of spheres of diameter 2 centred in f_1 and f_2, that is on a $3D$ sphere. With the help of the additional vertices d_1, d_2, d_3, the cube shape of $c_1, \ldots, c_8$ is forced. The same arguments may be ap-

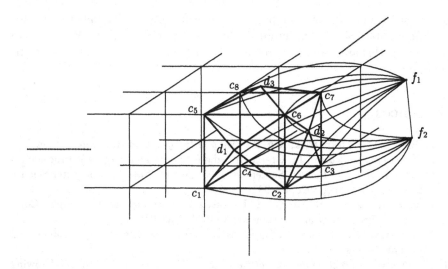

Fig. 7. A touching graph of $FRAME_4$ at the corner

plied to other boundary cubes, going from corners. Finally, the internal vertices are then tightly stretched inside the firm border.

The variables and clauses are represented within $FRAME_4$ in a similar way as previously, again connected together by paths. A problem is that we do not know exactly how many balls can touch one clause ball from one side of the frame layer, let this be a constant $t \geq 3$. Then we must connect the special vertex o (see the previous proof) by $t-2$ or $t-1$ paths to each vertex representing a clause.

The rest is very similar and we skip additional technical details here. □

5 Further work

If we want to proceed into higher dimensions, we get more and more difficult views of the situation. It is also related to the sphere-packing and kissing-number problems, that are very difficult in high dimensions and only little is known about them (except very special dimensions 8 and 24). We still do not have any idea how to construct a general reduction for unit-ball touching graphs, working in all dimensions. But since there is no obvious reason why the recognition problem should be easier in higher dimensions, we conjecture:

Conjecture. *The recognition of ball touching graphs and of unit-ball touching graphs is NP-hard in any fixed dimension.*

Another interesting question, which is not discussed in this work at all, is about the minimal size of the touching representation of a ball touching graph (related to the question whether their recognition belongs to *NP*). It seems that not all touching graphs have a representation using only small number of bits for coordinates, since we can easily force strange irrational distances. But we have no results in this area, and the problem may belong to *NP* using quite different certificate.

References

1. H. Breu, D.G. Kirkpatrick, *On the complexity of recognizing intersection and touching graphs of discs*, In: Graph Drawing (F.J. Brandenburg ed.), Proceedings Graph Drawing '95, Passau, September 1995, Lecture Notes in Computer Science 1027, Springer Verlag 1996, 88–98.
2. H. de Fraysseix, P.O. de Mendez, P. Rosenstiehl, *On triangle contact graphs*, Combinatorics, Probability and Computing 3 (1994), 233–246.
3. M.R. Garey, D.S. Johnson, Computers and Intractability, W.H. Freeman and Company, New York 1978.
4. P. Hliněný, *Contact graphs of curves (extended abstract)*, In: Graph Drawing (F.J. Brandenburg ed.), Proceedings Graph Drawing '95, Passau, September 1995; Lecture Notes in Computer Science 1027, Springer Verlag 1996, 312–323.
5. P. Koebe, *Kontaktprobleme der konformen Abbildung*, Berichte über die Verhandlungen der Sächsischen, Akad. d. Wiss., Math.–Physische Klasse 88 (1936), 141–164.
6. J. Kratochvíl, *Intersection graphs of noncrossing arc-connected sets in the plane*, Proceedings Graph Drawing '96, Berkeley, September 1996; Lecture Notes in Computer Science 1190, Springer Verlag, Berlin Heidelberg 1997.
7. C.B. Lekkerkerker, J.C. Boland, *Representation of finite graphs by a set of intervals on the real line*, Fund. Math. 51 (1962), 45–64.

Discrete Realizations of Contact and Intersection Graphs (Extended Abstract)

Jurek Czyzowicz[1][4] (czyzowicz@uqah.uquebec.ca)
Evangelos Kranakis[2][4] (kranakis@scs.carleton.ca)
Danny Krizanc[2][4] (krizanc@scs.carleton.ca)
Jorge Urrutia[3][4] (jorge@csi.uottawa.ca)

[1] Département d'Informatique, Université du Québec à Hull, Hull, Québec J8X 3X7, Canada.
[2] Carleton University, School of Computer Science, Ottawa, ON, K1S 5B6, Canada.
[3] University of Ottawa, Department of Computer Science, Ottawa, ON, K1N 9B4, Canada.
[4] Research supported in part by NSERC (National Science and Engineering Research Council of Canada) grant.

Abstract. Known realizations of geometric representations of graphs, like contact, intersection, etc., are "continuous", in the sense that the geometric objects are drawn in Euclidean space with real numbers as coordinates. In this paper, we initiate the study of dicrete versions of contact and intersection graphs and examine their relation to their continuous counterparts. The classes of graphs arising appear to have interesting properties and are thus interesting in their own right. We also study realizability, characterizations as well as intractability questions for the resulting new classes of graphs.
1980 Mathematics Subject Classification: 68R10, 68U05
CR Categories: F.2.2
Key Words and Phrases: Coin, Contact, Intersection, Interval graphs, Discrete, Planar graphs, NP.

1 Introduction

Classes of graphs having realizations either as intersection or contact graphs of given geometric objects have attracted the attention of several researchers in the literature, e.g. see the extensive literature cited in [9, 12]. Problems raised include realization, characterization, as well as intractability. Graphs thus realized have interesting applications ranging from scheduling theory [9] to motion planning [1].

Known studies of such graph-classes have only been concerned with "continuous" realizations of the concepts of "intersection" and "contact", in which the contact or intersection points have real numbers as coordinates. However, practical realizations of these concepts would seem to require the more natural concepts of discrete "intersection" and "contact" points, in the sense that the contact or intersection points have rationals or even "integers with a restricted range" as coordinates. By this we mean that the geometric objects have a predetermined set of points and only these points can serve as contact or intersection points among the objects.

A coin is a closed circular disc in the plane. Coin graphs are constructed from discs with disjoint interiors. Vertices of the graph are the coins. Two coins are adjacent if they touch each other. Such graphs are planar and have been considered previously in the literature [12]. A remarkable theorem attributed to P. Koebe by H. Sachs [12] states that every finite simple planar graph is a coin graph. The most recent proof of this theorem is given in [4]. It is intersesting to note that the discs required in the above representation of the planar graph are of arbitrary size. In addition, a related result in [7] shows that planar graphs are precisely the contact graphs of isosceles triangles.

Another class of graphs we will study are discrete intersection graphs. Among these are interval and unit square graphs which have been studied extensively in the literature [9]. Linear recognition algorithms for these graphs have appeared in [9, 5, 6], etc.

1.1 Preliminaries and notation

The class of coin graphs considered here are constructed from a single size coin.

Definition 1. The class of coin graphs, denoted by C, is defined as follows. Vertices are represented by unit discs such that the interiors of any pair of such discs have empty intersection. Two discs are adjacent if they touch.

Clearly the graphs in C are planar. The problem of recognizing whether a given graph is in the class C has been shown recently to be NP-complete [3].

Another class of graphs we will consider in this paper is the class of unit interval graphs (all intervals considered in the sequel are closed).

Definition 2. The class of unit interval graphs, denoted by I, is defined as follows. Vertices are represented by unit intervals. Two intervals are adjacent if they have a nonempty intersection.

It is known that unit interval graphs are precisely the interval graphs containing no induced copy of $K_{1,3}$ [9]. Thus the discrete unit interval graphs are a refinement of the unit interval graphs.

Notation.

From now on and for the rest of the paper if G denotes a class of graphs then G^n denotes the set of graphs in G with exactly n vertices. For example, we have the classes I^n, C^n, etc.

1.2 Results of the paper

In this paper we initiate the study of dicrete versions of the above mentioned classes of graphs. The classes of graphs arising appear to have interesting properties and are thus interesting in their own right. We also study realizability, characterizations as well as intractability questions for the resulting new graphs.

2   Discrete Coin Graphs

In the sequel we define new classes of planar coin graphs. For each integer $k \geq 3$ consider a regular k-gon P_k.

Definition 3. The class of polygonal coin graphs, denoted by $\mathcal{P}_k$ is defined as follows. Vertices are represented by isothets of the regular k-gon P_k such that the interiors of any pair of such k-gons have empty intersection. Two k-gons are adjacent if they touch at a polygon vertex.

Clearly the graphs in the class $\mathcal{P}_k$ are planar. We can prove the following theorem.

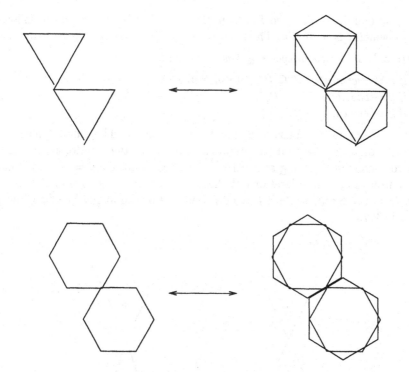

Fig. 1. The polygonal transformations required to prove that all polygonal coin graphs are coin graphs, for the case of equilateral triangles and hexagons.

Theorem 4. *All polygonal coin graphs are coin graphs, i.e. $\mathcal{P}_k \subseteq \mathcal{C}$, for all integers k.*

PROOF (OUTLINE) of Theorem 4. The theorem is easy for k even. Consider the layout of regular k-gons representing a given graph. Inscribe each polygon in a disc. Since k is even the resulting discs have mutually disjoint interiors. Moreover, the contact graph realized by these discs is identical to the contact graph realized by the k-gons. (See also Figure 1.) Therefore, we need only prove the assertion $\mathcal{P}_k \subseteq \mathcal{C}$ in the case k is odd.

In the sequel it will be convenient to use the notation $\mathcal{P}_k^{ve}$ as identical to the notation $\mathcal{P}_k$ in order to indicate that adjacency is determined by polygons touching on a plygon vertex. It will also be convenient to consider "adjacency" for polygonal layouts in which two k-gons are adjacent if they touch at a whole edge. More precisely we have.

Definition 5. The class $\mathcal{P}_k^{ed}$ is defined as follows. Vertices are represented by isothets of the regular k-gon P_k such that the interiors of any pair of such k-gons have empty intersection. Two k-polygons are adjacent if two polygon edges match.

Thus, by definition we have $\mathcal{P}_k^{ed} = \emptyset$, if k is odd. We can prove the following lemma which easily implies Theorem 4.

Lemma 6. *For each integer $k \geq 3$ we have that*

1. *if k is even then $\mathcal{P}_k^{ve} \subseteq \mathcal{P}_{kl}^{ve}$, for all integers l,*
2. *if k is odd then $\mathcal{P}_k^{ve} = \mathcal{P}_{2k}^{ed}$,*
3. *if k is even then $\mathcal{P}_k^{ve} = \mathcal{P}_k^{ed}$.*

PROOF of Lemma 6 Consider a side of the k-gon P_k. The side adjacent to it forms an angle $2\pi/k$ with it (see Figure 2). Let a, b be the exterior angles formed by two k-gons intersecting at a vertex. It is clear that $a + b = 4\pi/k$. If follows easily from angle considerations of the corresponding k-gons that if k is even then $a = b = 2\pi/k$, while if k is odd then one of the angles is π/k while the other is $3\pi/k$.

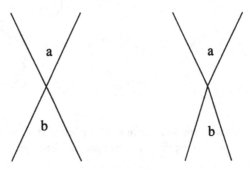

Fig. 2. The angles a, b of two touching polygons. In the picture to the left k is even, while in the picture to the right k is odd.

Now we can give the proof of the lemma. We transform the given polygon P_k into a new polygon Q_k. For given l, we inscribe P_k into a regular kl-gon Q_{kl}

such that the vertices of P_k are also vertices of Q_{kl}. Now suppose that P_k and P'_k are two touching k-gons. We consider separately the case where k is even and k is odd. It is easy to show that if k is even then the kl-gons Q_{kl} and Q'_{kl} touch at a vertex if and only if the k-gons P_k and P'_k touch at a vertex. This implies that if k is even then $\mathcal{P}^{ve}_k \subseteq \mathcal{P}^{ve}_{kl}$, for all integers l. If k is odd and the polygons P_k and P'_k touch at a vertex then the previous observations imply that one of the exterior angles at the contact point of the polygons is π/k while the other is $3\pi/k$. It follows that the corresponding polygons Q_{kl} and Q'_{kl} have nontrivial intersection. Moreover, this intersection is a polygon edge exactly when $l = 2$. It follows that if k is odd then $\mathcal{P}^{ve}_k \subseteq \mathcal{P}^{ed}_{2k}$.

To prove that $\mathcal{P}^{ed}_{2k} \subseteq \mathcal{P}^{ve}_k$ we essentially reverse the above argument. For a given $2k$-gon P_{2k} inscribe a k-gon as follows: form the k-gon Q_k with vertices the mid-points of the edges of P_{2k}. It follows that two $2k$-gons P_{2k} and P'_{2k} touch at an edge if and only if the k-gons Q_k and Q'_k touch at a vertex. Which completes the proof of the assertion.

The assertion $\mathcal{P}^{ve}_k = \mathcal{P}^{ed}_k$ for k even follows essentially the same ideas. One inscribes the given k-gon P_k into a k-gon Q_k such that the vertices of P_k are midpoints of edges of Q_k, and vice versa. Details of the proof are left to the reader.

This completes the proof of Lemma 6 and hence also the proof of Theorem 4. ∎

This raises the following interesting problem.

Problem 7. Is there a constant c such that $\mathcal{P}_k = \mathcal{C}$, for all $k \geq c$?

We show this is impossible by proving the following theorem.

Theorem 8. *For each k, there exist an n-vertex unit-disc contact graph with $n \in O(k)$ which is not representable by regular k-gons.*

PROOF (OUTLINE) We consider the contact graph depicted in Figure 3. It consists of the following components. An internal circle C, straight line double chains L around the internal circle and at equal angles, and an external circle. Both internal and external circles consist of unit circles. The external circle consists of circular chain segments S. These parts are connected in such a way that an imobile structure is formed. The corresponding parts can be constructed with unit discs thus forming a contact disc graph.

Let successive segments of type L be such that they are forming an angle of size $2\pi/n$. It is clear that we can achieve this by using a total of $O(n)$ unit discs in the graph. If $k \notin \Omega(n)$ then we note by using only regular k-gobs it will be impossible to form the graph depicted in Figure 3. ∎

2.1 Complexity of Recognition Problem

Theorem 9. *For each k, the problem of recognizing whether a given graph is in the class $\mathcal{P}_{4k}$ is NP-complete.*

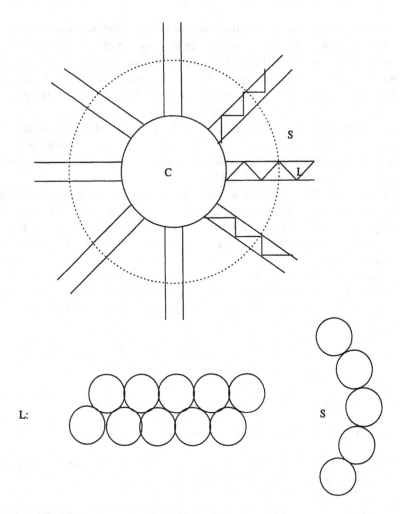

Fig. 3. Proving the lower bound in discretizing disc graphs. The graph is constructed from the components L, S and the circle C which is formed of a circular chain of unit discs.

PROOF (OUTLINE) We give only the argument for the class $\mathcal{P}_4$. Extensions to the classes $\mathcal{P}_{4k}$ are straightforward and will be given in the full paper.

We show how to reduce the problem of laying out a graph on a grid with dilation one to the above graph recognition problem. The former problem is known to be NP-complete [2, 11] even for the case when the graph is restricted to be a tree. For a given graph $G = (V, E)$ we define a new graph $G' = (V', E')$ such that

$$G' \in \mathcal{P}_4 \Leftrightarrow G \text{ is embedable on a grid with dilation 1.} \qquad (1)$$

It follows that the recognition problem for the class $\mathcal{P}_4$ is NP-complete. Since G is a tree if and only if G' is a tree it will also follow that the problem of recognizing whether a given tree is in the class $\mathcal{P}_4$ is NP-complete.

The graph G' is defined from G as follows. We replace each edge $\{u, v\}$ with the following caterpillar $C(u, v)$

- vertices: $u, v, u_1, u_2, u_3, u_4, u_5$ and x_1, x_2, y_1, y_2.
- edges:

$$\{u, u_1\}, \{u_1, u_2\}, \{u_2, u_3\}, \{u_3, u_4\}, \{u_4, u_5\}, \{u_5, v\}, \text{ and}$$

$$\{x_1, u_2\}, \{x_2, u_4\}, \{y_1, u_2\}, \{y_2, u_4\}.$$

The caterpillar itself is depicted in Figure 4. To prove the equivalence (1) observe

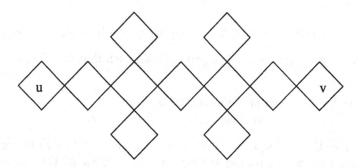

Fig. 4. The caterpillar $C(u, v)$ replacing the edge $\{u, v\}$.

that if the caterpillar $C(u, v)$ is represented in $\mathcal{P}_4$ then the centers of the squares representing the verices u, v must both lie either on a vertical or horizontal line. This completes the proof of the theorem. ∎

An interesting problem is the following.

Problem 10. For each k, is the problem of recognizing whether a given graph is in the class $\mathcal{P}_k$, NP-complete?

3   Discrete Intersection Graphs

Interval and unit square graphs have been extensively studied in the literature [9]. Linear recognition algorithms for these graphs have appeared in [9, 5, 6], etc. In the sequel we define new classes of unit interval graphs.

3.1 Discretizing intersection graphs

Intersection graphs of convex objects can be discretized by replacing intersection points with rational numbers which are sufficiently close to the original intersection points. To illustrate this point the following argument which is stated for the classes of intersection graphs of unit squares can be used for intersection graph representation by any type of convex objects.

Theorem 11. *Intersection graphs of unit squares can be realized as intersection graphs of unit squares all of whose corners have rational endpoints.*

PROOF We consider the Manhattan norm for points in the plane, $|(x,y)| = x+y$. Let $\{S_u : u \in V\}$ be a collection of closed unit squares representing the graph $G = (V, E)$, i.e. for all vertices u, v,

$$E(u,v) \Leftrightarrow S_u \cap S_v \neq \emptyset.$$

Let c_u be the center of the square I_u. Clearly, there exists a real $r > 0$ such that for all u, v

$$|c_u - c_v| < 1 \Rightarrow |c_u - c_v| < 1 - r, \text{ and } |c_u - c_v| > 1 \Rightarrow |c_u - c_v| > 1 + r.$$

Choose a sequence $\{r_u : u \in V\}$ of pairs of reals such that for all u, v,

- $|r_u| < r/2$,
- $c_u + r_u$ has both components rational, and
- if $c_u - c_v$ has both components integers then $r_u = r_v$.

Now consider the unit squares T_u with center $c_u + r_u$. It is easy to verify that this sequence of squares realizes the same graph G. Since all the corners of the squares are rationals it is starightforward that we can find an integer k such that $S^n = S^n_k$. ∎

One of the problems to be studied in this paper is the following.

Problem 12. What is the complexity of a discrete realization of a geometric graph when corresponding points are only allowed to have integer coordinates with "limited range"?

3.2 Discrete unit interval graphs

Definition 13. Let k be a given integer. To define the class of k-discrete unit interval graphs we consider intervals of identical size over the integers. A graph belongs to the class $\mathcal{I}_k$ if it can be represented by a collection of **pairwise distinct** intervals over the integers such that each interval has length exactly k and such that interval intersections can only occur at integer points.

Since all intervals are of equal length it is clear that $\mathcal{I}_k \subseteq \mathcal{I}$, for all k.

Theorem 14. *For all n there exists an integer $k \leq 2^{n-1}$ such that $\mathcal{I}^n = \mathcal{I}^n_k$.*

PROOF Let $\{I_u : u \in V\}$ be a collection of closed unit intervals representing the graph $G = (V, E)$, i.e. for all vertices u, v, $E(u, v) \Leftrightarrow I_u \cap I_v \neq \emptyset$.

We prove the existence of a unit interval realization of the graph such that no two endpoints of intervals of the new realization are identical.

Let $l(I), r(I)$ denote the left and right endpoints of the interval I. We prove the existence of a unit interval representation $\{J_u : u \in V\}$ of the graph in such a way that for all u, v,

- $l(I_u) < l(I_v) \Leftrightarrow l(J_u) < l(J_v)$,
- $|\{l(J_u), r(J_u) : u \in V\}| = 2n$,
- for any $x \neq y$ in the set $\{l(J_u), r(J_u) : u \in V\}$ we have that $|x - y| \geq 1$.

The proof is by induction on n. The inductive step $n = 1$ is trivial. Assume the induction hypothesis is true for interval graphs with $n - 1$ vertices. Remove the interval with the smallest left endpoint, say I_a. Consider all intervals $\{I_u : u \neq a\}$ intersecting I_a. Since all intervals have equal lengths, it is easy to show that the set $\{I_u : u \neq a, I_u \cap I_a \neq \emptyset\}$ forms a clique. By the induction hypothesis the set $\{J_u : u \neq a, I_u \cap I_a \neq \emptyset\}$ also forms a clique. In particular their left endpoints must lie within an interval of length 2^{n-2}. Since by the induction hypothesis no two endpoints can coincide, in follows that these left endpoints must lie within an interval of length $2^{n-2} - 1$. Let r be the rightmost left endpoint among the intervals $\{J_u : u \neq a, I_u \cap I_a \neq \emptyset\}$. Let J_a be the closed interval

$$J_a = \left[r - 2^{n-2} + \frac{1}{2}, r + \frac{1}{2} \right].$$

For each u define the intervals $\bar{J}_u = [2l(J_u), 2r(J_u)]$. It is clear that each interval has length 2^{n-1}. Moreover, it is easy to see that the resulting sequence $\{\bar{J}_u : u \in V\}$ of intervals represents the same graph G and satisfies all the conditions of the inductive steps. This proves the theorem. ∎

It is interesting to note that Theorem 14 improves on Theorem 3.2 of [5] which implies that for all n there exists an integer $k \leq n!$ such that $\mathcal{I}^n = \mathcal{I}^n_k$.

Theorem 15. *There exist graphs $G \in \mathcal{I}^n_{n/2-1} \setminus \mathcal{I}^n_{n/2-2}$, for $n \geq 2$.*

PROOF Let $n = 2m$. Consider the unit interval graph with vertex set $V = \{u_1, \ldots, u_m, v_1, \ldots, v_m\}$. Edges are defined as follows. Vertex u_i is adjacent exactly to the vertices

$$u_1, u_2, \ldots, u_m, v_1, v_2, \ldots, v_{i-1}$$

and vertex v_i is adjacent exactly to the vertices

$$v_1, v_2, \ldots, v_m, u_{i+1}, v_{i+2}, \ldots, u_m.$$

Since the degree sequence of the graph attains m different values it is easy to see that $G \notin \mathcal{I}^n_{n/2-2}$. The rest of the statement of the theorem is immediate. ∎

Definition 16. An interesting parameter for a (unit) interval graph G is $\mathcal{I}(G)$ which is the smallest integer m such that a discrete unit interval representation of the graph can be drawn with all integers of all interval endpoints in the range $[0, m]$.

Theorem 14 implies that for any unit interval graph G, $\mathcal{I}(G) \leq n2^{n-1}$. We can also prove the following lower bound.

Theorem 17. *There exist graphs $G \in \mathcal{I}^n$ such that $\mathcal{I}(G) \in \Omega(n \log n)$.*

PROOF (OUTLINE) Let $i(n)$ be the number of unit interval graphs on n vertices. Assume we could draw all unit interval graphs within the integer range $[1, m]$ with all (pairwise distinct) intervals of length exactly k. Since unit intervals are uniquely determined by a unique point (say, their cenrter) it is clear that there exist at most $\binom{m}{n}$ such graphs. Hence, $i(n) \leq \binom{m}{n}$. This implies that $n! \cdot i(n) \leq m^n$. Trivial calculations show that this implies that $m \in \Omega(n \log n)$. ■

An interesting problem is the following.

Problem 18. Give an algorithm for recognizing membership in the class $\mathcal{I}_k$. Is this an NP-complete problem? A related question is the following. For each graph $G \in \mathcal{I}^n$ compute $\text{disc}_{\mathcal{I}}(G) = \min_k \{G \in \mathcal{I}_k^n\}$. A similar question is the following. For each graph $G \in \mathcal{I}^n$ compute $\mathcal{I}(G)$. Can we tighten the bounds $cn \log n \leq \mathcal{I}(G) \leq n2^{n-1}$?

3.3 Discrete unit square intersection graphs

Definition 19. The class of unit square graphs, denoted by $\mathcal{S}$, is defined as follows. Vertices are represented by unit squares. Two vertices are adjacent if they have a nonempty intersection.

Definition 20. Let k be a given integer. To define the class of k-discrete unit square graphs we consider squares of identical side over the integer square lattice of integers. A graph belongs to the class $\mathcal{S}_k$ if it can be represented by a collection of squares over the square lattice of integers such that the side of each square is an interval of length exactly k.

Since all squares are of equal length it is clear that $\mathcal{S}_k \subseteq \mathcal{S}$, for all k. We can ask similar questions as with unit interval graphs.

Theorem 21. *For all n there exists an integer $k \leq 2^{n-1}$ such that $\mathcal{S}^n = \mathcal{S}_k^n$.*

PROOF (OUTLINE) Consider the two unit interval graphs formed by projecting the squares representing a given graph on the x and y axis. Then use the result of Theorem 14. ■

Problem 22. Give an algorithm for recognizing membership in the class $\mathcal{S}_k^n$. Is this an NP-complete problem? A related question is the following. For each graph $G \in \mathcal{S}^n$ compute $\text{disc}_{\mathcal{S}}(G) = \min_k \{G \in \mathcal{S}_k^n\}$.

We also state without proof the following two theorems.

Theorem 23. *For each graph $G \in S$ let G_x, G_y be the unit interval graphs obtained by projecting the squares representing G on the x- and y-axis. Then $\max\{\operatorname{disc}_I(G_x), \operatorname{disc}_I(G_y)\} \leq \operatorname{disc}_S(G) \leq \operatorname{disc}_I(G_x) \cdot \operatorname{disc}_I(G_y)$.* ∎

Theorem 24. *If $G_1, G_2 \in I^n$ such that $\operatorname{disc}_I(G_1) = \operatorname{disc}_I(G_2) = k$ then there exists a graph $G \in S^n$ such that $G_x = G_1, G_y = G_2$ and $\operatorname{disc}_S(G) = k$.* ∎

4 Conclusion

In this paper we considered discrete realizations of contact and intersection graphs. Contact graphs of regular polygons form a hierarchy of planar graphs and the recognition problem for such graphs is NP-hard. In contrast to this, intersection graphs are always descretizable and bounds on the complexity of the resulting discetizations are given.

Acknowledgements

Many thanks to Prosenjit Bose for useful converations that improved the presentation.

References

1. S. Abramowski, B. Lang, and H. Miller, "Moving Regular k-gons in Contact", In proceedings of International Workshop WG'88 on Graph-Theoretic Concepts in Computer Science, Jan van Leewen, editor, Springer Verlag Lecture Notes in Computer Science, vol. 344, pp. 229-242, 1989.
2. S. N. Bhatt and S. S. Cosmadakis, "The Complexity of Minimizing Wire Lengths in VLSI Layouts", Information Processing Letters 25 (1987) 263-267.
3. H. Breu and D. G. Kirkpatrick, "On the Complexity of Recognizing Intersection and Touching Graphs of Disks", In proceedings of Graph Drawing 95, pp. 88-98. Springer Verlag Lecture Notes in Computer Science.
4. G. R. Brightwell, and E. R. Scheinerman, "Representation of Planar Graphs", SIAM Journal on ???, Vol. 6, No. 2, pp. 214 - 229, May 1993.
5. D. G. Corneil, H. Kim, S. Natarajan, S. Olariu, and A. P. Sprague, "Simple Linear Time Recognition of Unit Interval Graphs", Information Processing Letters 55(1995) 99-104.
6. C. M. Herrera de Figueiredo, J. Meidanis, C. Picinin de Mello, "A Linear-Time Alforithm for Proper Interval Graph Recognition", Information Processing Letters 56 (1995) 179-184.
7. H. de Fraysseix, P. Ossona de Mendez, and P. Rosenstiehl, "On Triangle Contact Graphs", Combinatorics, Probability and Computing (1994) 3, 233 - 246.
8. D. R. Fulkerson and O. A. Gross, "Incidence Matrices and Interval Graphs", Pacific Journal of Mathematics, 15, 835-855, 1985.
9. M. C. Golumbic, "Algorithmic Graph Theory and Perfect Graphs", Academic Press, New York, 1980.

10. N. Korte and R. H. Möhring, "A Simple Linear-Time Algorithm to Recognize Interval Graphs", In proceedings of International Workshop WG'86 on Graph-Theoretic Concepts in Computer Science, G. Tinhofer and G. Schmidt, editors, Springer Verlag Lecture Notes in Computer Science, vol. 246, 1987.

11. Z. Miller and J. B. Orlin, "NP-Completeness for Minimizing Maximum Edge Length in Grid Embeddings", Journal of Algorithms 6, 10-16 (1985).

12. H. Sachs, "Coin Graphs, Polyhedra, and Conformal Mapping", Discrete Mathematics 134 (1994) 133-138.

Minimum-Area h-v Drawings of Complete Binary Trees

(Extended Abstract)

P. Crescenzi * and P. Penna

Dipartimento di Scienze dell'Informazione
Via Salaria 113, 00198 Roma
e-mail: {piluc, penna}@dsi.uniroma1.it

Abstract. We study the area requirement of h-v drawings of complete binary trees. An h-v drawing of a binary tree t is a drawing of t such that (a) nodes are points with integer coordinates, (b) each edge is either a rightward-horizontal or a downward-vertical straight-line segment from a node to one of its children, (c) edges do not intersect, and (d) if t_1 and t_2 are immediate subtrees of a node u, the enclosing rectangles of the drawings of t_1 and t_2 are disjoint. We prove that, for any complete binary tree t of height $h \geq 3$ and with n nodes, the area of the optimum h-v drawing of t is equal to (a) $2.5n - 4.5\sqrt{(n+1)/2} + 3.5$ if h is odd, (b) $2.5n - 3.25\sqrt{n+1} + 3.5$ otherwise. As far as we know, this is one of the few examples in which a closed formula for the minimum-area drawing of a graph has been explicitly found. Furthermore this minimum-area h-v drawing can be constructed in linear time. As a consequence of this result and the result of Trevisan (1996), we have that h-v drawings are *provably* less area-efficient than strictly upward drawings when we restrict ourselves to complete binary trees. We also give analogous results for the minimum-perimeter and the minimum-enclosing square area h-v drawings.

1 Introduction

Trees are one of the most common used structures in computer science and many techniques for the visualization of trees have been proposed. These techniques usually aim to find a layout satisfying specific aesthetic criteria. One of these criteria is the strictly upward grid straight-line (in short, *strictly upward*) requirement that imposes to map each node into a point with integer coordinates and each edge into a single straight-line segment, to place each node *below* its parent, and to not intersect two edges. This paper deals with a slight different criteria, that is, the h-v requirement.

---

* Starting from November 1, 1997, the author's new affiliation will be: Dipartimento di Sistemi ed Informatica, Università di Firenze, Via Cesare Lombroso 6/17, 50134 Firenze, Italy (e-mail: piluc@dsi2.dsi.unifi.it).

The definition of h-v drawing has been introduced in [9] and successively used in [2] as a tool to obtain strictly upward drawings for binary trees. Formally, an *h-v drawing* of a binary tree t is a drawing of t such that (see Fig. 1 and 6):

1. Nodes are points with integer coordinates.
2. Each edge is either a rightward-horizontal or a downward-vertical straight-line segment from a node to one of its children (that is, an h-v drawing is not strictly upward).
3. Edges do not intersect.
4. If t_1 and t_2 are immediate subtrees of a node u, the enclosing rectangles of the drawings of t_1 and t_2 are disjoint.

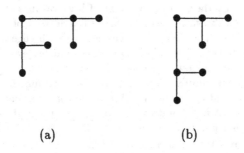

(a)   (b)

Fig. 1. Two h-v drawings of the complete binary tree of height 3.

The goal of this paper is to find minimum-area h-v drawings of complete binary trees. The width (respectively, height) of an h-v drawing is the width (respectively, height) of the smallest enclosing rectangle[2]. In the following we will denote the height and the width of any h-v drawing Δ with H_Δ and W_Δ, respectively. The *area* of Δ is then defined as $H_\Delta \cdot W_\Delta$.

1.1   Previous Results

In [9] a linear-time algorithm to construct an $O(n \log n)$-area h-v drawing of any binary tree with n nodes has been proposed. In [2] the authors show that such an algorithm is optimal since an infinite class of binary trees requiring $\Omega(n \log n)$ area exists. Moreover they proved that complete and Fibonacci binary trees admit a linear-area h-v drawing. This latter result has been first extended to AVL trees in [5] and subsequently to k-unbalanced, red-black and BB[α] trees in [3]. A polynomial-time (respectively, efficient parallel) algorithm to find the h-v drawing of any binary tree that minimize any " reasonable size function"

---

[2] We adopt the convention that both the width and the height are measured by the number of grid points

is given in [6] (respectively, [8]). In [6] two applications of this algorithm that use general binary trees (not the special case of complete binary trees) to visual interfaces are also presented.

The h-v drawing criterion is a restriction of the orthogonal straight-line one in which each edge of the tree is mapped to either a vertical (not necessarily downward) or an horizontal (not necessarily rightward) straight-line segment. Algorithms to obtain area-efficient orthogonal straight-line drawings of trees can be found in [1, 7, 10].

1.2 Our Results

The algorithm given in [2] for complete binary trees produces an h-v drawing whose area is at most $3n$. It is then natural to ask whether such an area is the minimum one. On the other hand, even if the algorithm in [6] shows that the minimum-area h-v drawing of a complete binary tree can be computed in polynomial time, no closed formula for this area function is known.

In this paper we both show that the algorithm in [2] is not the optimum one and *give a closed formula for the minimum-area requirement to h-v draw a complete binary tree*. In particular, for any such tree t of height $h \geq 3$ and with n nodes, the area of the optimum h-v drawing of t is equal to (a) $2.5n - 4.5\sqrt{(n+1)/2} + 3.5$ if h is odd, (b) $2.5n - 3.25\sqrt{n+1} + 3.5$ otherwise.

This result suggests an interesting comparison with the result of [11]. Indeed, in that paper it is shown that any complete binary tree admits an $(n+o(n))$-area strictly upward drawing. Our result thus shows that *h-v drawings are provably less area efficient than strictly upward drawings when we restrict ourselves to complete binary trees*. Indeed, it is not difficult to define other less "natural" infinite class of binary trees for which h-v drawings are less area efficient than strictly upward drawings. On the other hand, it is also possible to define an infinite class of binary trees for which strictly upward drawings are less area efficient than h-v drawings (the proof of these two last statements are here omitted: they will be included in the final version of the paper).

Finally, we also observe that analogous results hold for two other size functions defined in [6]: the perimeter and the enclosing-square area.

1.3 Preliminaries

In the following C_h denotes the complete binary tree of height h and n_h denotes the number of nodes of C_h.

For any $h > 1$, given two h-v drawings Δ_1 and Δ_2 of C_{h-1}, we denote by $\Delta_1 \ominus \Delta_2$ and $\Delta_1 \oslash \Delta_2$ the h-v drawings of C_h obtained by combining Δ_1 and Δ_2 as shown in Fig. 2(a) and 2(b), respectively. More precisely, the h-v drawing $\Delta_1 \ominus \Delta_2$ is obtained by translating to the right Δ_2 by as many grid points as W_{Δ_1} and by translating Δ_1 to the bottom by one grid point. The semantic of $\Delta_1 \oslash \Delta_2$ is defined similarly.

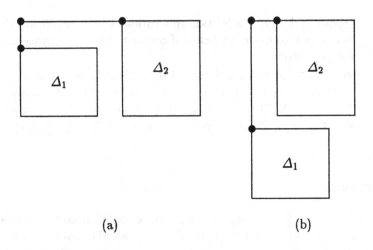

Fig. 2. The two h-v drawing operations.

2 The Algorithm

In this section we describe the algorithm to obtain an h-v drawing for a complete binary tree and we analyze its area requirement. In the next one we prove that this algorithm is optimal.

The algorithm constructs the h-v drawing following a bottom-up approach. More precisely, for each tree C_h, two different h-v drawings are produced: the optimum and the "useful" one. Intuitively, the basic property satisfied by these two drawings is that the width (respectively, height) of the useful drawing is equal to the width (respectively, height) of the optimum drawing minus (respectively, plus) one. On the ground of this property, the two drawings, denoted by O_h and U_h, respectively, are then combined in order to obtain O_{h+1} and U_{h+1}.

Let us first observe that, since we are dealing with complete binary trees, the following fact holds.

Lemma 1. *For any h and for any h-v drawing Δ of C_h, a "reverse" h-v drawing Δ^r of C_h exists such that $W_{\Delta^r} = H_\Delta$ and $H_{\Delta^r} = W_\Delta$.*

Proof. The proof is by induction on h. If $h = 1$ then we simply let Δ^r be equal to Δ. Otherwise, if $\Delta = \Delta_1 \ominus \Delta_2$ (respectively, $\Delta = \Delta_1 \oslash \Delta_2$) then we define $\Delta^r = \Delta_2^r \oslash \Delta_1^r$ (respectively, $\Delta^r = \Delta_2^r \ominus \Delta_1^r$). □

The algorithm to construct the two h-v drawings O_h and U_h for C_h (with $h \geq 3$) is described in Fig. 3 where Δ_3 and Δ_3^r denote the h-v drawings shown in Fig. 1(a) and 1(b), respectively.

Theorem 2. *For any $h \geq 3$, the drawing O_h of C_h produced by function MACBT with input h has area equal to*

```
function MACBT(h:integer): h-v drawings Oₕ and Uₕ of Cₕ;
begin
    if h = 3 then
        MACBT.Oₕ:=Δ₃;
        MACBT.Uₕ:=Δ₃ʳ;
    else
        MACBT.Oₕ:=(MACBT(h − 1).Uₕ₋₁)ʳ ⊖ (MACBT(h − 1).Oₕ₋₁)ʳ;
        MACBT.Uₕ:=(MACBT(h − 1).Oₕ₋₁)ʳ ⊖ (MACBT(h − 1).Oₕ₋₁)ʳ;
end;
```

Fig. 3. Algorithm to construct O_h and U_h.

$$\begin{cases} 2.5n_h - 4.5\sqrt{(n_h + 1)/2} + 3.5 \ \text{if } h \text{ is odd,} \\ 2.5n_h - 3.25\sqrt{n_h + 1} + 3.5 \quad \text{otherwise.} \end{cases}$$

Proof. The proof is by induction on h. For $h = 3$, the proof is straightforward. Let $h \geq 3$ and let us assume that the theorem is true for any height less than $h + 1$. Let us first observe that, if we denote with L_h and l_h the longer and the shorter side of O_h, respectively, then the longer and the shorter side of U_h is equal to $L_h - 1$ and $l_h + 1$, respectively (see Fig. 4).

Furthermore by the definition of function MACBT we obtain the following recurrence:

$$\begin{cases} L_h = 4 & \text{if } h = 3, \\ l_h = 3 & \text{if } h = 3, \\ L_h = 2l_{h-1} + 1 & \text{if } h > 3, \\ l_h = L_{h-1} & \text{if } h > 3. \end{cases}$$

The solution of this recurrence is:

$$L_h = \begin{cases} 2.5 \cdot 2^{\frac{h-1}{2}} - 1 \ \text{if } h \text{ id odd,} \\ 2^{\frac{h+2}{2}} - 1 \quad \text{if } h \text{ is even,} \end{cases}$$

and

$$l_h = \begin{cases} 2^{\frac{h+1}{2}} - 1 \quad \text{if } h \text{ id odd,} \\ 2.5 \cdot 2^{\frac{h-2}{2}} - 1 \ \text{if } h \text{ is even.} \end{cases}$$

It is then easy to see that the area of O_h, that is $L_h l_h$, is equal to

$$\begin{cases} 2.5n_h - 4.5\sqrt{(n_h + 1)/2} + 3.5 \ \text{if } h \text{ is odd,} \\ 2.5n_h - 3.25\sqrt{n_h + 1} + 3.5 \quad \text{otherwise.} \end{cases}$$

$\square$

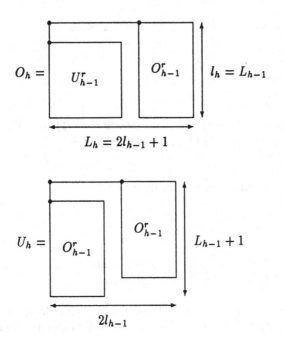

Fig. 4. The composition of O_{h-1} and U_{h-1} to obtain O_h and U_h.

3 The Proof of Optimality

In this section we prove that the h-v drawing O_h is the minimum-area one. To this aim we first need the following definition.

Definition 3 [6]. An h-v drawing Δ for C_h is an *atom* if, for any other h-v drawing Δ' of C_h, either $W_{\Delta'} > W_\Delta$ or $H_{\Delta'} > H_\Delta$.

It is easy to see that the minimum-area h-v drawing of C_h can be obtained only by combining atoms. We shall therefore suppose in the following that all the drawings are atoms.

Lemma 4. Let Δ be an h-v drawing for C_h such that $W_\Delta \leq H_\Delta$. Then an h-v drawing Δ' for the same tree exists such that

1. $W_{\Delta'} \leq W_\Delta$.
2. $\Delta' = \Delta'_1 \oslash \Delta'_2$, for some Δ'_1 and Δ'_2.
3. The area of Δ' is less than or equal to the area of Δ.

Proof. The proof is by induction on h. For $h = 1$ the theorem is obviously true. Let $\Delta = \Delta_1 \ominus \Delta_2$ (otherwise we can simply define $\Delta' = \Delta$). We distinguish the following two cases.

1. $\Delta_1 = \Delta_2$. From the inductive hypothesis we can assume that $\Delta_1 = \Delta_{1,1} \oslash \Delta_{1,2}$, for some $\Delta_{1,1}$ and $\Delta_{1,2}$ (see Fig. 5(a)).
   Let Δ' be the h-v drawing defined as (see Fig. 5(b))

$$\Delta' = (\Delta_{1,1} \ominus \Delta_{1,1}) \oslash (\Delta_{1,1} \ominus \Delta_{1,2}).$$

If $\Delta_{1,1} = \Delta_{1,2}$, then

$$H_{\Delta'} = H_{\Delta_{1,1}} + 1 + H_{\Delta_{1,1}} + 1 = H_{\Delta_{1,1}} + 1 + H_{\Delta_{1,2}} + 1 = H_\Delta + 1$$

and

$$W_{\Delta'} = 2W_{\Delta_{1,1}} + 1 = 2W_{\Delta_{1,2}} + 2 - 1 = W_\Delta - 1.$$

Hence the area of Δ' is equal to

$$(H_\Delta + 1)(W_\Delta - 1) = H_\Delta W_\Delta + W_\Delta - H_\Delta - 1 < H_\Delta W_\Delta$$

where the last inequality is due to the fact that $W_\Delta \leq H_\Delta$.
Otherwise, if $\Delta_{1,1} \neq \Delta_{1,2}$, it is easy to see that $H_{\Delta_{1,1}} \leq H_{\Delta_{1,2}} - 1$. Indeed, since Δ_1 is an atom we have that $W_{\Delta_{1,1}} > W_{\Delta_{1,2}}$ (otherwise, $\Delta_{1,2} \oslash \Delta_{1,1}$ would include Δ_1). From the fact that $\Delta_{1,1}$ and $\Delta_{1,1}$ are atoms it follows that $H_{\Delta_{1,1}} \leq H_{\Delta_{1,2}} - 1$. We then have that

$$H'_\Delta = H_{\Delta_{1,1}} + 1 + H_{\Delta_{1,2}} = H_\Delta$$

and

$$W_{\Delta'} = 2W_{\Delta_{1,1}} = W_\Delta.$$

Thus the area of Δ' is equal to that of Δ.
2. $\Delta_1 \neq \Delta_2$. Observe that, since Δ is an atom we have that $H_{\Delta_2} > H_{\Delta_1}$ (otherwise, $\Delta_2 \ominus \Delta_1$ would include Δ). From the fact that Δ_1 and Δ_2 are atoms it follows that $W_{\Delta_2} < W_{\Delta_1}$. Let us then replace Δ with $\tilde{\Delta} = \Delta_2 \ominus \Delta_2$. The height of $\tilde{\Delta}$ is at most $H_\Delta + 1$ while its width is less than W_Δ. That is, the area of $\tilde{\Delta}$ is less than or equal to that of Δ. We can now deal with $\tilde{\Delta}$ as in the previous case.

In both the above cases we have shown that the area of Δ' is less than or equal to that of Δ and the lemma thus follows. □

Informally, the next lemma states that if we shorten the longer side of O_h by c units, then its shorter side increases by at least c units. Conversely, if we shorten its shorter side by c units, then the other side increases by at least $2c$ units. More formally the following fact holds.

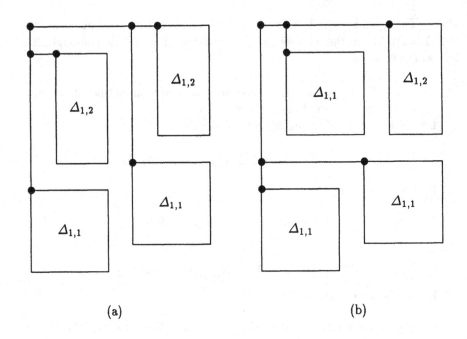

(a) (b)

Fig. 5. The transformation of Δ into Δ'.

Lemma 5. *For any h-v drawing Δ of C_h (with $h \geq 3$), whose longer and shorter side are L and l, respectively, either*

$$(L \leq L_h) \wedge (l \geq l_h + L_h - L)$$

or

$$(l \leq l_h) \wedge (L \geq L_h + 2(l_h - l)).$$

Proof. The proof is by induction on h. For $h \leq 3$ the proof is straightforward. Let us suppose that the lemma holds for any complete tree of height at most h with $h \geq 3$ and let Δ be an h-v drawing for C_{h+1}.

From Lemmas 1 and 4 we can assume, without loss of generality, that the longer side of Δ corresponds to its width and that $\Delta = \Delta_1 \ominus \Delta_2$, for some Δ_1 and Δ_2.

Since Δ is an atom, either $L = L_{h+1}$ and $l = l_{h+1}$ (in which case the lemma clearly follows) or one of the following two cases must hold.

1. $L < L_{h+1} = 2l_h + 1$. In this case we have that

$$W_{\Delta_1} \leq l_h.$$

Indeed, if $W_{\Delta_1} > l_h$ then $W_{\Delta_2} < l_h$ (since $W_{\Delta_1} + W_{\Delta_2} = L \leq 2l_h$). From the inductive hypothesis it follows that $H_{\Delta_2} \geq L_h + 2(l_h - W_{\Delta_2}) > L_h$. Replacing Δ_1 with O_h yields an h-v drawing for C_{h+1} which is included in

Δ thus contradicting the fact that Δ is an atom. Similarly, we can prove that

$$W_{\Delta_2} \leq l_h.$$

From the inductive hypothesis, it then follows that

$$H_{\Delta_1} \geq L_h + 2(l_h - W_{\Delta_1})$$

and

$$H_{\Delta_2} \geq L_h + 2(l_h - W_{\Delta_2}).$$

Hence

$$
\begin{aligned}
l &= \max(H_{\Delta_1} + 1, H_{\Delta_2}) \\
&\geq L_h + 2l_h + \max(1 - 2W_{\Delta_1}, -2W_{\Delta_2}) \\
&= 2l_h + L_h - \min(2W_{\Delta_1} - 1, 2W_{\Delta_2}) \\
&\geq 2l_h + L_h + 1 - L = L_{h+1} + l_{h+1} - L,
\end{aligned}
$$

where the last inequality follows from the fact that $L = W_{\Delta_1} + W_{\Delta_2} \geq \min(2W_{\Delta_1} - 1, 2W_{\Delta_2}) + 1$.

2. $l < l_{h+1}$. In this case, the shorter sides of both Δ_1 and Δ_2 are less than l_h. From the inductive hypothesis, it follows that

$$L^i \geq L_h + 2(l_h - l^i)$$

where L^i and l^i denote the longer and the shorter side of Δ_i, respectively, for $i = 1, 2$. Since $l^i < l_h$, the above inequality implies that

$$L^i + l^i > L_h + l_h.$$

Hence, we have that

$$W_{\Delta_1} > l_h + L_h - H_{\Delta_1}$$

and

$$W_{\Delta_2} > l_h + L_h - H_{\Delta_2}.$$

Since $H_{\Delta_1} + 1 \leq l$ and $H_{\Delta_2} \leq l$, we finally have that

$$L = W_{\Delta_1} + W_{\Delta_2} > 2l_h + 2L_h - 2l + 1 = L_{h+1} + 2(l_{h+1} - l).$$

The lemma thus follows. □

We are now in a position to prove the main result of this section.

Theorem 6. *For any $h \geq 3$, O_h is the minimum area h-v drawing for C_h.*

Proof. The proof is by induction on h. For $h = 3$ the proof is straightforward. Let $h \geq 3$ and let us assume that the theorem is true for any tree of height less than $h + 1$.

Let Δ be an h-v drawing for C_{h+1}. Let us denote with L and l the longer and the shorter side of Δ, respectively. Since Δ is an atom, either $L = L_{h+1}$ and $l = l_{h+1}$ (in which case the theorem clearly follows) or one of the following two cases must hold.

1. $l < l_{h+1}$.

   (a) $l \leq l_h$. From Lemmas 1 and 4, we can assume, without loss of generality, that $W_\Delta = L$ and $H_\Delta = l$ and that $\Delta = \Delta_1 \ominus \Delta_2$, for some Δ_1 and Δ_2. Because of the inductive hypothesis, we have that

   $$W_{\Delta_1} \geq \frac{L_h l_h}{l - 1} > L_h.$$

Thus

$$lL \geq H_{\Delta_1} W_{\Delta_1} + H_{\Delta_2} W_{\Delta_2} + W_{\Delta_1} \geq 2L_h l_h + L_h = L_{h+1} l_{h+1}.$$

   (b) $l > l_h$. Because of Lemma 5 we have that

   $$L \geq L_{h+1} + 2l_{h+1} - 2l.$$

Thus

$$lL \geq l(L_{h+1} + 2l_{h+1} - 2l) = L_{h+1} l_{h+1} + (l_{h+1} - l)(2l - L_{h+1}) > L_{h+1} l_{h+1},$$

   where the last inequality follows from the fact that $l \geq l_h + 1 > L_{h+1}/2$.

2. $L < L_{h+1}$. From Lemma 5 it follows that

   $$l \geq l_{h+1} + L_{h+1} - L > l_{h+1}.$$

Thus

$$lL \geq (l_{h+1} + L_{h+1} - L)L = l_{h+1} L_{h+1} + (L_{h+1} - L)(L - l_{h+1}) > l_{h+1} L_{h+1}.$$

Hence, we have proved that the area of Δ is at least equal to the area of O_{h+1}. The theorem thus follows. $\qquad\square$

In Fig. 6(a) and 6(b) the h-v drawing obtained by the algorithm proposed in [2] and the minimum-area h-v drawing, respectively, for C_5 are shown[3]. Observe that the area of the first h-v drawing is 70 while the area of the minimum one is 63.

---

[3] An "animation" of our algorithm with input C_5 is available from the WEB home page of the first author: http://www.dsi.uniroma1.it/~piluc/

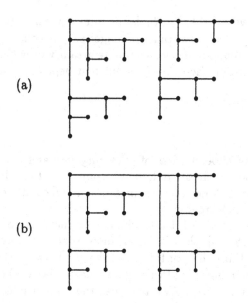

(a)

(b)

Fig. 6. Two h-v drawings for C_5.

4 Two More Size Functions

In this section we consider two other size functions among those defined in [6]. In particular, for any h-v drawing Δ, we define

1. $perimeter(\Delta) = H_\Delta + W_\Delta$.
2. $square(\Delta) = \max(H_\Delta, W_\Delta)$.

Let us first observe that from Lemma 5, we have that $perimeter(\Delta) \geq l_h + L_h$, for any drawing Δ of C_h. It thus follows that O_h also minimizes the $perimeter(\cdot)$ function.

Let us now consider the $square(\cdot)$ function. Again using Lemma 5 we have that, for any h-v drawing Δ of C_h, $square(\Delta) \geq \frac{l_h + L_h}{2}$. In order to find the h-v drawing for which $square(\cdot)$ is minimized, we define

$$S_h = \begin{cases} U_h & \text{if } h \leq 5, \\ (U_{h-2} \ominus U_{h-2}) \oslash (U_{h-2} \ominus U_{h-2}^r) & \text{if } h = 6, \\ (S_{h-2} \ominus S_{h-2}) \oslash (\bar{U}_{h-2} \ominus S_{h-2}) & \text{otherwise,} \end{cases}$$

where

$$\bar{U}_h = \begin{cases} O_h & \text{if } h \leq 5, \\ (O_{h-2} \ominus U_{h-2}) \oslash (U_{h-2} \ominus U_{h-2}) & \text{if } h = 6, \\ (\bar{U}_{h-2} \ominus S_{h-2}) \oslash (\bar{U}_{h-2} \ominus S_{h-2}) & \text{otherwise.} \end{cases}$$

It is easy to see that, for any h with $3 \leq h \leq 5$, S_h is the minimum enclosing square drawing. Moreover, for $h \geq 6$, we have that its longer side is equal to $\frac{l_h + L_h}{2}$, which implies that S_h is the minimum-enclosing square area h-v drawing. It is interesting to note that the constant factor of S_h is different depending on the fact that h is odd or even.

5 Open Questions

The first problem left open by this paper is that of obtaining similar results for other balanced trees (e.g. Fibonacci trees). Indeed, we conjecture that the algorithm to construct h-v drawings of Fibonacci trees presented in [2] computes the minimum-area h-v drawings of such trees.

Secondly, the upper bound known for AVL trees, whose class includes both complete and Fibonacci trees, is $18n$ [4], that is, much more than the lower bound we proved in this paper. We think that for this class, the gap between the upper and the lower bound could be reduced by focusing on a particular kind of AVL tree, requiring more area than any other AVL tree with the same height.

Finally, it may also be interesting to find a linear-time algorithm computing the minimum area h-v drawing of any binary tree. Observe that, for the complete and Fibonacci binary trees, this fact already holds.

References

1. T. Chan, M. T. Goodrich, S. R. Kosaraju, and R. Tamassia. Optimizing Area and Aspect Ratio in Straight-Line Orthogonal Tree Drawings. In *Proc. Graph Drawing 96*, LNCS 1190, 63-75, 1997.

2. P. Crescenzi, G. Di Battista, and A. Piperno. A note on optimal area algorithms for upward drawings of binary trees. *Computational Geometry: Theory and Applications*, 2:187-200, 1992.

3. P. Crescenzi and P. Penna. Upward Drawings of Search Trees. In *Proc. WG 96*, LNCS 1197, 114-125, 1997.

4. P. Crescenzi, P. Penna, and A. Piperno. Linear area upward drawings of AVL trees. *Computational Geometry: Theory and Applications*, to appear.

5. P. Crescenzi, and A. Piperno. Optimal-area upward drawings of AVL trees. In *Proc. Graph Drawing 94*, LNCS 894, 307-317, 1994.

6. P. Eades, T. Lin, and X. Lin. Minimum size h-v drawings. In *Advanced Visual Interfaces*, 386-394, World Scientific, 1992.

7. A. Garg, M. T. Goodrich, and R. Tamassia. Planar upward tree drawings with optimal area. *IJCGA*, 6:333-356, 1996.

8. P.T. Metaxas, G.E. Pantziou, and A. Symvonis. Parallel h-v drawings of Binary Trees. In *Proc ISAAC 94*, 487-496, 1994.

9. Y. Shiloach. Linear and planar arrangements of graphs. Ph.D. Thesis, Department of Applied Mathematics, Weizmann Institute of Science, Rehovot, Israel, 1976.

10. C.-S. Shin, S.K. Kim, and K.-Y. Chwa. Area-Efficient Algorithms for Upward Straight-Line Tree Drawings. In *Proc. COCOON 96*, LNCS 1090, 106-116, 1996.

11. L. Trevisan. A Note on Minimum-Area Upward Drawing of Complete and Fibonacci Trees. *Information Processing Letters*, 57:231-236, 1996.

Packing Trees into Planar Graphs

A. García[1], C. Hernando[2], F. Hurtado[3], M. Noy[3], J. Tejel[1]

[1] Dep. Métodos Estadísticos. Universidad de Zaragoza. Pl. San Francisco s/n. 50009 Zaragoza (Spain)
[2] Dep. Matemàtica Aplicada I. Universitat Politècnica de Catalunya. Diagonal 647. 08028 Barcelona (Spain)
[3] Dep. Matemàtica Aplicada II. Universitat Politècnica de Catalunya. Pau Gargallo 5. 08028 Barcelona (Spain)

Abstract. The main problem considered in this paper is the following: given two trees both with n vertices, whether it is possible to draw them on the plane with the same set of vertices without crossings and duplicated edges. We formulate this problem in terms of packing graphs and give a solution in several situations. We also solve some related problems on drawing trees and cycles.

1 Introduction

We say that the graphs $H_1, \ldots, H_k$ can be *packed* into a graph G if G contains subgraphs isomorphic to $H_1, \ldots, H_k$ and pairwise edge disjoint. The problem of packing graphs has been widely studied, especially when G is a complete graph. For instance, it is known that two trees with n vertices none of them equal to a star can be packed into K_n [2]. The packing of three trees presents more difficulties and has been completely solved recently [3]. A recent survey on packing graphs is [1].

We observe that when G is a planar graph, a packing into G is equivalent to a planar drawing of the graphs $H_1, \ldots, H_k$ without sharing edges. In this case we obtain a simultaneous drawing of all the graphs $H_1, \ldots, H_k$ on the same set of vertices. Coloring each of the graphs with a different color, one obtains a clear picture since there are no crossings and no multiple edges. This solution is a good option in some situations, since the information from a collection of k separate drawings can be difficult to integrate.

The main problem we discuss in this paper is the *tight planar packing* of two trees: given two trees T_1 and T_2 both with n vertices, none of them equal to a star, is it possible to pack them into some planar graph with n vertices? We conjecture that the answer is in the affirmative. We observe that although T_1 and T_2 can be packed into K_n as mentioned above, the union of the two graphs in K_n can certainly be non planar. We also observe that three trees with n vertices never admit a tight planar packing, since a planar graph has at most $3n - 6$ edges.

In this paper we solve the conjecture in several cases. When one of the two trees is a path we show that it can be packed with any other tree T different from a star. This provides a drawing of T together with a spanning path visiting all the vertices and not using edges of T. We also show that the problem can be solved when one of the two trees has maximum degree $n - 2$, and that any tree T different from a star can be packed together with a copy of T.

The technique used to prove these results is based on finding special drawings of trees on the set of vertices of a convex polygon. This is the content of Section 2. In Section 3 we present the results on the packing of two trees, and in Section 4 we study related problems, particularly planar packings of cycles.

2 Drawing trees on a convex polygon

Given any tree T of order n, it is clear that T can be drawn without crossings on a convex polygon with n sides and that, in general, it can be drawn in many different ways. A property of drawings that turns out to be significant is which edges are used from the boundary of the polygon. The next two lemmas and the theorem below deal precisely with this topic.

Lemma 1. *Let T be a tree with $n > 3$ vertices and let T' be the tree obtained from T by adding a new terminal node h connected to a node v of degree $r \geq 2$. If T can be drawn on a convex polygon using only m edges on the boundary , then T' also admits a drawing with only m edges on the boundary.*

Proof. Since $r \geq 2$, there exist two neighbors w and w' of v such that the counterclockwise path from w to w' along the boundary of the polygon does not contain neither v nor any neighbor of v. Then at least one boundary edge $v_i v_{i+1}$ between w and w' does not belong to T, since otherwise there would be a cycle. Moreover, it is easy to see that it can be chosen so that h can be inserted between v_i and v_{i+1} and connected to v without creating crossings.

Let T be a tree with $p > 1$ vertices and with m terminal nodes denoted $h_1, \ldots, h_m$. Let $\overline{T}$ be the tree obtained from T by adding a new terminal node $\overline{h}_i$ connected to h_i for every $i = 1, \ldots, m$. Clearly $\overline{T}$ has at least four vertices and as many terminal nodes as T. Then we have the following lemma.

Lemma 2. *In any drawing of $\overline{T}$ on a convex polygon, the m edges $h_i \overline{h}_i$ lie necessarily on non-consecutive boundary edges. Moreover, $\overline{T}$ can be drawn using exactly m boundary edges.*

Proof. If $h_i \overline{h}_i$ is drawn as an internal diagonal then the remaining vertices of the polygon are split into two regions that cannot be connected without crossing $h_i \overline{h}_i$, since the degree of h_i in $\overline{T}$ is equal to two. Hence $h_i \overline{h}_i$ is a boundary edge.

We prove now the second part of the statement, by induction on m, the number of terminal nodes. Let $v_1, \ldots, v_n$ be the vertices of the convex polygon, listed counterclockwise, where $n = |V(\overline{T})|$. If $m = 2$ then $\overline{T}$ is a path, which can be drawn in a zigzag way, $v_1 v_2 v_n v_3 v_{n-1} \ldots$, using exactly two boundary edges.

For $m > 2$, let us consider the path C starting at $\overline{h}_m$ an ending in the first encountered node w with degree greater than 2. Let $w_{k+1} = \overline{h}_m, w_k =$

$h_m, \ldots, w_2, w_1 = w$ be the nodes of C. By removing $w_2, \ldots, w_{k+1}$ from $\overline{T}$ we obtain a tree T' with $m-1$ terminal nodes, $\bar{h}_1, \ldots, \bar{h}_{m-1}$, among a total of $n-k$ nodes. T' can be drawn on the convex polygon $v_1 v_2 \ldots v_{n-k}$, by induction, with only $h_1 \bar{h}_1, \ldots, h_{m-1} \bar{h}_{m-1}$ becoming boundary edges. We can assume without loss of generality that w becomes v_{n-k} in that drawing. Now we can add the path C in a zigzag way as $v_{n-k} v_n v_{n-k+1} v_{n-1} \ldots$, and the only new boundary edge will correspond precisely to $\bar{h}_m h_m$.

The two lemmas above imply clearly the following result.

Theorem 3. *Any tree T different from a star can be drawn on a convex polygon without using consecutive edges of the boundary. Moreover, the minimum number of boundary edges in any drawing of T is equal to the number of terminal nodes of the tree obtained after removing all the terminal nodes of T.*

3 Drawing two trees

We say that the graphs $H_1, \ldots, H_k$ can be *packed* into a graph G if G contains subgraphs isomorphic to $H_1, \ldots, H_k$ and pairwise edge disjoint.

Observe that packing a family of planar graphs $H_1, \ldots, H_k$ into a planar graph G amounts to drawing simultaneously the graphs in the plane without any crossings and without multiple edges. Of course this is trivial if one does not bound the number of vertices of G, but it is a challenging problem if $|V(G)| = \max |V(H_i)|$. If this condition is fulfilled, we say that $H_1, \ldots, H_k$ admit a *tight planar packing*. The following lemma justifies the study of drawings on a convex polygon presented in the preceding section.

Lemma 4. *Suppose two graphs H_1 and H_2 both with n vertices can be drawn on a convex n-gon without sharing any edge. Then they admit a tight planar packing.*

Proof. Without loss of generality we can assume that the vertices of the convex polygon lie on the equator of a sphere. Then we can draw the edges of H_1 in one hemisphere and the edges of H_2 in the opposite hemisphere. Since the two drawings were planar and did not share edges, we obtain a simple planar graph drawn on the sphere.

Our first application is to show that one can always draw a given combinatorial tree in the plane together with a spanning path without crossings or multiple edges.

Theorem 5. *Any tree T with n vertices, different from a star, admits a tight planar packing with a path P_n with n vertices.*

Proof. Let $\widehat{T}$ the tree obtained from T by removing all the terminal nodes. Let us consider first the case in which $\widehat{T}$ is not a single edge joining two vertices, possibility only arising when T is the so called *double-star*.

We know from Section 2 that T can be drawn on a convex polygon $v_1, \ldots, v_n$, described counterclockwise, using only sides of the polygon corresponding to the edges $h_i \bar{h}_i$, $i = 1, \ldots m$, where $h_1, \ldots, h_m$ are the terminal nodes of $\widehat{T}$ and $\bar{h}_1, \ldots, \bar{h}_m$ are terminal nodes of T respectively connected to $h_1, \ldots, h_m$. Observe that this drawing does not contain edges between $\bar{h}_i$ and h_j or $\bar{h}_j$, for $j \neq i$, because $\bar{h}_i$ is a terminal node of T. Also, there is no edge between h_i and h_j as they are terminal nodes of $\widehat{T}$, only adjacent if T were a double-star.

We draw now P_n using all the boundary edges different from the $h_i \bar{h}_i$, whose extremes are connected in a zigzag way by diagonals not shared with T, as proved in the preceding paragraph (see Figure 1).

If T is a double star, having two adjacent vertices with degrees $n - k$ and k, we draw T on the convex polygon using the edges $v_1 v_2, v_1 v_3, \ldots, v_1 v_{k+1}$ and $v_n v_{k+1}, v_{n-1} v_{k+1}, \ldots, v_{k+3} v_{k+1}, v_{k+2} v_{k+1}$. Now we can draw P_n as $v_{k+1} v_k \ldots v_3 v_2 v_{k+2}\ v_{k+3} \ldots v_n v_1$.

As a second application of our technique we show how to draw simultaneously two copies of the same tree.

Theorem 6. *Let T be any tree with n vertices different from a star. Then T and a copy of T admit a tight planar packing.*

Proof. Let P be a convex polygon with vertices $v_1, \ldots, v_n$, numbered as they appear counterclockwise on its boundary. According to Theorem 3, T can be drawn on P without using consecutive edges of the boundary. A copy of this drawing of T is now obtained by rotating the first drawing one unity counterclockwise. By construction no two edges on the boundary are shared by these two drawings. On the other hand, no edge $v_i v_j$ with $j \neq i + 1$ can be shared, as this would mean that both edges $v_i v_j$ and $v_{i-1} v_{j-1}$ were present in the first drawing, which is impossible as these two diagonals cross each other.

To conclude this section we mention that a tree with n vertices and maximum degree $n - 2$ admits a tight planar packing with any tree with n vertices different from a star. We omit the proof since it does not introduce new techniques.

4   Related results

In the study of planar packings it makes sense to consider graphs other than trees or the packing of more than two graphs. In this section we present some results about packing cycles and about packing three paths of the same length. We start with the following question: when two cycles of length n admit a tight planar packing, i.e. a packing into a planar graph with n vertices.

Theorem 7. *Two cycles of length n admit a tight planar packing if, and only if, $n = 6$ or $n > 7$.*

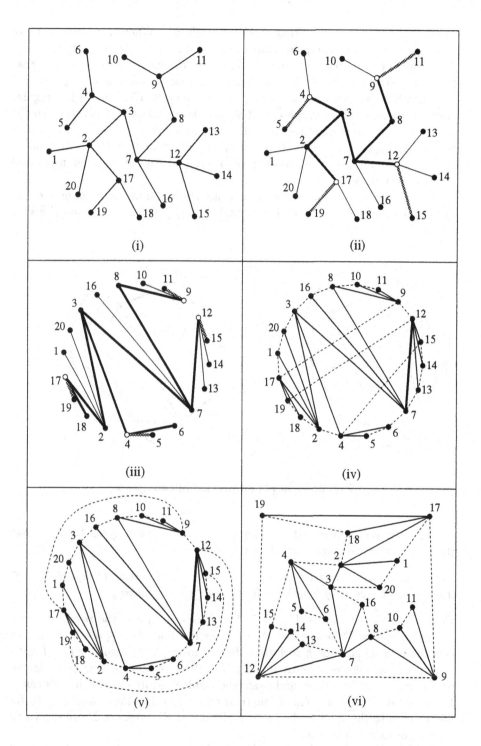

Fig. 1. Drawing a tree on a convex polygon (i,ii,iii) and a planar packing of the tree with a path (iv,v,vi).

Proof. If $n = 3, 4$ it is clear that there is no solution. When $n = 5$ the union of two disjoint C_5 is the complete graph K_5, which of course is not planar.

To show that there is no solution either in the case $n = 7$, we recall that a graph contractible to K_5 is non planar. Suppose G is a planar graph with 7 vertices whose edges are the disjoint union of two cycles C_7. Then G is 4-regular and the complementary graph $\overline{G}$ is 2-regular, hence equal to $C_3 \cup C_4$ or to C_7. If $\overline{G} = C_3 \cup C_4$ then clearly G contains $K_{3,3}$ as a subgraph and is not planar. If $\overline{G} = C_7 = v_1 v_2 \cdots v_7$, then G contains a graph contractible to K_5 that has as vertices v_1, v_2, v_4, v_6 and the contraction of $v_3 v_7$, where the path joining v_1 and v_2 goes through v_5.

Finally, an explicit construction provides a solution in the remaining cases. The constructions for $n \geq 6$ and even, and for $n \geq 9$ and odd are exemplified in Figure 2.

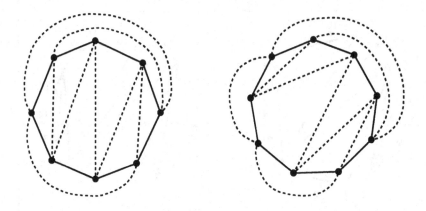

Fig. 2. Tight planar packing of two cycles.

Our next result deals with the packing of a tree and a cycle. Observe that the condition on the maximum degree is necessary.

Theorem 8. *Given a tree T with $n \geq 5$ vertices and maximum degree $\Delta(T) \leq n - 3$, T and a cycle of length n admit a tight planar packing.*

Proof. By induction on n. If $n = 5$ then T is a path and the result is easily checked. Let now T' be a tree with $n + 1$ vertices and $\delta(T') \leq n - 2$. T' always contains a terminal node h such that after removing h the resulting tree T with n vertices has degree at most $n - 3$ (the only exception to this claim is the double-star with 6 vertices, and again the result is easily checked in this case). By induction T and a cycle of length n admit a tight planar packing. To fix ideas, suppose the cycle is drawn as a circle and the edges of T are drawn as internal or external chords.

Let v be the vertex adjacent to h in T', let u and w be the two neighbors of v in the cycle, and let L be the arc of circle between u and w not containing v.

If v can be connected by an internal chord to a point x in L that is not a vertex of T without creating crossings, then we can put h in x and draw the edge vh to get a packing of T' and a cycle of length $n+1$. Otherwise there has to be a path in T from u to w through the interior of the circle (this situation is illustrated in Figure 3.) Now we can argue in the same way trying to connect v to a point x' in L by an external chord. If this were not possible there would be also a path from u to w through the exterior of the circle, and T would contain a cycle. This completes the proof.

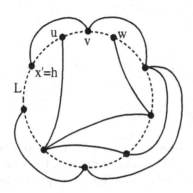

Fig. 3. Packing a tree and a cycle.

We observe that the above result together with the remark at the end of Section 3 actually provide an alternative proof to Theorem 5. Nevertheless the proof there was constructive instead of the inductive technique used here.

Our last results deal with the packing of three graphs. The proof of the first two is very similar to that of Theorem 7, and is omitted.

Theorem 9. *Three cycles of length $n-2$ can be packed into a planar graph with n vertices if, and only if, $n=6$ or $n>7$.*

Theorem 10. *Three cycles of lengths $n-3, n-2$ and $n-1$ can be packed into a planar graph with n vertices if, and only if, $n=6$ or $n>7$.*

An explicit construction for the case $n=9$ is exemplified in Figure 4. Observe that the union of the three cycles shown in the picture gives rise to a maximal planar graph (a triangulation), showing that the result is best possible. This is also the case in our final result, whose proof is omitted.

Theorem 11. *For all $n \geq 3$, there is a packing of three paths of length $n-2$ into a planar graph with n vertices.*

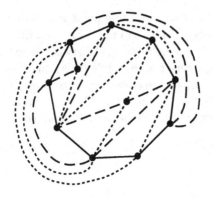

Fig. 4. Packing three cycles.

5   Final remarks

In this paper we have presented several results about planar packings of trees and cycles. The main tool is based on drawing trees on a convex polygon with specific properties. We believe this technique might be used to settle other cases, if not completely, of our conjecture:

Conjecture: Any two trees with $n \geq 4$ vertices different from a star admit a tight planar packing.

A case study shows that the conjecture is true for $n \leq 7$.

References

1. B. Bollobás, Extremal Graph Theory, in *Handbook of Combinatorics, vol. II*, R.L. Graham, M. Grötschel and L. Lovász (eds.), pp. 1231-1292, North-Holland (1995).
2. S.M. Hedetniemi, S.T. Hedetniemi and P.J. Slater, A note on packing two trees into K_n, *Ars Combinatoria* 11 (1981), 149-153.
3. M. Maheo, J.F. Saclé y M. Woźniak, Edge-disjoint placement of three trees, *European J. of Combinatorics* 17 (1996), 543-563.

The Three-Phase Method: A Unified Approach to Orthogonal Graph Drawing *

Therese C. Biedl[1] and Brendan P. Madden[2] and Ioannis G. Tollis[3]

[1] McGill University, 3480 University St. #318, Montreal, Quebec H3A2A7,
therese@cgm.cs.mcgill.ca
[2] Tom Sawyer Software, 804 Hearst Avenue, Berkeley, CA 94710,
bmadden@tomsawyer.com
[3] Dept. Computer Science, Univ. of Texas at Dallas, Richardson, TX 75083,
tollis@utdallas.edu

Abstract. In this paper, we study orthogonal graph drawings from a practical point of view. Most previously existing algorithms restricted the attention to graphs of maximum degree four. Here we study orthogonal drawing algorithms that work for any input graph, and discuss different models for such drawings. Then we introduce the three-phase method, a generic technique to create high-degree orthogonal drawings. This approach simplifies the description and implementation of orthogonal graph drawing, and can be applied to global as well as interactive and incremental settings.

1 Background

In recent years, graph drawing has created intense interest due to its numerous applications. In networking and database applications, graph drawings serve as a tool to help display large diagrams efficiently. Different drawing styles have been investigated (see [4, 5, 12, 21] for overviews). One drawing technique, called orthogonal drawing, routes edges along the rows and columns of an underlying rectangular grid. Specific uses of orthogonal graph drawings include Entity Relationship (ER) Diagrams and large industrial schematics. The goal is to obtain aesthetically pleasing drawings. Common objectives include small area, few bends, and few crossings.

A drawing cannot be understood clearly if two edges overlap. Therefore, a feasible orthogonal drawing, with nodes drawn as points, is possible only if the maximum degree of the graph is at most four. Many orthogonal drawing heuristics have been developed for such graphs, see for example [2, 16, 18].

---

* This work was performed while the first author was working at, and the third author was consulting with Tom Sawyer Software. It was, in part, funded by the NIST under grant number 70NANB5H1162. This paper is part of a series about the orthogonal library of the Graph Layout Toolkit produced by Tom Sawyer Software. Patent on these and related results is pending.
These results are part of a Ph.D thesis of the first author at Rutgers University under the supervision of Prof. Endre Boros.

If the degree of a node v is larger than four, then more than one grid-point must be assigned to v. To maintain the semblance to a point, one uses a box which should not be too large. We discuss different models of high-degree drawings in Section 2.

There are different approaches to orthogonal graph drawing with high degree. One approach is to modify the input graph until it has maximum degree 4, by splitting nodes into chains or cycles of nodes. Then an algorithm for 4-graphs is applied, possibly with some modification to ensure that split nodes are drawn as boxes. This approach was taken in the GIOTTO-system [19] based on the algorithm by Tamassia [18], and in the algorithm by Biedl and Kant [2]. Its main disadvantage is that we have to decide on the ordering of the edges around the node to split it, which imposes unwanted structure. Additionally, the boxes are often rather large relative to the degree of a node.

Another approach is to directly assign boxes to nodes. Again there arises the difficulty to determine which edge should attach where along a side of the box. Also, to increase space for adding a box, previously existing boxes may be stretched, so control over the dimensions of the boxes is hard to achieve. Most algorithms for visibility representations, i.e., orthogonal drawings in which all edges are drawn as straight lines, work in this fashion, see for example [17, 20]. Not every graph has a visibility representation.

Only recently have there appeared algorithms for orthogonal drawings of any input-graph [7, 15]. These papers present algorithms where each edge has at most one bend. The first paper achieves a minimum number of bends (under the assumption that each edge may bend at most once) for planar graphs in a specialized model. The second paper presents an algorithm for general graphs with area guarantees of at most $m \times m/2$.

We present a third approach to create high-degree drawings, called the *three-phase method*. The main difference from the other approaches is that it first creates an infeasible drawing, called a *sketch*: we draw nodes as points and route edges with overlaps. Once all nodes and edges have been placed, we increase the nodes to boxes and determine the assignment of edges to particular port locations. This choice can now be made while avoiding crossings, since the routing of edges is known. We study this method in Section 3.

The three-phase method enables us to easily create an interactive framework, i.e., we allow the user to change the resulting drawing by adding or deleting nodes or edges, such that these changes result in only a small distortion of the picture. We study interactive drawings in Section 4, and end in Section 5 with conclusions and open problems.

We have implemented the three-phase method as part of the Graph Layout Toolkit, a family of general-purpose graph visualization libraries developed by Tom Sawyer Software. This toolkit is available with a license agreement for use in commercial applications, and may be obtained for a very small fee by academic institutions.[4]

---

[4] For information, see *http://www.tomsawyer.com* or contact *info@tomsawyer.com*.

2 Different models for high-degree drawings

In an orthogonal drawing, nodes should resemble points, hence it is undesirable that node dimensions are arbitrarily large. We present here three different models of orthogonal drawings, and discuss their advantages and disadvantages.

2.1 The Unlimited-growth model

In the first model, which we call the *unlimited-growth model*, no restrictions are placed on the dimensions of the boxes of the nodes. The GIOTTO system works in the unlimited-growth model [19], and so do all algorithms for visibility representations [17, 20]. The main advantage of the unlimited-growth model is that frequently we can stretch a node to cover the bend of an incident edge (see e.g. in [6]). In particular, planar graphs can be drawn without bends [17, 20]. The main disadvantage of this model is that nodes are not recognizable as points, and that we have no means of influencing the width and the height of the nodes.

2.2 The Kandinsky model

Fößmeier and Kaufmann introduced a model called the *Kandinsky model* [7]. In such a drawing, there are two different types of grid-lines. The grid-lines of a coarse grid are used to place the nodes. The grid-lines of a finer grid serve to allow edges to attach on each side of a node.

The Kandinsky model has many appealing features. It is possible to draw the nodes such that the boxes are aligned and have the same size, thus the concept of points has been generalized. The Kandinsky model is probably the best choice for drawings with uniform node dimensions. A disadvantage of the Kandinsky model is that it is somewhat wasteful in terms of bends and grid-size. There exists a graph that must be drawn with $m - 4$ bends in the Kandinsky model, but that can be drawn without bends in another model.

2.3 The Proportional-growth model

The unlimited-growth model permits drawings with few bends, but may result in large node dimensions. The Kandinsky model permits control over the node dimensions, but may do so at the cost of introducing bends. Therefore we developed a third model for high degree orthogonal drawings, which we refer to as the *proportional-growth model*. In this model, we allow nodes to grow in size, but the growth must be reasonable.

Precisely, assume that we are given a drawing Γ. For each node, let $r(v)$ be the number of edges that attach on the right side of the box of v in Γ. Similarly, we define $l(v), t(v), b(v)$ for the other three sides. Our condition can then be formulated as follows:

Definition 1. A two-dimensional drawing is said to be in the *proportional-growth model*, if for each node v, the width of the box of v is $\max\{1, t(v), b(v)\}$, and the height of the box of v is $\max\{1, r(v), l(v)\}$.

We have found the proportional-growth model to be a good model for practical purposes. The growth of a node is related to its degree, and therefore nodes are not too distorted. In fact, some users have remarked on the degree-related node-growth as a positive feature of a graph visualization library, since it immediately displays influential nodes in diagrams.

3 The three-phase method

In this section, we introduce the three-phase method. Besides the three main phases, which are illustrated in Fig. 1, there are pre-processing and post-processing steps. In Fig. 5 on Page 10, we show a flow chart of the process of creating an orthogonal drawing.

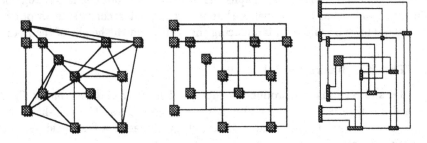

Fig. 1. A drawing after the three main phases.

The three-phase method is not an algorithm in itself, but it is a framework that unifies different approaches. It can be used for either of the three models. It is easily changeable, since an algorithm for a phase can be replaced by a better algorithm, if one becomes available in the future.

3.1 Pre-processing steps

Before starting the layout process, we apply the transformations needed to convert the graph into a *normalized graph*, i.e. a connected graph without reflexive edges and without nodes of degree 1. If the input graph is not connected, then we draw each connected component separately. Each reflexive edge is removed and added into the finished drawing close to the node. Each node of degree 1 is removed, and added into the finished drawing close to its neighbor.

3.2 Node placement

The first main phase of the algorithm is called *node placement*. In this and the following phase, we draw the nodes as points, not as boxes. During the node placement phase, we assign each node to a point in an $n \times n$-grid such that no two nodes are placed at the same point.

Any node placement is feasible, even though some can be achieved only by allowing more than one bend per edge. A node placement is said to be in *general position* if no two nodes are placed on the same grid-line. If G is a normalized graph, then for any node placement in general position, we can achieve an $m \times m$-grid and one bend per edge (see Section 3.6).

A node placement is an assignment of nodes to an $n \times n$-grid. Thus, it can be formulated as a 0-1 integer program with n^3 variables, and adapted to a number of objectives, such as for example the edge length. Solving this problem to optimality is prohibitively slow, therefore one has to develop heuristics.

Median placement We present here one heuristic for node placement, called the *median placement* strategy. To add a node v into a given drawing, let $w_1, \ldots, w_r$ be the embedded neighbors of v. A natural place for v is "in the center of its neighborhood", that is, roughly equidistant from $w_1, \ldots, w_r$.

Let w_j be placed at (x_j, y_j), and sort the neighbors of v such that $x_{i_1} \leq \ldots \leq x_{i_s}$ and $y_{j_1} \leq \ldots \leq y_{j_s}$. Let k be the median of $\{1, \ldots, s\}$ and define the *median center* as (x_{i_k}, y_{j_k}). Place v_{i+1} in a newly added row and column next to the median center. To compute the median center we need the relative order of x_j and y_j, but not the absolute values, and it can be found in $\mathcal{O}(deg(v))$ time.

The median-placement should be used in an interactive setting if we want to add a new node into an existing drawing. A similar technique is one of the options used in the interactive technique described in [13].

3.3 Edge routing

The second phase of the algorithm is called *edge routing*. Assume that the node placement has been chosen. Now we want to decide on a route for each edge. During this phase, edge routes may intersect nodes or overlap each other, such conflicts will be removed during port assignment. For each edge, there are at most two possible routes with one bend.

Computing an edge routing with at most one bend per edge corresponds to a 0-1 integer program with at most $m + 2n$ variables, which can be adapted to objectives such as the desired dimensions of the nodes and the half-perimeter of the drawing. Solving this problem to optimality is prohibitively slow. A simple heuristic, using a Eulerian circuit, yields the following result. Details are omitted.

Lemma 2. *For any node placement in general position, there is an edge routing with one bend per edge such that at most $\lceil \frac{deg(v)}{2} \rceil$ edges attach on each side of a node v. It can be found in $\mathcal{O}(m)$ time.*

Another heuristic, based on randomized rounding of the optimal fractional solution of the 0-1 program yields the following result. Details are omitted.

Lemma 3. *For any normalized graph and any node placement in general position, if HP_{IP} is the optimal half-perimeter (among the edge routings with one bend per edge), and HP_{RR} is the randomized rounded half-perimeter, then*

$$P\left(HP_{RR} > HP_{IP} \cdot \left(1 + \sqrt{4\ln(4n)/n}\right)\right) < \frac{1}{n}.$$

Routes with more than one bend Sometimes we want to route edges with more than one bend. In this case, we *triple the grid* before edge routing, i.e., in both orientations, we add one new grid-line before and after each used grid-line. Define the *off-grid-lines* of a node v as the grid-lines before and after the grid-line of v. To route an edge $e = (v, w)$ with more than one bend, we use the off-grid-lines at v and w, see Fig. 2. An off-grid-line may be used by more than one edge route, such overlap will be removed during port assignment.

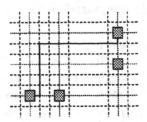

Fig. 2. An edge route with three bends. Grid-lines are dotted, off-grid-lines are dashed.

With three bends per edge, we can achieve the minimal possible node dimensions, again using a Eulerian circuit. Details are omitted.

Lemma 4. *For any node placement there exists an edge routing with three bends per edge such that at most $\lceil \frac{\deg(v)}{4} \rceil$ edges attach on each side of a node v. It can be found in $\mathcal{O}(m)$ time.*

3.4 Port assignment

In the third phase, called *port assignment*, we increase node dimensions, adding new rows and columns if needed. Afterwards, each node v has a number of intersections with grid-lines, these places are called the *ports* of v. We assign a port to each endpoint of an edge such that no two edges overlap.

Such an assignment is not always possible without adding bends to edges. We will study in the following sufficient conditions for the existence of a port assignment without additional bends. Then we study how to achieve these conditions by adding bends and changing the node placement. Port assignment is done for one grid-line at a time, after the feasibility of port assignment has been assured for all grid-lines. Thus, during port assignment for a row, the port assignments in the columns stay unchanged, and vice versa.

One node per grid-line Assume row r contains only one node v. We add $\max\{1, r(v), l(v)\} - 1$ new rows after r, and extend v to cover these rows. We assign the edges to ports of v in such a way that no two incident edges on one side intersect. The resulting drawing is in the Kandinsky model.

Hamiltonian paths If the graph induced by the nodes in one row contains a hamiltonian path, then port assignment is possible such that the resulting drawing is in the proportional-growth model. Details are omitted, see Fig. 3.

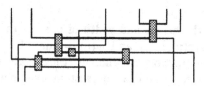

Fig. 3. Port assignment if the row graph has a hamiltonian path (shown in thick lines).

Non-overlapping trees If the graph of the nodes and bends in one row contains a rooted spanning tree such that for any two siblings v and w, the column-intervals spanned by the descendants of v respectively w are disjoint, then port assignment is possible such that the resulting drawing is in the proportional-growth model. Details are omitted, see Fig. 4.

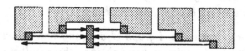

Fig. 4. Port assignment if the row graph contains a non-overlapping tree.

Resolving conflicts Port assignment is not always possible, and we also may not want to spend the time to find out, since testing some of the conditions is $\mathcal{NP}$-complete. Hence, we proceed as follows: Pick a condition that we would like to satisfy. Apply a simple heuristic to determine whether the condition is satisfied (it may erroneously report that it is not). If the condition is not satisfied, determine edges and nodes that prevent the condition, these are the *conflicts*.

Conflicts can be resolved in two possible ways. One simple solution is to re-route a conflicting edge with two or three bends. Alternatively, we can split the nodes in one grid-line into two groups, and move one of the groups to a newly added grid-line. This maintains the property that every edge is routed with at most one bend. Details are omitted.

3.5 Post-processing steps

There are two types of post-processing steps. First, we have to undo the changes of the pre-processing step, i.e., add previously removed trees and reflexive edges and combine the drawings of connected components. Secondly, we improve the drawing, by applying compaction techniques from VLSI-design, see the book by Lengauer [9].

3.6 Bounds

Let $h(v)$ and $w(v)$ be the height and width of v, so in the proportional-growth model, $h(v) = \max\{1, r(v), l(v)\}$ and $w(v) = \max\{1, t(v), b(v)\}$. We use the values as updated after the port assignment (they might change if we re-route an edge or re-locate nodes). In a drawing with at most one bend per edge, each used grid-line contains a node, therefore the total height cannot be more than $\sum_{v \in V} h(v)$. If (v, w) is an edge drawn without bend, then its row is used for both v and w. So if t_h is the number of horizontal straight edges, then the height is $\sum_{v \in V} h(v) - t_h$. Similarly one shows a bound for the width.

Lemma 5. *If Γ is a drawing with at most one bend per edge, and with t_h and t_v straight horizontal and vertical edges, respectively, then the height is at most $\sum_{v \in V} h(v) - t_h$, and the width is at most $\sum_{v \in V} w(v) - t_v$.*

Two important results follow from this lemma and Lemmas 2 and 4.

Theorem 6. *For any normalized graph G and any node placement in general position, we can find an edge routing and port assignment such that the resulting grid-size is $m \times m$ and each edge has one bend.*

Theorem 7. *For any normalized graph G and any node placement, we can find an edge routing and port assignment such that the resulting side-length of the grid is at most $\frac{3}{2}m + \frac{3}{4}n$, the half-perimeter is at most $3m + n$, and each edge has at most three bends.*

4 Interactive drawings

In this section we show that the three-phase scheme is helpful for developing efficient interactive high-degree orthogonal drawing algorithms. In an interactive setting we are given a legal orthogonal drawing, and an operation such as insert a node, insert an edge, delete a node, delete an edge, or move a node. The objective is to perform the desired change without major disturbance to the existing drawing, in order to preserve the "mental map" [11]. Algorithms for interactive orthogonal drawing were presented in [10, 13, 14].

The main problem in interactive drawing is to detect sufficient existing space or increase the space for adding edges and nodes, and to delete superfluous space. We attack this problem by uncompressing and then again compressing the drawing with every interactive change. Also, to find a suitable place to add grid-lines, if needed, we revert the valid drawing back into the sketch-status with every change. See Fig. 5 for a flow-chart of the interactive process.

Thus, we need only describe the changes to the sketch in the five possible interactive operations. For each, there are only a few affected grid-lines of the sketch, and only for these grid-lines do we undo the port assignment.

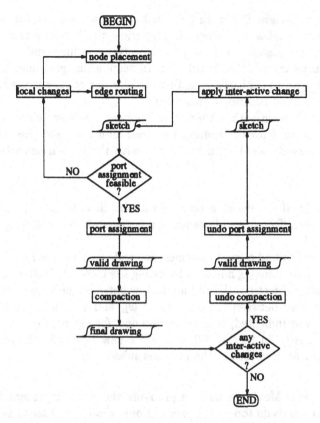

Fig. 5. The flow-chart of the three-phase method.

Delete an edge If we delete an edge $e = (v, w)$, the affected grid-lines are the ones that contain a segment of e. We delete all segments of e, and test for each affected grid-line whether it still contains elements; if not, we delete it. Then we try to join each affected grid-line with another grid-line, if this is feasible for port assignment.

Delete a node If we delete a node v with incident edges $e_1, \ldots, e_d$, the affected grid-lines are those that contain some segment of some edge $e_1, \ldots, e_d$, and both grid-lines of v. We delete the edges $e_1, \ldots, e_d$ as described above. Then we delete v, and check whether the grid-lines containing v are now empty; if so, we delete them. Then we merge each affected grid-line with another grid-line, if this is feasible for port assignment.

Insert an edge If we insert an edge $e = (v, w)$, the affected grid-lines are both grid-lines of both v and w. We test whether we need a bend for e, and if so, where to place it.

We first try to route e without bend. Thus, if v and w share a grid-line, then we draw e as a straight line, if this is feasible for port assignment. Otherwise, we try whether v and w are in neighboring grid-lines. If so, we test whether it is feasible for port assignment if we join these two grid-lines and add edge e to it. If not, then we try whether v and w are in neighboring grid-lines in the other direction and proceed similarly. If neither of these options is successful, then we conclude that e must be drawn with a bend.

If e must be drawn with a bend, then we have two possibilities to route e. Possibly a preference for the routing has been indicated (see later in "Moving a node"). Otherwise, we determine a route with the chosen heuristic for edge routing.

Insert a node If we insert a node v with incident edges $e_1, \ldots, e_d$, where $e_i = (v, w_i)$, the affected grid-lines are the horizontal and vertical grid-lines of w_i, $i = 1, \ldots, d$.

We first need to find a good placement for node v. This could have been indicated already (see "Moving a node") by listing the closest neighbors w_S, w_N, w_E and w_W in either direction. We add a new row between the rows of w_S and w_N, and a new column between the columns of w_E and w_W, and place v in it. If no placement was indicated, then we choose one, for example using the median placement discussed in Section 3.2, and add a new row and column there. We route $e_1, \ldots, e_d$ as described in the previous subsection.

Moving a node Moving a node is probably the most important interactive change, and it builds on top on the previous operations. In order to move a node v, we determine the nodes w_N, w_S, w_E and w_W that are closest to the desired new place of v in the four directions. We also store for each incident edge of v whether it attached to v horizontally or vertically.

We delete v and all its incident edges, and then re-insert v, using the information about the four closest neighbors. Then we insert the incident edges of v with the same bend placement as before, if possible.

Worst-case bounds At any time, every edge has at most one bend, so we use at most one row and one column for each edge. For some nodes, all incident edges may attach in one orientation, say horizontally, thus one column is not accounted for by any edge. Consequently, the total grid-size may be at most $m + n$ in either orientation.

Theorem 8. *Let G be a graph that is changed interactively, through insertions, deletions, and move operations. Under any sequence of operations, we can maintain a drawing of G with only local changes. At any time, the grid-size is at most $(m + n) \times (m + n)$, and there is at most one bend per edge, where n and m are the current number of nodes and edges.*

5 Conclusion

In this paper, we introduced the three-phase method for creating orthogonal drawings of graphs with high degrees. This method breaks the layout problem into three separate phases, and thus permits detailed study of each. With this method, we can create drawings in a newly defined model, called proportional-growth model, which allows us to save bends and at the same time maintains box dimensions of nodes within reasonable bounds.

The three-phase method is general and flexible, and has numerous applications, including the following results for normalized graphs:

- Any simple graph can be embedded in an $\frac{m+n}{2} \times \frac{m+n}{2}$-grid with at most 1 bend per edge [3].
- Any simple graph can be embedded with half-perimeter $2m-n$ with $m-n+1$ bends and at most one bend per edge [1].
- Any simple triconnected planar graph can be embedded without crossings in an $(m - n + 1) \times \min\{m - n, \frac{m}{2}\}$-grid with $m - n$ bends and at most 1 bend per edge [3].
- For any given node order $\{v_1, \ldots, v_n\}$, we can build a drawing incrementally by adding nodes in this order, such that with any addition there are only local changes. At any point in time, the grid-size is at most $(\frac{m}{2}+n) \times (\frac{2}{3}m+n)$ and there is one bend per edge [3].
- *Relative node placement constraints*: For any pair of nodes v, w, we can specify whether v should be below w, or whether v should be to the left of w, or both, as long as these constraints are not contradictory in themselves. The drawing has grid-size $m \times m$ and at most 1 bend per edge [1].
- *Port specifications*: We can specify constraints of the form "edge (v, w) should attach at the top side of node v, and it should be the kth of the edges attaching there" [1].

One other advantage of the three-phase method is its adaptability: the algorithm for a phase can be replaced by a better algorithm, when one becomes available in the future. We expect further improvements on the area bounds and number of bends, based on this method.

As for open problems, we plan to develop improved node placement schemes. In theory, a node placement in general position gives the best known worst-case bounds (compare Theorem 6 with Theorem 7). But if we place more than one node into a grid-line, and if this does not create a conflict for port assignment, then we save one bend and therefore also one grid-line (Lemma 5). So a node placement scheme is best in terms of area if it permits many nodes in one grid-line, while satisfying a necessary condition for port assignment. How can such a node placement be found? How can node placements be found that ensure a small number of crossings?

References

1. T. Biedl. *Orthogonal Graph Visualization: The Three-Phase Method With Applications.* PhD thesis, RUTCOR, Rutgers University, May 1997.
2. T. Biedl and G. Kant. A better heuristic for orthogonal graph drawings. In *2nd European Symposium on Algorithms*, volume 855 of *Lecture Notes in Computer Science*, pages 124–135. Springer Verlag, 1994.
3. T. Biedl and M. Kaufmann. Static and incremental high-degree orthogonal drawings. In *5th European Symposium on Algorithms*, Lecture Notes in Computer Science. Springer Verlag, 1997. To appear.
4. F. Brandenburg, editor. *Symposium on Graph Drawing 95*, volume 1027 of *Lecture Notes in Computer Science*. Springer Verlag, 1996.
5. G. Di Battista, P. Eades, R. Tamassia, and I. Tollis. Algorithms for drawing graphs: an annotated bibliography. *Comp. Geometry: Theory and Applications*, 4(5):235–282, 1994.
6. U. Fößmeier, G. Kant, and M. Kaufmann. 2-visibility drawings of planar graphs. In North [12], pages 155–168.
7. U. Fößmeier and M. Kaufmann. Drawing high degree graphs with low bend numbers. In Brandenburg [4], pages 254–266.
8. G. Kant. Drawing planar graphs using the canonical ordering. *Algorithmica*, 16:4–32, 1996.
9. T. Lengauer. *Combinatorial Algorithms for Integrated Circuit Layout.* Teubner/Wiley & Sons, Stuttgart/Chicester, 1990.
10. K. Miriyala, S. Hornick, and R. Tamassia. An incremental approach to aesthetic graph layout. In *6th Intl. IEEE Workshop Computer-Aided Software Eng.*. 1993.
11. K. Misue, P. Eades, W. Lai, and K. Sugiyama. Layout adjustment and the mental map. *J. Visual Languages and Computing*, pages 183–210, June 1995.
12. S. North, editor. *Symposium on Graph Drawing 96*, volume 1190 of *Lecture Notes in Computer Science*. Springer Verlag, 1997.
13. A. Papakostas, J. M. Six, and I. Tollis. Experimental and theoretical results in interactive orthogonal graph drawing. In North [12], pages 371–386.
14. A. Papakostas and I. Tollis. Issues in interactive orthogonal graph drawing. In Brandenburg [4], pages 419–430.
15. A. Papakostas and I. Tollis. High-degree orthogonal drawings with small grid-size and few bends. In *5th Workshop on Algorithms and Data Structures*, Lecture Notes in Computer Science. Springer Verlag, 1997. To appear.
16. A. Papakostas and I. Tollis. A pairing technique for area-efficient orthogonal drawings. In North [12], pages 355–370.
17. P. Rosenstiehl and R. E. Tarjan. Rectilinear planar layouts and bipolar orientation of planar graphs. *Discrete Computational Geometry*, 1:343–353, 1986.
18. R. Tamassia. On embedding a graph in the grid with the minimum number of bends. *SIAM Journal of Computing*, 16(3):421–444, 1987.
19. R. Tamassia, G. Di Battista, and C. Batini. Automatic graph drawing and readability of diagrams. *IEEE Trans. Systems, Man and Cybernetics*, 18(1), 1988.
20. R. Tamassia and I. Tollis. A unified approach to visibility representations of planar graphs. *Discrete Computational Geometry*, 1:321–341, 1986.
21. R. Tamassia and I. Tollis, editors. *DIMACS International Workshop, Graph Drawing 94*, volume 894 of *Lecture Notes in Computer Science*. Springer Verlag, 1995.

NicheWorks - Interactive Visualization of Very Large Graphs

Graham J Wills

Room 1u334, Lucent Technologies (Bell Laboratories),
1000 E. Warrenville Road, Naperville IL 60566, USA
Email: gwills@research.bell-labs.com, Web: www.bell-labs.com/~gwills

Abstract

The difference between displaying networks with 100–1000 nodes and displaying ones with 10,000–100,000 nodes is not merely quantitative, it is *qualitative*. Layout algorithms suitable for the former are too slow for the latter, requiring new algorithms or modified (often relaxed) versions of existing algorithms to be invented. The density of nodes and edges displayed per inch of screen real estate requires special visual techniques to filter the graphs and focus attention. A system for investigating and exploring such large, complex data sets needs to be able to display both graph structure and node and edge attributes so that patterns and information hidden in the data can be seen. We describe a tool that addresses these needs, the NicheWorks tool. We describe and comment on the available layout algorithms and the linked views system, and detail an example of the use of NicheWorks for analyzing web sites.

1. Introduction

NicheWorks is a visualization tool for the investigation of very large graphs. By 'very large' we mean graphs for which we cannot look at the complete set of labeled nodes and edges on one static display. Typical analyses performed using NicheWorks have between 20,000 and 1,000,000 nodes. On current mid-range workstations a network of around 50,000 nodes and edges can be visualized and manipulated in real time with ease. NicheWorks allows the user to examine a variety of node and edge attributes in conjunction with their connectivity information. Categorical, textual and continuous attributes can be explored with a variety of one-way, two-way and multi-dimensional views.

Originally designed for telephony applications, a number of other data sources have been analyzed NicheWorks has been adapted and modified into a general purpose tool. It has been applied to a variety of different problem areas, including:

- **Relationships in a large software development effort**. The goal is to understand functional relationships between pieces of a large software system and see how changes to one part impact other parts. By examining modification history we can create links between files indicating their degree of 'co-modification'.
- **Web site analysis**. Navigation is one goal of web analysis; another is simply to allow the user to understand how a site is laid out.

- **Correlation analysis in large databases.** In looking for patterns in a database consisting of many variables, it is helpful to have some way of summarizing the variables' relationships to each other. We display standardized correlations to get a first look at the variables and see how they are related.

2. Overview

The NicheWorks tool is part of a suite of visualization views that have been created by the Bell Labs Visualization Group for interactive analysis of large data sets. It shares a number of features with its sibling tools (e.g. SeeSoft [Ei94]), including:

- Ability to hide parts of a graph via manipulation of node/edge attribute views.
- Tools to color nodes and edges based on their attributes.
- Drag and drop mapping of attributes to shapes and labels.
- Selective labeling of nodes under user control;
- Interactive data interrogation via the mouse.

Methods specific to graph analysis that are incorporated into NicheWorks include:

- Automatic selection propagation from nodes to edges and vice versa.
- Selection propagation within a graph by following edges (one step or connected component).
- Interactive Pan/Zoom and Rotate facility.

This paper describes the methodology behind NicheWorks and our general approach to visualizing large complex data sets; in this case, weighted graphs. Section 3 details layout algorithms and section 4 the interactive interface. Section 5 presents an example of the tool for some real-life data and section 6 summarizes our findings.

3. Layout

Among others, Coleman [Co96] gives a list of properties towards which good graph layout algorithms should strive. We want our algorithms to lay out very large general weighted graphs, producing a straight-edge layout that reflects the edge weightings and that places nodes close to other nodes to which they are similar. Di Battista [Ba94] gives four aesthetics that are important for the general graph case. Since our layout is straight-edge, we trivially satisfy the aesthetic of avoiding bends in edges. Instead of keeping edge lengths uniform, we wish them to reflect the edge weights; our algorithms try to set the edge lengths inversely proportional to the weights, so that the strongest linked nodes are closest together in the ideal layout. Due to the computational cost of multiple edge-crossings detection, we elect to ignore the criterion that edge crossings should be minimized, trusting that our algorithms will produce good results without directly involving a measure of edge-crossings. Since we wish to show clusters of nodes and discriminate nodes far from such clusters, the aesthetic of distributing nodes evenly is not obviously useful, and we relax it considerably. There are two types of algorithms used for laying out graphs in NicheWorks. First, an initial layout (section 3.1) is chosen. The user may then chose to improve the graph layout with one or more incremental algorithms (section 3.2).

In the following discussion, algorithms are run on each connected component of the graph. The final layout is achieved by placing the components close to each other, with the largest in the center. The algorithm currently in place for this stage is rather naïve: Components are represented by a circle sufficiently large to encompass the component. The circles are then laid out using a greedy algorithm that places the circles as close as possible to the center of the display in decreasing order of size. Figure 8(a) shows the limitations of this approach. On an implementational note, we use any available parallel machine architecture to process each component separately. This is trivial to implement as no synchronization is necessary until all the components are placed together at the end.

3.1 Initial Layout

Initial layouts are not yet a strong point of our work; suggestions by this paper's referees should prove helpful for future work. Currently, the available initial component layout algorithms are fairly simple, comprising:

- **Circular Layout**. Nodes are placed on the periphery of a single circle
- **Hexagonal Grid**. Nodes are placed at the points of a regular hexagonal grid.
- **Tree Layout**. Nodes are placed with the root node in the center, then each connected node is put in a circle around that.

The first two layout algorithms simply place the nodes at random locations either on the circle or on the grid. The tree layout algorithm was inspired by the cone-tree 3D visualization method for large hierarchies [Ro91], but avoids the occlusion problem induced by a 3D visual system while still coping with the size graphs typically displayed in cone tree examples. [Ba94] gives other examples of radial layout algorithms of which this is an example. The tree layout algorithm also works well for DAGs and has proved to be useful for both general directed and undirected graphs. In the latter cases we find the source of the graph by working inward from the leaves until we find the center-most node(s). If there are multiple sources, the algorithm creates a fake root node as parent to all the real root nodes and lays out this enhanced graph.

The algorithm creates a tree from the enhanced graph by creating a subgraph G', initially consisting of just the root node. An iterative scheme is performed whereby all nodes that are one step away from G' are added to G', along with the strongest weighted edge from that node to G'. A naive implementation of this algorithm runs in $O(DE)$, where D is the tree depth and E is the number of edges. Each node is then labeled with the size of its subtree. A variable indicating subtree angle is also attached to each non-leaf node. The root node is positioned at the center and given an angle of 360 degrees. This indicates the angular span of its subtree. We then perform the following iterative layout method:

For each of the leaf nodes of the positioned graph, we divide up the angular span available to it subtree using the size of each of its children's subtrees as weights. The children are placed on a circle with radius proportional to their distance from the root node and are placed at the midpoint of their individual angular ranges, with their parent in the center of the overall range (complying with a common criterion for hierarchical layouts mentioned in, for example, [Co96]). An example is shown in figure 1. The root node (R) is drawn at the center, with its children on a circle centered at R of radius 1. R has a subtree of size 20 and its child S has a subtree of size 10, so S acquires an angular span of 360*10/20 = 180 degrees. Its child T with subtree of size 5 gets a span of 180*5/10 = 90 degrees, U gets 180*2/10 = 36 degrees and V gets 180*3/10 = 54 degrees.

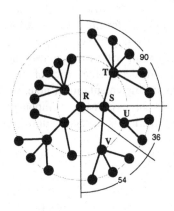

Figure 1. Radial placement

3.2 Incremental algorithms

There are three incremental algorithms available. In each case, the user defines a potential function which describes the disparity between a weighted edge and the length of that edge. The edge length should be inversely proportional to its weight, so that strongly tied nodes are close together. Two of the more useful functions are a sum of terms of the following form:

a) $(1-dw)^2$ b) $|1-dw|$

where d is the edge length and w is the edge weight. Each potential contribution is minimized when $d = 1/w$. The difference between (a) and (b) can be seen if we add a small perturbation to the optimal solution, making it instead $d = 1/w + \varepsilon$

Then we get   a) $\left(1-w\left[\frac{1}{w}+\varepsilon\right]\right)^2 = w^2\varepsilon^2$ b) $|1-w(1/w+\varepsilon)| = w\varepsilon$

so for a small absolute perturbation of the distance, (a) is more forgiving of minor variations than (b), assuming a transformation of the weights.

An important point to note is that the potential is a function *only* of the graph edges - if two nodes do not have an edge between them, then the distance between them is irrelevant to the potential function. This characteristic ensures that the potential calculations are fast (it is the primary reason for the efficiency of our approach), but has the drawback that there is no force repelling nodes from each other.

3.2.1 Steepest Descent

For this method we consider the potential of the graph to be a function of the 2N-dimensional vector of locations of its nodes. Moving the location of the graph in this high dimensional space is equivalent to moving every node in the graph simultaneously. We calculate the gradient of this vector and move in that direction a suitable amount. Then we can move the configuration in 2N-space to the specified point along the gradient direction. The basic method is described for one dimensional functions in [BuFa85].

This method is a relatively slow method, with each step requiring the calculation of several gradient potential functions for offsets from the current location in 2N-space. Although each calculation is of order $O(E)$ so the order of the whole process is $O(E)$, the constant multiplier is quite high and our informal experience suggests that the number of iterations required to achieve good results is around $O(\sqrt{E})$ giving an overall order of $O(E\sqrt{E})$.

3.2.2 Simulated Annealing Swapping Algorithm

This algorithm randomly picks a pair of nodes and calculates the difference in potential if the nodes were swapped. If the potential increase is allowed by the annealing algorithm, then the nodes' positions are swapped. Details of annealing algorithms in a graph layout context can be found in Davidson and Harel's paper [DaHa96]. They use an annealing approach to decide whether to move a node to a new position and we use annealing to decide whether to swap nodes, but the process is conceptually very similar.

3.2.3 Repelling algorithm

The descent algorithm of (3.2.1) can produce layouts with nodes placed very close to each other since it only uses inter-node distances if there is a edge between them. To solve this problem, we introduced a last-stage algorithm to be run a few times only which calculates the nearest neighbors for all nodes and then moves the closest ones apart a small distance. Running this a few times will move overlapping nodes apart.

This algorithm uses a quad-tree with an implementation as described in [NiHi93] that is $O(logN)$ for all three operations of adding, deleting and calculating nearest neighbors. Thus each step of the algorithm is $O(N\ logN)$, which is acceptable.

4. Interactive interface

The interface to NicheWorks is an instance of a linked views environment, described in [Wi90], [Wi97] and [EiWi95]. In this paradigm, each view of the data displays both the data themselves and a state vector that is attached to the data. This vector dictates how each datum should contribute to the view appearance. In our implementation the possible states are:

- *Deleted*   Treat the data point as if it were not present
- *Normal*   Show the data
- *Highlighted* Show the data so it will stand out against *normal* data
- *Focused*   Show as much detail as possible on the data

Furthermore the user should be allowed to modify the state vector by interacting with the data views. For example, selecting a specific bar from a bar chart view and highlighting it will change the data state vector for items represented by that bar, causing other views of the data immediately to update their representation.

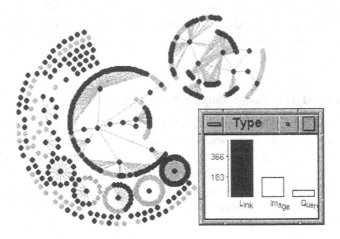

Figure 2. Web site with nodes of type 'link' highlighted

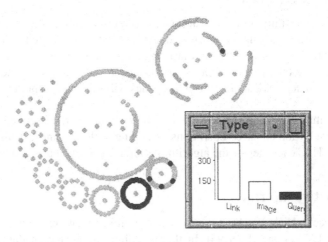

Figure 3. Nodes with degree zero have been *deleted* and of the rest, nodes with type 'query' have been *highlighted*.

In NicheWorks there several options for displaying the graph using the state vector, and for interacting with the graph. We use a small data set consisting of a few hundred web pages and links between them to exemplify the approach. This data set was collected by listing all the pages near the top level of the author's directory and feeding the references to them to MOMspider [Fi94] which uses references in those documents to search out new pages on the web.

Figure 2 shows the results of selecting only nodes labeled as 'link' (a standard web page). Selected nodes are drawn in a highlight color, with unselected nodes in gray. Edges are only drawn from selected nodes to other selected nodes. To create figure 3, we created a histogram of the degree of each node (not shown) and then used the mouse to select those of degree zero. We then set their state to 'deleted' as we are not interested in these degenerate components. In the 'Type' bar chart we then select the 'query' type to see which links called query routines.

There is an interesting component where all the child nodes are queries. We move the pointer over those nodes so that the labels appear and disappear rapidly. We see that all the query nodes are searches into a film database for various films, and the central node is called 'films96.html' - it looks like a page of film reviews.

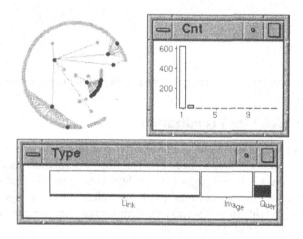

Figure 4. Using edge statistics to highlight nodes and show the distributions of statistics for those selected nodes

In figure 4, we show how fairly complex queries can be posed naturally through the linked views metaphor. We have changed the 'Type' bar chart to a spineplot, where each bar has a fixed height and the width of the bar indicates its count. Within each bar the darker area shows the percentage of selected cases within the bar as a height. We have also created a bar chart of edge counts, showing the number of times a URL refers to another URL. We have selected all counts above one; i.e. all those links to a URL from a URL that occur multiple times. This selection defines a subset of highlighted edges which in turn highlights those nodes that are endpoints of the edges. The type bar chart and the NicheWorks view both immediately show this result; we can see that images never have multiple edges to them, regular pages sometimes do and queries do about half the time.

The state vector can be useful when laying out large graphs. If we set a node's state to *deleted*, then it plays no part in the layout process, nor do any edges involving it. Thus we can use the deletion mechanism to look at subsets, trying layouts only for them, or using partial layouts to speed up positioning a very large graph. An example of the former is shown in figures 5(a) and 5(b). In the former we have selected an important web page (a bibliography) and use the *one step* menu option twice to expand the selection to nodes up to two steps away. The result is not very clear, so we *delete* the unselected points and choose the tree layout option to arrive at figure 7(b), which shows the layout more clearly.

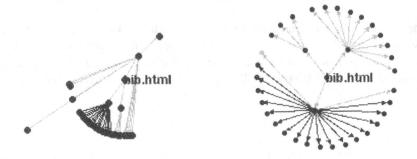

Figure 5. A subset of the web site positioned (a) as part of the whole site (b) by itself

5. Example: Web Site Visualization

The MOMspider web crawler was used to search and index all web pages accessible locally from the author's home page. The resulting information totaled 733 pages (nodes) and 758 links between pages (edges). Statistics were collected on both nodes and edges, including number of times a link was referred to in a page. This statistic was used for the weight. In this section we demonstrate how we use NicheWorks to understand the structure of this small site.

Figure 6. Circle, Hexagonal and Tree layout methods for web site data

Figure 6 shows the layouts for each of our three methods. We ran the swapping algorithm for ten seconds on the hexagonal grid to achieve the above layout. Each view shows nine separate components of differing sizes. The circle layout does not look very promising as it shows the size of the clusters well, but not their structure. The tree layout hides the size to an extent (for example consider the very dense cluster towards the bottom right) to show structure better. The Hexagonal grid method shows a bit of each. We run the move algorithm for 10 seconds on the most promising two, the hex and tree layouts, to give figures 7(a) and 7(b). As might be expected, the two layouts appear fairly similar as far as individual components are concerned. The tree layout followed by move appears best; we'll use it from now on.

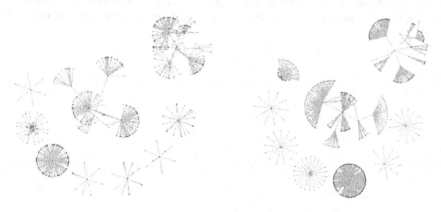

Figure 7. Results of the move algorithm for (a) Hex layout (b) Tree layout

There is an immediately noticeable pattern in several of the components; a central node with connections to every other node in the component and no other edges. These are collections of information with one index page referencing many others. Although most of these 'central node' components in figure 7(b) are symmetrical, there is one that is very asymmetrical at the top left. We zoom in on it and selectively label the center node to produce figure 8. By interrogating, we see that they are all queries for a database server, each query being a request for a film name. The different line lengths indicate that some films are referred to more frequently than others from the central page.

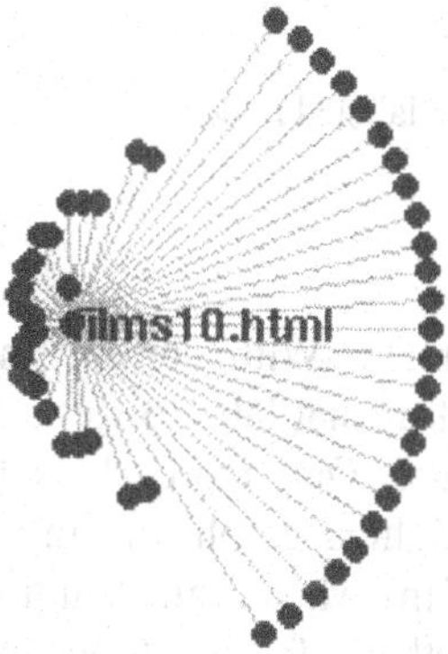

This component was created to index a list of best films of 1996. Some films were mentioned only once (the ring of far away circles on the right), others more often. Since simple components like this can be solved for a zero potential, the distance from the central node is exactly inversely proportional to the number of times the film is mentioned in the article. We focus on the innermost nodes

Figure 8. An asymmetrical component

and note that "The English Patient" and "Fargo" are the most commonly mentioned.

We zoom in on the large central component in figure 7(b) and label some representative nodes to give figure 9. The central page here is the home page for the author's department, with a ring of general purpose pages around it, most of which are not accessed since they go off-site. Two interesting exceptions are the 'who.html' page, with its list of images of people and links to their home pages, and a set of pages for ordering books on-line using the local 'bookbot' system.

Exploring web networks is an emerging field. It is important for both site administrators and for users navigating a site. The size of the networks and their ad-hoc complexity make it a natural candidate for this form of network visualization.

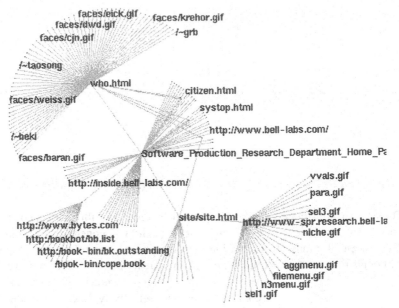

Figure 9.. Department home page component with user-labeled nodes

6. Conclusions and Future Work

Displaying and navigating large networks is a hard problem. With current and foreseeable limitations on display size and resolution, it is clear that labeled views of complete large networks are impossible with static layouts.. Our approach is to provide a tool that allows the user to interact with the weighted graph, making it possible to position and focus rapidly on different subsets of the whole, thus building up knowledge about the entire graph. In our opinion, methods for the following operations are essential:

- Defining a subset of the graph based *both* on graph structure and on values of any available node/edge variables.
- Providing a range of robust layout tools suitable for different types of graph.
- Laying out subgraphs.
- Giving immediate and reversible control over mappings from data attribute to node and edge attributes (labelling, olor, size, shape, line style, etc.)
- Rapid ability to pan, zoom and rotate the graph in the viewing window. The speed must be sufficient to make the action appear truly interactive.
- Means for the user to retrieve full details on nodes and edges.

We have described the NicheWorks tool and given an overview of the linked windows environment in which it is embedded. The NicheWorks layout algorithms have been designed to work well for large weighted graphs and have been implemented so as to be robust against slight data irregularities. A number of selection tools and methods are available which have been described elsewhere [WI96]. The goal of NicheWorks is to allow the user to interact with large graphs; to allow them to try ideas, focus on different aspects of the data and to create views that spark intuition. We have used perceptually good attribute encodings [ClMc84, ClMc88] and expanded principles of Tufte [T83, Tu90] and Bertin [Be83] into the interactive domain (via the linked windows environment, section 4). One of the most pleasing aspects of the project is that domain experts with little or no graph-theoretic or statistical background can use it to gain knowledge about graphs in their own area.

NicheWorks allows users to visualize weighted networks with hundreds of thousands of nodes and edges. It combines statistical data views with graph layouts and visualization methods from the computer science disciplines using an interactive linked views environment. It is currently being used for a number of tasks including software analysis, fraud detection and document correlations. We welcome all comments and suggestions as we continue to improve it to be better, faster and ultimately more *informative*.

7. References

Ba94 Di Battista, G., Eades, P., Tamassia, R. and Tollis, I. (1994) *Algorithms For Drawing Graphs: An Annotated Bibliography* Computational Geometry 4 (1994) 235-282

Be83 Bertin, J. (1983) *Semiology of Graphics* University of Wisconsin Press

BuFa85 Burden, R. and Faires, J.D. (1985) *Numerical Analysis (3rd ed.)* PWS publishers, Duxbury Press, Boston MA 02116

ClMc84 Cleveland, W. S. and McGill, R. (1984) Graphical Perception: Theory, experimentation, and application to the development of graphical methods Journal of the American Statistical Association, 79 pp 531-554

ClMc88 Cleveland, W. S. and McGill, R., eds. (1988) Dynamic Graphics for Statistics Wadsworth & Brooks, California

Co96 Coleman, M. K. (1996) *Aesthetics-Based Graph Layout For Human Consumption* Software Practice And Experience, 1996, Vol 26(12), Pp 1415-1438

DaHa96 Davidson, R. and Harel, D. (1996) *Drawing Graphs Nicely Using Simulated Annealing* ACM Transactions On Graphics, Vol 15, No. 4, 1996, Pp 301-331

EiWi95 Eick, S. G. and Wills, G. (1995) *High Interaction Graphics* European Journal of Operations Research #81 (1995) pp. 445-459

Ei94 Eick, S.G. (1994) *Graphically displaying text* Journal of Computational and Graphical Statistics, 3(2), pp. 127-142

EiWi93 Eick, S. and Wills, G. (93) *Navigating Large Networks with Hierarchies* Proceedings of IEEE Visualization '93

Fi94 Fielding, R. (1994) Maintaining Distributed Hypertext Infostructures: Welcome to MOMspider's Web Proceedings of 1st intl. conf. on the World-Wide-Web, Geneva

NiHi93 Nievergelt, J. and Hirichs, K. (1993) *Algorithms and Data Structures with Applications to Graphics and Geometry* Prentice Hall, Englewood Cliffs, NJ 07632

Ro91 Robertson, G. G., Mackinlay J. D., and Card, S. K. *Cone Trees: Animated 3D Visualizations of Hierarchical Information.* In Proceedings of the ACM Conference on Human Factors in Computing Systems (CHI'91), pp. 189-194. ACM Press, 1991.

Tu83 Tufte, E.R. (1983) *The Visual Display of Quantitative Information* Graphics Press, PO Box 430, Cheshire, Connecticut 06410

Tu90 Tufte, E.R. (1990) *Envisaging Information* Graphics Press, PO Box 430, Cheshire, Connecticut 06410

Wi97 Wills (1997) *Visual Exploration of Large Structured data Sets* New Techniques and Technologies for Statistics II, IOS Press, Washington DC

Wi96 Wills, G. J. (1996) *Selection: 524,288 Ways to say "This is Interesting"* Proceedings of IEEE InfoVis '96, pp 54-60

Wi90 Wills, G. Unwin, A., Haslett, J. and Craig, P. (1990) *Dynamic Interactive Graphics For Spatially Referenced Data* Softstat '89 Fortschritte Der Statistik-Software 2, Gustav Fischer Verlag, Stuttgart, Pp 278-287

Extending the Sugiyama Algorithm for Drawing UML Class Diagrams: Towards Automatic Layout of Object-Oriented Software Diagrams

Jochen Seemann

Institut für Informatik, Am Hubland, 97074 Würzburg,
seemann@informatik.uni-wuerzburg.de

Abstract. The automatic layout of software diagrams is a very attractive graph drawing application for use in software tools. Object-oriented software may be modelled using a visual language called the Unified Modeling Language (UML). In this paper we present an algorithm for the automatic layout of UML class diagrams using an extension of the Sugiyama algorithm together with orthogonal drawing. These diagrams visualize the static structure of object-oriented software systems and are characterised by the use of two main types of edges corresponding to different relationships between the classes. The graph drawing algorithm accounts for these concepts by treating the different edge types in different ways.

1 Introduction

Object-oriented modeling techniques such as Booch [1] or OMT (Object Modeling Technique) [10] and their graphical representations of object-oriented software design have become very popular in recent years. The Unified Modeling Language (UML) [14] is an up-coming standard for specifying and visualizing various aspects of object-oriented software systems. UML is a graphical language derived from several existing notations commonly used to specify the design of object-oriented software. Some important diagrams are those representing the architecture of the software. In UML these are known as static structure diagrams. Class diagrams form a subset of these which is used to represent the static structure of classes and the relationships between them.

In this paper we present a technique for the automatic layout of UML class diagrams. Our algorithm is based on a combination of an extension of the well-known Sugiyama algorithm and orthogonal drawing techniques. We have implemented the algorithm as a part of a tool called UML workbench, which is used to demonstrate new techniques for software engineering tools.

The paper is organised as follows. Section 2 gives an overview of the diagrams and the related graphs that should be drawn. Section 3 describes the phases of our layout algorithm. In section 4 we show an example of the output from our drawing algorithm. Finally, we discuss the application of the algorithm and give some ideas for further work in this field.

2   UML Class Diagrams

Figure 1 shows an example of a UML class diagram. The classes are depicted by rectangles containing a list of the attributes and operations (methods) of the class. There are two main categories of relationships between classes:

- inheritance relationships (Fig. 1: Expression / TypeExpression or ModelElement / Instance): these represent generalization-specialization relationships allowing a hierarchical classification of classes. These relationships are known from object-oriented programming languages.
- associations (Fig. 1: Type / TypeExpression): these represent a more general relationship between classes as for example in ER-diagrams. An aggregation (Fig. 1: Instance / Value) is a special association representing a containment of classes and is drawn with a special symbol.

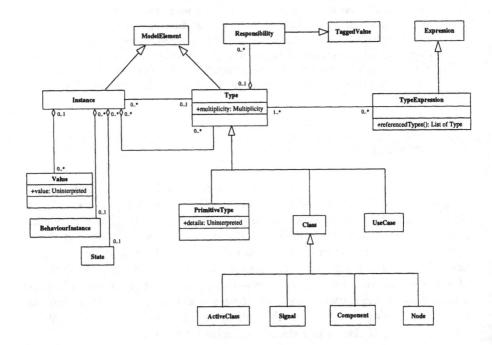

Fig. 1. Example class diagram in UML notation (as printed in [14])

A diagram can be seen as a graph whose nodes represent classes and whose edges represent relationships. Because of the semantic richness of UML both the nodes and the edges may have attributes (Fig. 1). These attributes are represented textually or in some cases graphically.

We have identified some important conditions for the drawing of UML class diagrams that lead to the idea of our algorithm:

- Nodes do not have a fixed size.
- The main difference to usual graph drawing problems is that we have two types of edges between the nodes, that must be treated by the layout algorithm in two different ways.
- The most important condition for the placement of classes is the inheritance structure of the graph. The subclasses (specializations) should always be placed below their superclasses (generalizations). This leads to a hierarchical structure with the classes placed on several levels.
- The subgraph of a diagram, which contains only the classes and the inheritance relationships, is a directed acyclic graph.
- The placement of classes connected to other classes by association edges is free, but should lead to short edges and few crossings.
- The subgraph of a diagram, which contains the classes and the association relationships, is a general graph that may be cyclic.
- The software engineer expects a layout of the diagram that satisfies these constraints even if there are alternative layouts for the graph with, for example, fewer crossings. This means the diagrams drawn by our algorithm may be far from an edge-crossing minimal solution.

We conclude the following drawing strategy from these characteristics:

- First place the classes involved in inheritance relationships in a hierarchical structure.
- Then place the remaining classes preserving the basic structure.

3 Algorithm

The Sugiyama algorithm is a widely used technique for drawing directed graphs [13], which has been well-analyzed [4, 5] and improved in many ways during recent years [6, 9].

Our approach is to use the Sugiyama algorithm with some modifications for the placement of all classes involved in inheritance relations. We then place the remaining classes using our extension to the basic algorithm. Then the other classes are placed using an incremental placement algorithm. This placement may influence the first Sugiyama placement again.

We represent a UML class diagram as a graph G, whose nodes are classes and whose edges are relationships among these classes.

Our algorithm consists of several phases:

Phase 1 - Preparation

1. First we remove direct cycles in G. These edges are removed and stored as attributes of the nodes involved in the cycle.
2. Then we compute the subgraph I of G that contains only the classes related through inheritance together with their inheritance connections. If I does not form a single graph, we add a new (hidden) node as root of all the different partial graphs. I is a directed acyclic graph.

Phase 2 - Sugiyama Layout

1. For the graph I we compute a first layering as described in [13]. The nodes are assigned to different layers according to the structure of the inheritance graph.
2. Reduction of crossings: this phase reorders the nodes in each layer to reduce the number of crossings. As in [13], we use barycentric ordering [15] as a heuristic method for this purpose.

   We also take into account that there may be other edges between nodes of graph I due to association relationships between the classes. If there are such edges, the nodes should be placed next to each other where possible. We compute the set O of these additional edges between nodes of I. From O we compute priorities for the placement of the nodes on each layer. If O contains an edge between two nodes on the same layer, these nodes should be placed as neighbours on the layer.

   In some cases, the drawing of an association edge between nodes placed in different layers can be improved. For example, if a node X placed in the layer L_i has an association edge to a node Y placed in layer L_j $(i < j)$, and there are no inheritance edges from X to nodes in layers between L_i and L_j, this node X can be placed in layer L_j. If X has several such edges, it is placed in the layer with the lowest index.
3. After the previous step is completed, mark all nodes that are neighbours in a layer and an edge of O exists between them. Remove this edge from O.

Phase 3 - Incremental Extension

In this phase, we extend the layout incrementally, until all nodes have been placed in the diagram.

1. Compute the set S, which contains the nodes of Graph I.
2. Select nodes from G that are not in S and which are connected to nodes of graph S due to association relationships. We construct sets S_i for each node N_i of S in two steps as follows:
   - First, for each node X in G, not in S, select X for S_i if it has one or more connections to exactly one node N_i.
   - Next, for each node X in G, not in S and not in any S_i, choose X if it has more than one connection to nodes in S. The node X is added to the set S_i that relates to a node N_i connected to X and that has the minimum number of elements.

   Now we are able to extend the existing layout in the neighbourhood of node N_i with each of these sets S_i. If there are one or two nodes in S_i we simply place those nodes in the layer to the right and to the left of N_i. If there are more than two nodes in S_i we insert another sublayer into the diagram. The nodes of S_i are placed in that layer with directed edges to N_i. If there is already a new sublayer above or below the layer of node N_i, we use this existing sublayer. If N_i is marked in the last step of Phase 2, we have to use a sublayer even if there are less than three nodes in S_i, because we cannot add a new node to the right or/and to the left of N_i on the layer.

3. Add all nodes of all sets S_i to S.
4. Repeat steps 2. and 3. until all nodes of the original graph G are placed.

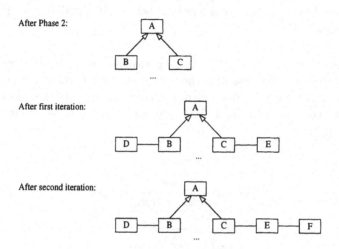

Fig. 2. Incremental extension

5. Optimize the node positions in each layer. In this step we try to reduce the number of crossings and bends of the association edges between classes on a layer. This step also improves the aspect ratio of the drawing. The nodes involved in inheritance relationships are not moved anymore. Fig. 3 shows an example of such a transformation.

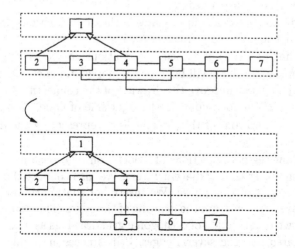

Fig. 3. Optimization using sublayers

Phase 4 - Orthogonal Drawing of Association Connection Edges

In UML diagrams, association relationships are usually drawn as orthogonal line segments and inheritance relationships as straight lines of any angle. As shown in Fig. 1 inheritance relationships may also be drawn as orthogonal segments. Straight line drawing is preferred however, because it is easier to distinguish the different edge types.

1. Compute the node sizes (Fig. 4). We have to take into account that our nodes consist of the class with their attributes and parts of the attributes of the relationships. The node size depends also on the position of other nodes that are connected to the node. For this reason the node size is calculated as in Fig. 4.

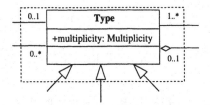

Fig. 4. Calculation of the nodesize

2. Add hidden nodes to construct the drawing of the edges remaining in set O and the edges handled in the second part of Phase 3, Step 2. We add a hidden node next to (right or left) each of the two nodes we have to connect. If the edges cross at least one layer, we add hidden nodes on all layers that are between these two nodes.
3. Now we are able to do the fine tuning of the node positions on the layers as in other Sugiyama implementations [6, 7].
4. Compute line positions of association relationships between adjacent nodes on the same layer. These will be drawn as horizontal lines. The hidden nodes may be expanded to improve the drawing of the connecting lines (For example in Fig. 5. the hidden node on the right side of node 'Shape').
5. Compute line positions of the straight lines representing the inheritance relations.
6. Compute line segment positions of association relationships between classes on the same layer or on adjacent layers. We also connect the hidden nodes added in Step 2 of this phase. We insert horizontal gridlines between the existing layers to construct the drawing of the edges. This technique is similar to the known construction of orthogonal grid drawings as in [3]. The gridlines can be shared between several edges. If it is necessary, additional gridlines are inserted. We use a sweep line algorithm as in [11] for this task.

4   Implementation

We have developed a tool called UML workbench. Using this tool we have investigated the drawing algorithms for UML class diagrams described by an underlying scripting language [12]. The following example (Fig. 5), with a graph taken from [10], shows the layout produced by our algorithm. Further examples are shown in Fig. 6 and Fig. 7.

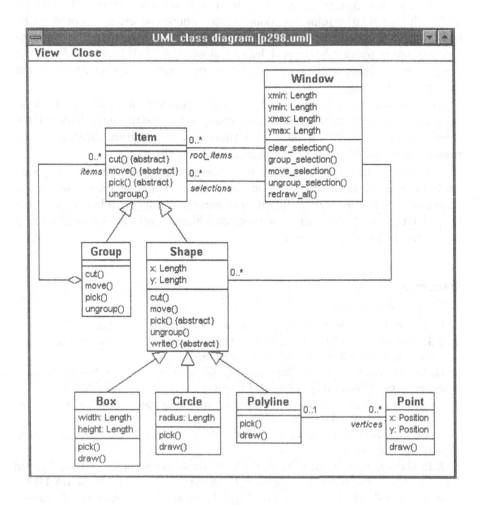

Fig. 5. Example from [10] drawn by the algorithm

5   Conclusion

We have presented a technique for the automatic generation of UML class diagrams. It is fundamentally the adaptation and combination of various algorithms developed during recent years among the graph drawing community. The growing use of visual languages in the field of object-oriented software engineering will lead to interesting graph drawing applications such as graphical browsers and CASE-Tools.

This work is a first step towards the automatic layout of object-oriented software diagrams. We have implemented the most important subset of the UML static class diagram notation.

We have good results for object-oriented architectures with considerable use of inheritance relationships but poor results where the architecture is based heavily on associations. In order to overcome this problem we plan to investigate an extension to planar hierarchical drawing algorithms [3]. It is expected that the graph characteristics can be used to select the most appropriate layout algorithm for a particular diagram.

For interactive applications such as CASE tools our algorithm has a useful property; when the software developer changes the relationships between the classes, we preserve the fundamental structure of the diagram. In this way we preserve the class hierarchy as a "mental map" for the user.

The algorithm will be used in a reverse-engineering tool for C++ software projects. Further work will focus on clustering and folding techniques in the diagrams, because software diagrams may consist of a few hundred classes. For large diagrams node clustering and diagram folding algorithms are as important as the graph layout algorithm itself.

References

1. G. Booch: *Object-Oriented Design*, Benjamin/Cummings Publishing, 1991
2. F.J. Brandenburg, editor: Proceedings of Graph Drawing '95, Vol. 1027 of *Lecture Notes in Computer Science*, Springer Verlag, 1996
3. G. Di Battista, P. Eades, R. Tamassia, I.G. Tollis: Algorithms for drawing graphs : an Annotated Bibliography, *Comput. Geometry Theory Appl.*, 4:235-282, 1994
4. Peter Eades, Kozo Sugiyama: How to draw a directed graph, *Journal of Information Processing*, 14(4):424-437, 1990
5. A. Frick: Upper bounds on the Number of Hidden Nodes in the Sugiyama Algorithm, in [8], pp. 169-183
6. E.R. Ganser, E. Koutsofios, S. North, K.-P. Vo: A technique for drawing directed graphs, *IEEE Transactions on Software Engineering*, 19(3): 214-230, March 1993
7. F. Newbery-Paulisch, W. F. Tichy: Edge: An extendible graph editor, *Software - Practice and Experience*, 20(1): 63-88, June 1990
8. S. North, editor: Proceedings of Graph Drawing '96, Vol. 1190 of *Lecture Notes in Computer Science*, Springer Verlag, 1997
9. P. Mutzel: An Alternative Method to Crossing Minimization on Hierarchical Graphs, in [8], pp. 318-333
10. J. Rumbaugh, M. Blaha, W. Premerlani, F. Eddy, W. Lorenson: *Object-Oriented Modeling and Design*, Prentice-Hall, 1991
11. G. Sander: A Fast Heuristic for Hierarchical Manhattan Layout, in [2], pp. 447-458

12. J. Seemann, J. Wolff von Gudenberg: *OMTscript - eine Programmiersprache für objekt-orientierten Software-Entwurf*, Technical Report, Department of Computer Science, Würzburg University, 1997
13. Kozo Sugiyama, Shojiro Tagawa, and Mitsuhiko Toda: Methods for visual understanding of hierarchical system structures, *IEEE Transactions on Systems, Man, and Cybernetics* SMC-11(2): 109–125, February 1981
14. Rational Software Corporation: *The Unified Modeling Language 1.0*, only available via WWW: http://www.rational.com, January 1997
15. J. Warfield: Crossing Theory and Hierarchy Mapping, *IEEE Transactions on Systems, Man, and Cybernetics* SMC-7(7): 505–523, July 1977

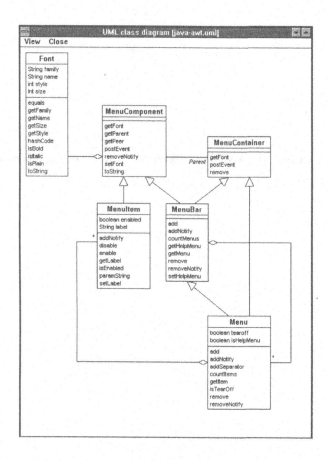

Fig. 6. Example from the AWT class library for Java (© by Sun Microsystems Inc.)

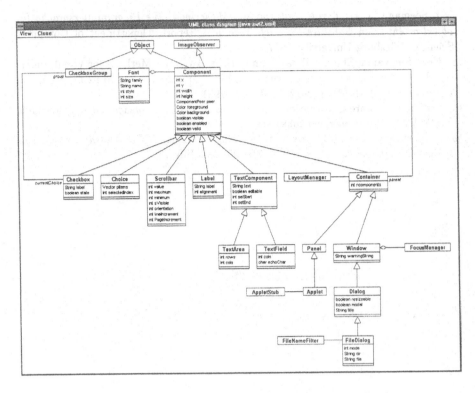

Fig. 7. Example from the AWT class library for Java (© by Sun Microsystems Inc.)

Graph Drawing and Manipulation with *LINK*

Jonathan Berry[1] Nathaniel Dean[2] Mark Goldberg[3]
Gregory Shannon[4] Steven Skiena[5]

[1] Elon College (berryj@numen.elon.edu)
[2] Bell Laboratories Innovations (nate@research.bell-labs.com)
[3] Rensselaer Polytechnic Institute (goldberg@cs.rpi.edu)
[4] Milkyway Technologies (shannon@milkyway.com)
[5] SUNY Stony Brook (skiena@cs.sunysb.edu)

Abstract. This paper introduces the *LINK* system as a flexible tool for the creation, manipulation, and drawing of graphs and hypergraphs. We describe the basic architecture of the system and illustrate its flexibility with several examples. *LINK* is distinguished from existing software for discrete mathematics by its layered interface, including a graphical user interface tied into an object-oriented Scheme language interface with access to Tk, and an extensible underlying set of C++ libraries. We conclude by briefly discussing roles *LINK* has played in research and education.

1 Background

Over the past several years, there have been several efforts to construct software systems for discrete mathematics, and in particular, for the manipulation of graphs. None, however, has resulted in a product with influence comparable to the familiar symbolic mathematics packages.

Some notable existing systems for discrete mathematics are *Combinatorica* [14], Steven Skiena's extension package for *Mathematica*, *NETPAD* [11] due to Nathaniel Dean and others at Bellcore, *SetPlayer* [1], due to Mark Goldberg and his students at Rensselaer Polytechnic Institute, and Gregory Shannon et. al.'s *GraphLab* [13]. For various reasons, none of these systems has the potential to be a widely-useful environment for both graph manipulation and computation. The authors of these systems recognized this and proposed the development of *LINK*, which was to be a freely-available and portable software system for discrete mathematics overcoming the various shortcomings of existing systems. After three years of development, the system is now freely available from the *LINK* web site:

http://dimacs.rutgers.edu/Projects/LINK.html.

LINK features a 200 page on-line manual and an on-line tutorial.

LINK 's design philosophy placed flexibility as the highest priority, and this led to the selection of STk, Erick Gallesio's object-oriented Scheme language interface to John Ousterhout's portable, interpretive Tk graphics system. [12][6]. Tk enables involved graphics programming without any knowledge of the X-window system, and offers the advantages of interpretation and portability at the cost of speed. This means that the system is not appropriate for viewing massive data sets. Graph views with a few thousand objects have been used, but these took several minutes to load on a Sparcstation 5.

Several other graph manipulation systems have been designed using Tk, including Graphlet [6], which is built on top of the LEDA C++ library [10]. However, these systems rely on Tcl, a language more similar to operating system shell scripting languages than high-level programming languages. *LINK*'s command interface, on the other hand, offers the mathematician or computer scientist a high-level Scheme language interface with an object-oriented front-end to Tk. Scheme is a compact, standardized dialect of Lisp, a functional language useful for symbol manipulation.

The remainder of the paper is broken into sections describing *LINK*'s layered interface, then illustrating its flexibility with examples, and finally giving examples of its roles in research and education.

2 *LINK*'s Templated C++ Libraries

Underlying the *LINK* system is a set of object-oriented C++ libraries which designed to offer a rich and coherent set of graph and collection objects to support programming in the pursuit of research.

2.1 *Collections* and *Containers*

The basis of the *LINK* system is a set of classes grouped into two hierarchies: *Collection* and *Container*. The *Collection* hierarchy consists of multisets (bags), sets, and sequences, while the *Container* hierarchy is subdivided into data structures such as lists and arrays and keyed dictionary structures such as binary heaps, red-black trees, etc. As in the Standard Template Library, the glue that binds all container and collection objects is the *Iterator*. *LINK*'s *Iterator* object can be used to retrieve the elements of any *Collection* or *Container*, and has been used to create a robust set of constructors and assignment operators which allow the easy transfer of data between any two collection or container objects.

The hierarchies of *Collection* and *Container* classes give library programmers a common interface for performing simple operations such as copying, assignment, insertion, deletion, extraction, comparison, display of values, and the set primitive operations. Also available are standard queries such as membership and whether or not the structure is empty, sorted, a permutation, or a subset of another.

Collection objects are templated by both element type and implementation structure (a *Container* object) to allow programmers to experiment with different set and sequence implementations. This close relationship between *Collections* and *Containers* supports a reference counting scheme which allows *Collection* objects to be passed around efficiently without the unnecessary copying of elements.

Extension classes of the *Collection* hierarchy offer functionality including the manipulation of power sets, combinations, cartesian products, and permutations.

2.2 The *Graph* Hierarchy

The *Collection* hierarchy allows the definition of a rich variety of graph objects. These are also arranged into a hierarchy so that objects from different graph classes can interact. As in the *Collection* hierarchy, objects of the *Graph* hierarchy can be copied and assigned gracefully. The classes of the *Graph* hierarchy will be introduced below.

---

[6] See http://fmi.uni-passau.de/himsolt/Graphlet.

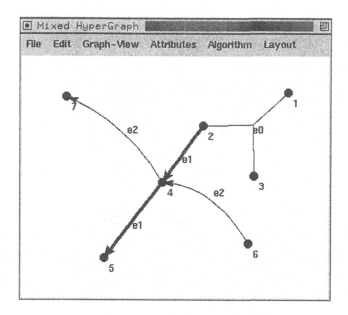

Fig. 1. A "mixed" hypergraph

Graph **Types** Central to *LINK*'s design philosophy are the goals that many different graph types should be available to the user and that algorithms need only be written once to work on many types of graph. The *Collection* hierarchy has been used extensively to meet the both goals, with the result that *LINK* users may now select between 12 types of graph by specifying the edge type. An edge is a collection of vertices, and the *Collection* hierarchy enables us to offer graph types, classified by the following pertinent questions:

- **Multigraph or not?** Multiple edges between the same vertices can be allowed or not.
- **Binary Graph or Hypergraph?** Edges can be defined as groups of two vertices or not. Relaxing this restriction results in hypergraph objects.
- **Directed, Undirected, or Mixed Graph?** Each edge is either a multiset or a sequence of vertices. Graphs can either limit their edge sets to undirected edges or directed edges, or they can make no such restriction. The result of the latter is a set of graph types in which a single graph object instance might contain both directed and undirected edges.

A *directed* hyperedge has been defined to be a set of vertices in which one vertex is specified to be a "sink," and the other vertices are assumed to precede that vertex [8]. We give a more general definition: a directed hyperedge is simply a sequence of vertices. A "mixed" hypergraph is shown in Figure 1, and a "mixed" binary graph with multiple edges is shown in Figure 2. Figure 1 is particularly interesting since it illustrates the two different modes of displaying hyperedges. Edge *e1* has been thickened for clarity, and edge *e0* is drawn using *star draw* mode, in which edge segments radiate from a central, draggable edge label. The other two edges are displayed using *path draw* mode, in which identically-labeled edge segments connect the vertices of the hyperedge. The

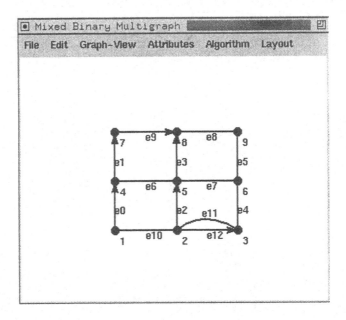

Fig. 2. A "mixed" binary multigraph

complete order of vertices within a *directed* hyperedge, however, will only be visible if the path draw display mode is used. The graphical user interface described below in Section 3.2 allows the user to change the edge display mode for the whole graph or some selected subset of the edges.

Graph **Methods** All graph objects have the same core functionality, and the member functions implementing this functionality can be broken roughly as follows:

- Vertex and edge manipulation routines such as the insertion and deletion of vertices and edges, either as individuals or in groups.
- Vertex and edge access routines which return specified individual or groups of vertices and edges.
- Queries to determine the order and size of the graph, whether or not it is directed, binary, simple, etc., whether or not two vertices are adjacent, and whether or not the graph is isomorphic to another. The latter test uses *nauty*, Brendan McKay's well-known and practical isomorphism testing tool. [9].
- Routines to find the neighbors and incident edges of vertices. These routines use *Collection* functionality to return appropriate answers for undirected, directed, and mixed graphs. Also included are routines to return the sequence of edge objects associated with a path of vertices and vice versa.
- Input and output operations for graphs, vertices, and edges, both to and from files and the screen.
- Conversion and construction operations which take sets of vertices and edges and produce graph objects. This set of routines also can convert into adjacency or incidence matrix representation.

– Edge comparison routines – two edges are comparable if they contain the vertices of the same name in the same order. Inequalities are resolved using lexicographic ordering.

Multigraphs also feature subgraph collapsing and extracting methods which are supported by the graphical user interface. These are useful when studying properties such as the chromatic polynomial, which are defined in terms of graph contractions.

2.3 Attributes

LINK provides a mechanism for creating and manipulating attributes of graphs, vertices, and edges. The default attributes for all graph objects (including vertices and edges) are currently *name, direction, width, size, weight, x, y, color, label, mark, type, starttime, finishtime, back, low, distance, pred,* and *forefather*. To save space, the attribute mechanism stores a single copy of each attribute for the entire graph until individual attributes are changed (at which point an individual copy is made for the affected object). Some of the default attributes are used by the graphical user interface, and some are used by fundamental graph algorithms. If different attributes are desired, however, defining new attributes is simple, both for the library programmer and the interface user.

```
STk>   (describe (graph (current-graph-view)))
#[<dbingraph*> #p63f5cc] is an instance of class <dbingraph*>
Slots are:
     val = {[1 2 3 4 5 6]  {<1 2> <1 6> <2 5>  <3 4> <5 3> <5 6>}}
#f
STk>   (map color (vertices (current-graph-view)))
("black" "black" "black" "black" "black"
 "black")
STk>   (define vg(car(vertices(current-graph-view))))
#[undefined]
STk>   (set! (color vg) "green")
#[undefined]
STk>   (map color (vertices (current-graph-view)))
("green" "black" "black" "black" "black"
 "black")
```

Fig. 3. This Scheme code segment retrieves and manipulates the graph of the current *graph-view* (window) after the user has constructed a graph in it.

3   The STk Interface

STk is a complete programming environment in itself, and *LINK* inherits all of its functionality. In addition to a standard R^4RS Scheme interpreter, STk provides an object-oriented extension based on the Common Lisp Object System called STklos, as well as operating system shell access, regular expression processing, and Unix socket handling. The STklos extension enables the scheme programmer to define classes and generic functions, and *LINK*'s interface takes full advantage of this power. All of the basic *LINK* objects have been "wrapped" into the STk interpreter so that users can create, manipulate, and destroy them, and *LINK*'s graphical user interface consists exclusively of new STklos classes so that users may take advantage of high-level, object-oriented functionality to manipulate their data.

3.1 *LINK*'s STklos Objects

```
STk>    (define gv (show-graph (graph '(1 2 3) '((1 2) (2 3) (1 3)))))
#[undefined]
STk>    (define g (graph gv))
#[undefined]
STk>    (define eg (car (edges gv)))
#[undefined]
STk>    (define e (edge eg))
#[undefined]
STk>    eg
#[<edge-item> #p64ba68]
STk>    e
#[<edge*> #p64c7a0]
STk>    (set! (weight eg) 3.21)
#[undefined]
STk>    (find-double-attribute 'weight e)
3.21
STk>    (set-double-attribute! 'weight 2.1 e)
#[undefined]
STk>    (weight eg)
2.1
```

Fig. 4. Fundamental attribute operations

The *LINK* interface user has access to graph objects at both the graphical user interface level and the Scheme command language level. *LINK*'s manual contains dozens of Scheme programming examples, and we will include some below. Figure 3 contains Scheme code to retrieve the colors of the vertices that a user has created using the graphical interface. The example subsequently changes the color of the first vertex, a change reflected graphically on the screen. Note that *vg* is an STklos object which contains both a graphics field (displayed on the screen) and a reference to an underlying *LINK* vertex object. STklos gives users the considerable convenience of setting fields (or "slots") using the *set!* primitive. The expression (color vg) is a shorthand way of extracting the "color" field from the vertex.

The fundamental STklos graph objects of the *LINK* system correspond to the graph types described in Section 2.2: *vertex, edge, graph, bingraph, dbingraph, ubingraph, hypergraph, uhypergraph, dhypergraph mbingraph, mdbingraph, mubingraph, mhypergraph, muhypergraph, and mdhypergraph.* These and their most important methods are available to the *LINK* interface user, and detailed in the manual.

```
STk>    (define gv (show-graph (graph '(1 2 3)
                    '((1 2) (2 3) (1 3)))))
#[undefined]
STk>    (map weight (edges gv))
(1.0 1.0 1.0)
STk>    (random-edge-weights (graph gv))
#[<ubingraph*> #p63c0c0]
STk>    (map weight (edges gv))
(38.0 58.0 13.0)
STk>    (define mst (kruskal (graph gv)))
#[undefined]
STk>    (describe mst)
#[<set<edge*>> #p6f1c98] is an instance of class <set<edge*>>
Slots are:
    val = {{1 2} {2 3}}
#f
STk>    (map (lambda (x) (find-double-attribute 'weight x))
                (set-edge->list mst))
(38.0 13.0)
```

Fig. 5. Scheme code to call an algorithm and examine the results.

The *LINK* objects mentioned above are mirrored by special STklos graphics objects. The most important correspondences are those between classes representing the fundamental graph objects. The three most important STklos classes in *LINK* are:

- *graph-view*: a window which contains a *LINK graph* object.
- *vertex-item*: a class containing an STklos *oval* graphics object and a *LINK vertex* object.
- *edge-item*: a class containing (potentially many) STklos *line* graphics objects and a *LINK edge* object.

Figure 4 illustrates the difference between STklos graphics objects and *LINK* objects. In this example, a graph is defined and displayed in a *graph-view* window called *gv*. The method *(graph gv)* returns the *LINK* graph object associated with this graph-view. The example then extracts the first *edge-item* from the graph-view and extracts the *edge* associated with that edge-item. STklos supports "virtual" data within a class, and the *LINK* interface takes advantage of this by inseparably linking the attributes of the edge to those of the edge-item. When the user evaluates or modifies an attribute of the edge-item, such as its weight, that request is translated into an evaluation or assignment to the corresponding attribute of the underlying edge. For example, In figure 4, changes to the edge-item's weight attribute are reflected in the edge's weight attribute and vice versa.

Figure 5 shows a more detailed example in which a graph is created and displayed, its edges are assigned random weights, and a minimum spanning tree is computed, extracted, and manipulated. This example illustrates two different levels of attribute retrieval: directly from a graphics object, and via a *LINK* object. The former method is used in the command *(map weight (edges gv))*, which extracts the *weight* slot from each STklos edge object in the graph window called *gv*. Note that *mst*, the variable used to store the result of the spanning tree algorithm, is a *LINK set* object. This is converted into a Scheme list using the *set-edge→list* method (provided with all collection objects).

```
(define (forefather-binding graph-view)
 (strongly-connected-components (graph graph-view))
 (bind (slot-ref graph-view 'graph-toplevel) "<KeyPress-f>"
      (lambda (x y)
          (flash (vertex-item
                  (find-vertex-attribute 'forefather
                      (slot-ref (car *link:selected-vertex-items*) 'vertex)) graph-view)))))
```

Fig. 6. This complete code segment binds the "f" key so that the forefather of a selected vertex is flashed. The strongly-connected-components algorithm sets an attribute called *forefather* that is subsequently retrieved to decide which vertex to flash.

It is legitimate to ask why *LINK*'s *Collection* objects are at all useful in a list-based Scheme interpreter (why not just have all algorithms return Scheme lists?). Figure 7 provides an answer. Many of *Collection*'s methods, including the copying, insertion, deletion, query, and set primitive operations are available from the STklos interface. The example in Figure 7 is a simple graph sum computation using the set primitive operations.

```
STk>    (define g (graph '(1 2 3) '((1 2) (2 3) (3 1))))
#[undefined]
STk>    (define h (graph '(2 3 4) '((2) (2 3 4) (3 4))))
#[undefined]
STk>    (define ng (graph (+ (vertices g) (vertices h))
                          (+ (edges g) (edges h))))
#[undefined]
STk>    (describe ng)
#[<uhypergraph*> #p63e848] is an instance of class <uhypergraph*>
Slots are:
    val = {[1 2 3 4] {{1 2} {1 3} {2} {2 3} {2 3 4} {3 4}}}
#f
```

Fig. 7. An example which uses *Collection* methods

3.2 Graphical User Interface

STklos provides a core set of graphics classes (windows, labels, buttons, etc.) which make interface customization a high-level operation, and *LINK*'s graphical interface consists of a set of classes which inherit from these. The result is that standard graph

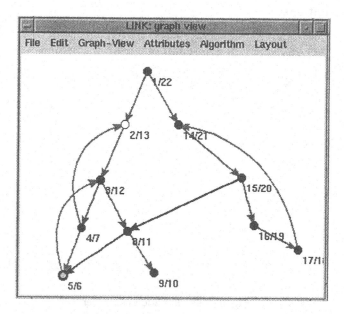

Fig. 8. Forefather finding with finishing times depicted

operations such as graph creation, insertion and removal of vertices and edges, execution of algorithms, and animation viewing are both point and click features and high-level STklos operations. Multiple graph windows can be viewed at once, and multiple algorithm animations can be stepped through side-by-side for comparison. Graphics attributes such as world coordinates, size, arrow shape, label, color, text-color, stipple, outline width, outline color, and font can be evaluated and changed easily.

It is important to note that any command in the graphical user interface corresponds to a STklos command that could have been typed into the command line prompt. This makes *LINK* a powerful environment for systematically constructing, executing, viewing, modifying, and rerunning experiments. This interface has been used in several research projects, and two will be abstracted in Section 5.

3.3 Flexibility

Consider the following example of the interface's flexibility. When describing the strongly connected components of a directed graph, it is important to relate the concept of the *forefather* of a vertex. Simply stated, the forefather $\phi(v)$ of a vertex v with respect to a depth-first search is the vertex reachable from v which has the maximum finishing time in the depth-first search. *LINK*'s interface can easily be tailored to illustrate the concept intuitively. Figure 6 shows, in its entirety, the STKlos code necessary to "bind" the f key on the keyboard to a function which will flash the forefather of a vertex selected with the mouse. Figure 8 shows a graph-view in which a depth-first search has been run and the discovery and finishing times of the vertices are displayed. Selecting a vertex, then pressing the f key highlights a vertex's forefather by flashing it several times.

3.4 Animations

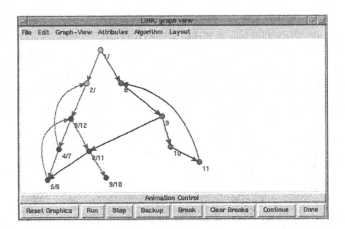

Fig. 9. The animation controller

When *LINK* algorithms are added to the C++ libraries, they can be augmented with special animation commands which modify the attributes of the graph's vertices and edges. These commands are executed if the algorithm is run from the interface (as opposed to being run from a standalone C++ program). Algorithms selected from a graph-view bring up an animation controller/debugger which allows the user to step through the algorithm forwards and backwards, set breakpoints, continue, and restart. An example animation of a depth-first search is shown in Figure 9.

4   Algorithms, Generators, and Layouts

LINK's libraries include several fundamental algorithms for manipulating, generating, and drawing graphs, including depth and breadth-first search, Kruskal's and Prim's minimum spanning tree algorithms, Goldberg & Tarjan's maximum flow algorithm, strongly connected components, generators for random graphs, cycles, complete graphs and grid graphs, and circular, random, grid, spring, and component-wise layout algorithms. This algorithms library will certainly grow as the system develops and new versions are distributed. If the reader is interested in contributing to this effort, please contact Jonathan Berry at berryj@numen.elon.edu.

5   Research Examples

LINK has already demonstrated its usefulness in research, and its role in two recent projects will be summarized below.

5.1 Latka Tournaments

In the first project, Brenda Latka, while visiting DIMACS from Lafayette College, used *LINK* to assist in her study of infinite antichains of tournaments (complete directed graphs). An antichain of tournaments is a set for which it is impossible to embed any tournament in the set within any other. Latka is interested in the construction of infinite antichains of tournaments. [5, 7]. Arguments to prove that a given set of tournaments is an antichain typically show that a specific sub-tournament of any given tournament cannot be mapped to any sub-tournament of any other tournament in that class. A crucial element of these arguments is the use of a non-trivial edge attribute: the number of directed 3-cycles in which an edge participates. Sub-tournaments [tbh]

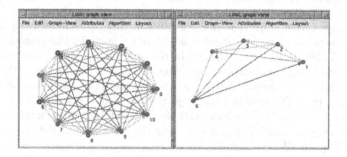

Fig. 10. A Latka tournament and an induced subgraph extracted with *LINK*'s graphical user interface

and these edge attributes are easily visualized using *LINK*'s features: the former can be extracted by clicking on vertices and selecting a menu option, while the latter can be computed (and the edges colored) by small programs written in *LINK*'s command language. Figure 10 shows an instance of a special class of tournaments defined by Latka (see [5, 7] for details), and an induced sub-tournament extracted by pointing and clicking (see the *LINK* web page for a color image).

Once *LINK* functionality had been used to generate Latka Tournaments, compute the edge attributes, and color their edges accordingly, Latka and Jonathan Berry used *LINK* to visualize dozens of these tournaments and variations upon them. Soon patterns began to emerge, leading Latka to conjecture that adding an extra parameter to her initial tournament construction would reveal the first known *infinite* set of infinite antichains of tournaments. The conjecture, still unproven, will be detailed in a future paper.

During this time, Jonathan Berry was also serving as a mentor for Chris Burrows, a participant in the the *NSF Research Experience for Undergraduates (REU)* at DIMACS. He used *LINK* to study isomorphisms of certain sub-tournaments by implementing some special invariants described by Latka and using *LINK*'s point and click access to *nauty*. During this process, we observed that one of the Latka tournaments also happened to be a Paley tournament. [7] Burrows then used *LINK* to find two ad-

---

[7] The well-studied Paley tournaments consist of p vertices, where p is prime and congruent to 3 mod 4. Arc (i, j) exists iff $j - i$ is a quadratic residue mod p.

ditional Latka-Paley tournaments, then develop the conjecture that no other Latka tournaments are Paley tournaments.

5.2 Market Basket Analysis

In another project, a set of supermarket data compiled and studied previously at Bell Laboratories was analyzed in a more meaningful way with *LINK*. Considering each type of item in a shopper's "basket" (e.g., bananas, 2% milk, skim milk) to be a vertex, "market-basket analysis" attempts to identify customer buying patterns by examining receipts. Correlations between purchases identified by the analysis can be used, for example, to arrange products more advantageously on the shelves or manipulate prices.

When a shopper purchases a set of items at once, we must represent this grouping somehow. An obvious approach to the problem is to place an edge between each pair of vertices in a basket, thus producing a graph where each purchase is represented by a clique. Given such a graph, however, the original "baskets" cannot by reconstructed. Nathaniel Dean and Jonathan Berry used *LINK* to re-model the problem using hypergraphs (graphs where each edge might contain more or fewer than two vertices) where each hyperedge represents a single shopper's basket. In addition to the considerable space savings inherent in this solution, more real-world information is preserved. Furthermore, the STk command language used by *LINK* makes it possible to pose interactive queries such as: "find all purchases in which both a snack food item and a beverage were purchased." A paper describing this work in more detail is available at the *LINK* web site [3].

6 *LINK* as an educational tool

LINK's flexible interface makes it an valuable educational tool, both in the classroom and as a vehicle for interesting assignments. The key binding example discussed above (see Figure 6) enables the instructor to present the forefather concept as a puzzle to engage students. This was done recently with encouraging success in an algorithms course at Elon College. The instructor also made extensive use of *LINK*'s graphical user interface and interactive algorithm animations in class. A discussion of *LINK*'s role in computer science education is found in [2].

7 Conclusion

The early development and primary designers and developers of *LINK* are detailed and acknowledged, respectively, in [4], while the current system is described in the manual available from the web site. Jonathan Berry took over the direction of the project in June, 1995, and spent a year at DIMACS preparing the public release.

Currently, *LINK* runs only on Unix systems (including Linux), but there is no major obstacle preventing a port to Windows, since the graphics system upon which *LINK* relies, STk [6], has already been ported. *LINK* for Windows is anticipated before the end of 1997.

With further development, *LINK* can become a formidible tool for prototyping, teaching, and experimentation. Development directions in the near future include improving the documentation, extending the algorithms library, and improving the STklos interface.

8 Acknowledgments

We would like to acknowledge the support of DIMACS and the *LINK* grant: CCR-9214487. DIMACS is a cooperative project of Rutgers University, Princeton University, AT&T Laboratories, Lucent Technologies/Bell Laboratories Innovations, and Bellcore. DIMACS is an NSF Science and Technology Center, funded under contract STC-91-19999; and also receives support from the New Jersey Commission on Science and Technology. The original primary investigator of the *LINK* project was Daniel Gorenstein, the founding director of DIMACS.

We would also like to acknowledge the contributions of Patricia K. Fasel of Los Alamos National Laboratory, who was the original project leader and who helped design the graph hierarchy and implemented many system fundamentals. Many students have helped with the *LINK* effort as well, and we acknowledge their effort.

References

1. D. Berque, R. Cecchini, M. Goldberg, and R. Rivenburgh. The setplayer system for symbolic computation on power sets. *Journal of Symbolic Computation*, 14:645–662, 1992.

2. J. Berry. Improving discrete mathematics and algorithms curricula with LINK. In *SIGCSE/SIGCUE Conference on Integrating Technology into Computer Science Education*, pages 14–20, 1997.

3. J. Berry and N. Dean. Market basket analysis with LINK. submitted to Congressus Numerantium, 1996.

4. J. Berry, N. Dean, P. Fasel, M. Goldberg, E. Johnson, J. MacCuish, G. Shannon, and S. Skiena. LINK: A combinatorics and graph theory workbench for applications and research. Technical Report 95-15, Center for Discrete Mathematics and Theoretical Computer Science (see also: http://dimacs.rutgers.edu), Piscataway, NJ, 1995.

5. G. Cherlin and B. Latka. A decision problem involving tournaments. Technical Report 96-11, Center for Discrete Mathematics and Theoretical Computer Science, 1996.

6. E. Gallesio. The stk reference manual. Technical Report RT 95-31a, I3S CNRS, Université de Nice - Sophia Antipolis, France, 1995.

7. B. Latka. Finitely constrained classes of homogeneous directed graphs. *The Journal of Symbolic Logic*, 59(1):124–139, March 1994.

8. E. Mäkinen. How to draw a hypergraph. *International Journal of Computer Mathematics*, 34:177–185, 1990.

9. B. McKay. Nauty user's guide. Technical Report TR-CS-90-02, Australian National University, 1990.

10. K. Mehlhorn and S. Nähger. Leda: A platform for combinatorial and geometric computing. *CACM*, 38(1):96–102, Jan 1995.

11. M. Mevenkamp, N. Dean, and C. Monma. NETPAD user's guide and reference guide, 1990.

12. J. Ousterhout. *Tcl and the Tk Toolkit*. Addison-Wesley, 1994.

13. G. Shannon, L. Meeden, and D. Friedman. SchemeGraphs: An object-oriented environment for manipulating graphs, 1990. Software and documentation.

14. S. Skiena. *Implementing Discrete Mathematics: Combinatorics and Graph Theory with Mathematica*. Addison-Wesley, 1990.

Graph-Drawing Contest Report

Peter Eades[1], Joe Marks[2], and Stephen North[3]

[1] Department of Computer Science
University of Newcastle
University Drive – Callaghan
NSW 2308, Australia
E-mail: eades@cs.newcastle.edu.au
[2] MERL–A Mitsubishi Electric Research Laboratory
201 Broadway
Cambridge, MA 02139, U.S.A.
E-mail: marks@merl.com
[3] AT&T Research
180 Park Ave., Bldg. 103
Florham Park, NJ 07932-0971, U.S.A.
E-mail: north@research.att.com

Abstract. This report describes the Fourth Annual Graph Drawing Contest, held in conjunction with the 1997 Graph Drawing Symposium in Rome, Italy. The purpose of the contest is to monitor and challenge the current state of the art in graph-drawing technology [2, 3, 4].

1 Introduction

Text descriptions of the three graphs for the 1997 contest can be found on the World Wide Web (WWW) at URL www.inf.uniroma3.it/calendar/gd97/-contest/rules.html. Graph A is an artificial graph that was designed as a special challenge for orthogonal graph-drawing algorithms. Graph B represents a collection of WWW pages and similarity measures between some of them. For both of these graphs an effective drawing had to communicate not only the edge connections between vertices, but also any vertex- or edge-attribute values peculiar to the graph. Graph C represents the calls made between a set of telephone numbers. Participants were required to submit a video of an interactive exploration through which they answered several specific questions about the graph.

Approximately 13 submissions were received by the contest deadline. The winners were selected by the organizers, and are shown below.

2 Winning submissions

2.1 Graph A

This directed graph contains 84 nodes and 333 edges. It was contrived without reference to a real-world application. The contest rules stated that only orthogonal drawings would be acceptable.

This graph was designed to present a special challenge for standard techniques. Nevertheless, most contestants were able to produce excellent drawings of it. In fact, five of the six submissions were essentially the same ideal drawing. All five entries were jointly awarded first prize. The winning authors were:

- B. Bascary, B. Cattan, A. Cohen-Solal, M. Philip, and H. Szigeti, (szigeti@eurecom.fr), Telecom Paris, France.
- Vladimir Batagelj and Andrej Mrvar ([Vladimir.Batagelj, Andrej.Mrvar]@uni-lj.si), University of Ljubljana, Slovenia.
- Michael Kaufmann, (mk@informatik.uni-tuebingen.de), Universität Tübingen, Germany.
- Gunnar Klau, (guwek@mpi-sb.mpg.de), MPI für Informatik, Saarbrücken, Germany.
- Thomas Ziegler and Petra Mutzel, ([tziegler, mutzel]@mpi-sb.mpg.de), MPI für Informatik, Saarbrücken, Germany.

Figure 1 contains a representative drawing from the set of winners. Remarkably, the similar drawings seem to have been created using a variety of techniques. For example, Ziegler and Mutzel based their drawing on a planar embedding of a maximal subgraph [7]. The embedding was augmented by hand, and then edges were drawn using Tamassia's bend-minimization algorithm [8]. In contrast, Kaufmann used a spring embedder [6] and direct manipulation to unfold the original graph and uncover symmetries. He then used the "Kandinsky" approach [5], augmented with some ideas from [8], to draw the edges with a minimum number of bends.

2.2 Graph B

This graph contains 47 nodes and 264 edges. It represents a collection of WWW pages in which pages of similar content have been linked automatically. Each page is represented by a graph node with one attribute, the page title. Associated with each edge are the degree of similarity between the connected pages, and a list of words indicating the basis of commonality.

Few graph-drawing systems handle text labels well, and fewer still incorporate clustering or partitioning algorithms for grouping similar vertices. We therefore expected Graph B to be a tough challenge. Nevertheless, the winning drawing, shown in Figure 2, is an excellent visualization of the graph data. It was produced by Vladimir Batagelj and Andrej Mrvar ([Vladimir.Batagelj, Andrej.Mrvar]@uni-lj.si) from the University of Ljubljana, Slovenia, using the "Pajek" system (see http://vlado.fmf.uni-lj.si/pub/networks/pajek/). They used Ward's hierarchical clustering technique to compute vertex clusters [1], which were then rearranged automatically to minimize edge crossings. A final refinement of vertex locations was performed manually.

2.3 Graph C

This graph contains 452 nodes and 768 edges. It was extracted from a large telephone-call database. Graphs like this are used in fraud investigations. Asso-

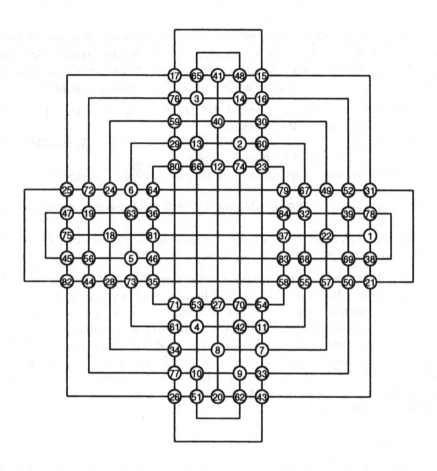

Fig. 1. Representative winner, Graph A.

ciated with the graph are several questions. Contestants were required to submit a video of their graph-drawing system being used to answer these questions.

There were two joint winners for this graph. The first video was submitted by Cristian Ghezzi (xtian@co.umist.ac.uk), of UMIST, Manchester, England. His Virtual Data Browser (VDB) is a 3D graph-drawing system (see `http://www.-co.umist.ac.uk/xtian/VRML/vdb/vdb.html`). It is written in Java, and uses SGML and VRML as the input and output file formats. A screen snapshot is shown in Figure 3.[4] Much use is made of 3D navigation, color, and interactive exploration to answer the posed questions.

The other winning video was submitted by Falk Schreiber and Carsten Friedrich

---

[4] Of course, a screen snapshot does not do justice to the system, which is why video submissions were required in the first place.

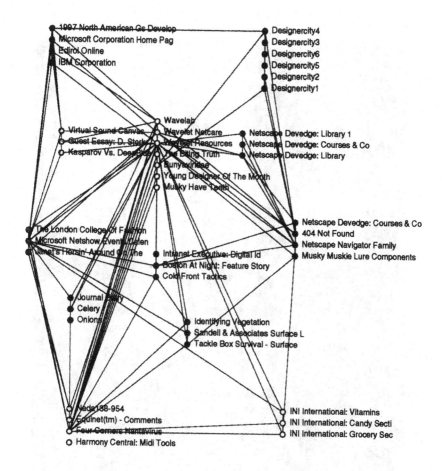

Fig. 2. Winner, Graph B.

([schreibe, friedric]@fmi.uni-passau.de) of Universität Passau, Germany. Their system is a more conventional 2D graph-drawing system, based on the "ffGraph" library (see `http://www.fmi.uni-passau.de/ friedric/ffgraph/ffgraph/-ffgraph.html`). A screen snapshot is shown in Figure 4. It shows clusters that arose naturally from a spring-method layout. These clusters were useful in answering several of the questions. Color and interactive exploration were also used to good effect.

3   Observations and Conclusions

The number of outstanding drawings for Graph A was a real surprise, especially because they were produced using several different techniques and ideas. In previous years, such a graph might not have been drawn well by most systems,

442

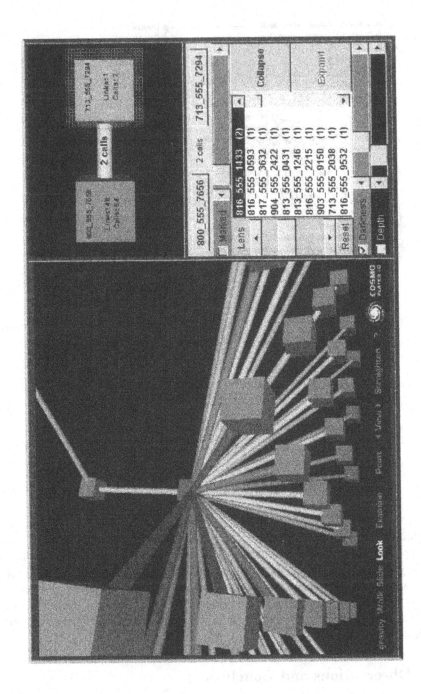

Fig. 3. Screen snapshot of joint winner, Graph C.

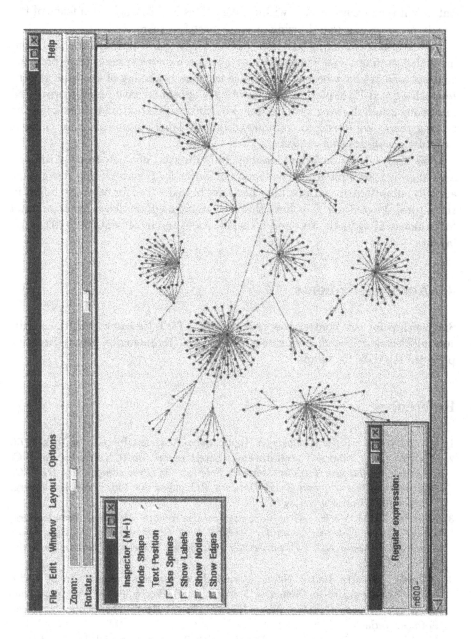

Fig. 4. Screen snapshot of joint winner, Graph C.

but now it seems comfortably within range. This is encouraging evidence of the steady progress in graph drawing over the past few years.

It was also encouraging to receive entries for Graph B that utilized partitioning and clustering ideas. These concepts have been underrepresented in previous contests, and yet have been studied since the very beginning of the field. Recent research on text-label placement does not yet appear to have been incorporated into many graph-drawing systems, but we expect this situation to change in the coming years. We intend to continue to include label placement as an element in future graph-drawing contests.

Soliciting video submissions was hopefully the start of a new trend. Although several groups attempted the production of videos for Graph C, only two groups actually submitted ones. Hopefully they have blazed a trail for others to follow in the future. We continue to believe that a dynamic medium like video is essential for demonstrating interactive graph visualization, and incremental and 3D graph layout.

4 Acknowledgments

Sponsorship for this contest was provided by AT&T Research, MERL–A Mitsubishi Electric Research Laboratory, and Lucent Technologies. Denny Bromley provided the data for Graph B.

References

1. M. Anderberg. *Cluster Analysis for Applications.* Academic Press, New York, 1973.
2. P. Eades and J. Marks. Graph-drawing contest report. In R. Tamassia and I. G. Tollis, editors, *Lecture Notes in Computer Science: 894 (Proceedings of the DIMACS International Workshop on Graph Drawing '94)*, pages 143–146, Berlin Heidelberg, October 1994. Springer-Verlag.
3. P. Eades and J. Marks. Graph-drawing contest report. In F. J. Brandenburg, editor, *Lecture Notes in Computer Science: 1027 (Proceedings of the Symposium on Graph Drawing GD '95)*, pages 224–233, Berlin Heidelberg, September 1995. Springer-Verlag.
4. P. Eades, J. Marks, and S. North. Graph-drawing contest report. In S. North, editor, *Lecture Notes in Computer Science: 1190 (Proceedings of the Symposium on Graph Drawing GD '96)*, pages 129–138, Berlin Heidelberg, September 1996. Springer-Verlag.
5. U. Fößmeier and M. Kaufmann. Drawing high degree graphs with low bend numbers. In F. J. Brandenburg, editor, *Lecture Notes in Computer Science: 1027 (Proceedings of the Symposium on Graph Drawing GD '95)*, pages 254–266, Berlin Heidelberg, September 1995. Springer-Verlag.
6. A. Frick, H. Mehldau, and A. Ludwig. A fast adaptive layout algorithm for undirected graphs. In R. Tamassia and I. G. Tollis, editors, *Lecture Notes in Computer Science: 894 (Proceedings of the DIMACS International Workshop on Graph Drawing '94)*, pages 388–403, Berlin Heidelberg, October 1994. Springer-Verlag.

7. M. Jünger and P. Mutzel. Solving the maximum weight planar subgraph problem by branch and cut. In G. Rinaldi and L. Wolsey, editors, *Proceedings of the Third Conference on Integer Programming and Combinatorial Optimization (IPCO)*, pages 479–492, 1993.

8. R. Tamassia. On embedding a graph in the grid with the minimum number of bends. *SIAM Journal on Computing*, 16(3):421–444, 1987.

Author Index

Springer
and the
environment

At Springer we firmly believe that an international science publisher has a special obligation to the environment, and our corporate policies consistently reflect this conviction.

We also expect our business partners – paper mills, printers, packaging manufacturers, etc. – to commit themselves to using materials and production processes that do not harm the environment. The paper in this book is made from low- or no-chlorine pulp and is acid free, in conformance with international standards for paper permanency.

Springer

Lecture Notes in Computer Science

For information about Vols. 1–1278

please contact your bookseller or Springer-Verlag